Francis-99 Workshop 1: Steady Operation of Francis Turbines 2014

Journal of Physics: Conference Series Volume 579

Trondheim, Norway
15 – 16 December 2014

Editors:

M. J. Cervantes
C. Trivedi
O. G. Dahlhaug
T. Nielsen

ISBN: 978-1-63439-912-8
ISSN: 1742-6588

Printed from e-media with permission by:

Curran Associates, Inc.
57 Morehouse Lane
Red Hook, NY 12571

Some format issues inherent in the e-media version may also appear in this print version.

Printed by Curran Associates, Inc. (2015)

For permission requests, please contact the Institute of Physics
at the address below.

Institute of Physics
Dirac House, Temple Back
Bristol BS1 6BE UK

Phone: 44 1 17 929 7481
Fax: 44 1 17 920 0979

techtracking@iop.org

Additional copies of this publication are available from:

Curran Associates, Inc.
57 Morehouse Lane
Red Hook, NY 12571 USA
Phone: 845-758-0400
Fax: 845-758-2634
Email: curran@proceedings.com
Web: www.proceedings.com

TABLE OF CONTENTS

PREFACE FRANCIS-99

Francis-99 is a set of upcoming workshops jointly organized by the Norwegian University of Science and Technology (NTNU), Norway and Luleå University of Technology (LTU), Sweden in the same spirit as the previous Turbine-99 workshops. The Francis-99 workshops aim during the coming years to determine the state of the art of high head Francis turbine simulations (flow and structure) under steady and transient operating conditions as well as promote their development and knowledge dissemination openly. Three workshops are initially planned:

- Workshop 1: steady operation of Francis turbines (December 2014)
- Workshop 2: transient operation of Francis turbines (December 2016)
- Workshop 3: FSI of Francis turbines (December 2018)

A high head Francis turbine model, named the Tokke model, has been designed and experimentally investigated at the Water Power Laboratory, NTNU. The complete geometry of the model and mesh are now freely available on the site www.francis-99.org together with a large set of experimental pressure and velocity measurements. The organisers expect this geometry to become with time a reference test case to the hydraulic community for research and development on high head Francis turbines and the workshops a meeting place to discuss developments, potentials, issues... on a common and open test case.

The present proceeding contains the papers presented at the first workshop at NTNU the 15th and 16th of December 2014. 50 participants were present at the workshop and a total of 14 papers were presented. A large variety of codes and models were used highlighting different issues in the simulation of high Francis turbines.

The editors:
Prof. Michel J. Cervantes (LTU, NTNU),
Dr. Chirag Trivedi (NTNU)
Prof O.G. Dahlhaug (NTNU)
Prof. T. Nielsen (NTNU)

Peer review statement

All papers published in this volume of *Journal of Physics: Conference Series* have been peer reviewed through processes administered by the proceedings Editors. Reviews were conducted by expert referees to the professional and scientific standards expected of a proceedings journal published by IOP Publishing.

Journal of Physics: Conference Series **579** (2015) 012001 doi:10.1088/1742-6596/579/1/012001

Prediction of dynamic blade loading of the Francis-99 turbine

J Nicolle, S Cupillard

Institut de recherche d'Hydro-Québec, Varennes, Qc, Canada, J3X 1S1

nicolle.jonathan@ireq.ca

Abstract. CFD simulations focusing on capturing dynamic fluctuations of the flow for three operating points were performed for a scale model of a high head Francis turbine. A mesh sensitivity study showed an influence of the near wall resolution, consequently a low Reynolds mesh with a SST turbulence model was used. Rotor/stator fluctuations are well reproduced with the full turbine simulation at all operating points. Velocity contours and average velocity profiles from LDV measurements in the draft tube confirm that the flow physics is generally well reproduced. Simplified approaches such as profile transform and Fourier transform underestimated the measured fluctuations. As full turbine simulations were time-consuming, a simulation with only the draft tube was performed at part load to predict the fluctuations in the draft tube cone. The SAS-SST turbulence model was able to capture the vortex rope behavior.

1. Introduction

While a large part of the design of hydraulic turbines still involves steady state simulations due to time constraints, more and more emphasis has recently been accorded to the dynamic behavior of the turbine blades. Greater flexibility of hydraulic machines is expected due to the integration of renewable energy but, at the same time, the competitive process also pushes for thinner blade design to achieve high efficiency while limiting material costs. Unit availability is also a concern as there is a strong push to move from systematic to conditional maintenance in order to limit downtime. All these factors bring new challenges regarding dynamic loadings and fluid structure interaction to power utilities as well as turbine providers.

The Francis-99 workshop series targets this problem very acutely by allowing an exchange platform to evaluate and improve numerical tools used by industry and universities. It is supported by experimental data from the scale model of a high head Francis turbine installed on a platform at NTNU in Norway. To avoid the pitfall of similar workshops held previously, the geometry of a complete turbine, including the runner, is freely available. This means that simulations can be performed on the whole machine, not just part of it. Work on this geometry should extend over a few years. The first workshop targets the dynamic loads under three steady state operating conditions. For a high head Francis turbine, significant rotor/stator interactions (RSI) are expected due to strong wakes of the guide vanes and close spacing between guide vanes and runner blades. This paper will focus mainly on the RSI and ways to capture it while limiting simulation costs.

2. Experimental dataset

Three operating points were suggested for the workshop; they are presented in table 1. To cover the wide operating range of the machine, one point is at part load (PL), one is at best efficiency point (BEP) and the last one is at high load (HL).

Table 1. Operating points.

	GV Angle (°)	Head (m)	Velocity (rev/s)	Flow rate (m^3/s)	Power (kW)	Efficiency (%)
Part Load	3.91	12.29	6.77	0.071	6.14	71.69
Best Efficiency Point	9.84	11.91	5.59	0.203	21.97	92.61
High Load	12.44	11.84	6.16	0.221	23.28	90.66

Pressure sensors were placed at various locations in the turbine, as can be seen in figure 1. In the rotating frame, a bandwidth filter had to be used on the experimental signal to remove the grid frequency of 50 Hz and its harmonics, which dominated the raw signal. After this treatment on the fast Fourier transform (FFT), shown in figure 1 for P42 at BEP, the normalized frequency of the RSI clearly appears as the dominant frequency. No more treatment was done on the signal, although it was still noisy in the rotating part. Experimental signal analysis shows that draft tube vortex rope is also present at PL and HL.

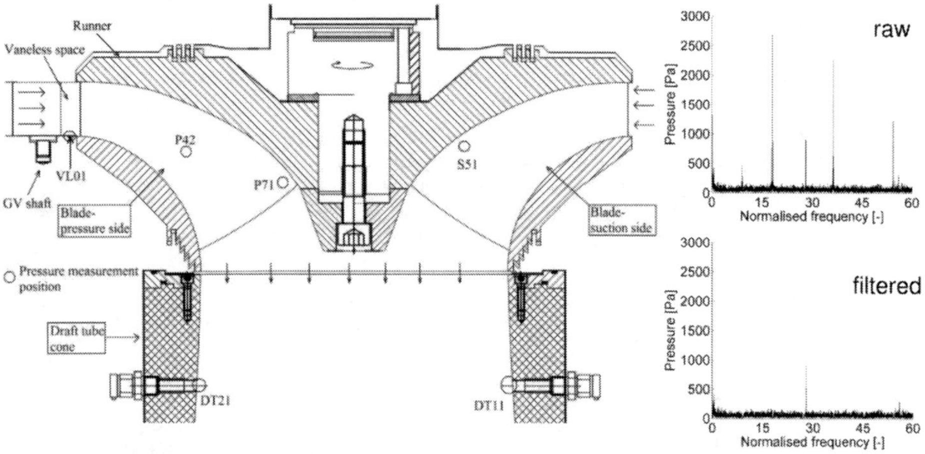

Figure 1. Pressure sensors on the Francis-99 model with the example of a raw (upper graph) and a filtered (lower graph) signal at P42 where the blade passing frequency is 28.

Laser Doppler velocimetry (LDV) measurements were also available on two horizontal planes in the draft tube below the runner to allow comparison of the axial and tangential velocities. Since the runner velocity and measured head (H) for the LDV campaign differ slightly from the proposed simulation points, experimental velocities were scaled assuming the normalized quantities (n_{ed} and Q_{ed}) were kept constant using (1). This translated into correction factors of 0.96574 and 0.96899 for the BEP and HL velocities. No correction was required for PL.

$$\frac{U_{OP}}{U_{LDV}} = \left(\frac{H_{OP}}{H_{LDV}}\right)^{1/2} \quad (1)$$

3. Numerical setup and sensitivities

Most of the geometry was reconstructed in Catia either from plans or by using curves and surfaces from the ANSYS ICEM CFD files provided [1]. This step was necessary in order to use our meshing tools afterwards. Small modifications were made to the runner cap, runner channel, distributor height and draft tube sections in order to better reflect the actual geometry. Further checks, not presented here, showed very limited impact of these changes.

3.1. Meshes

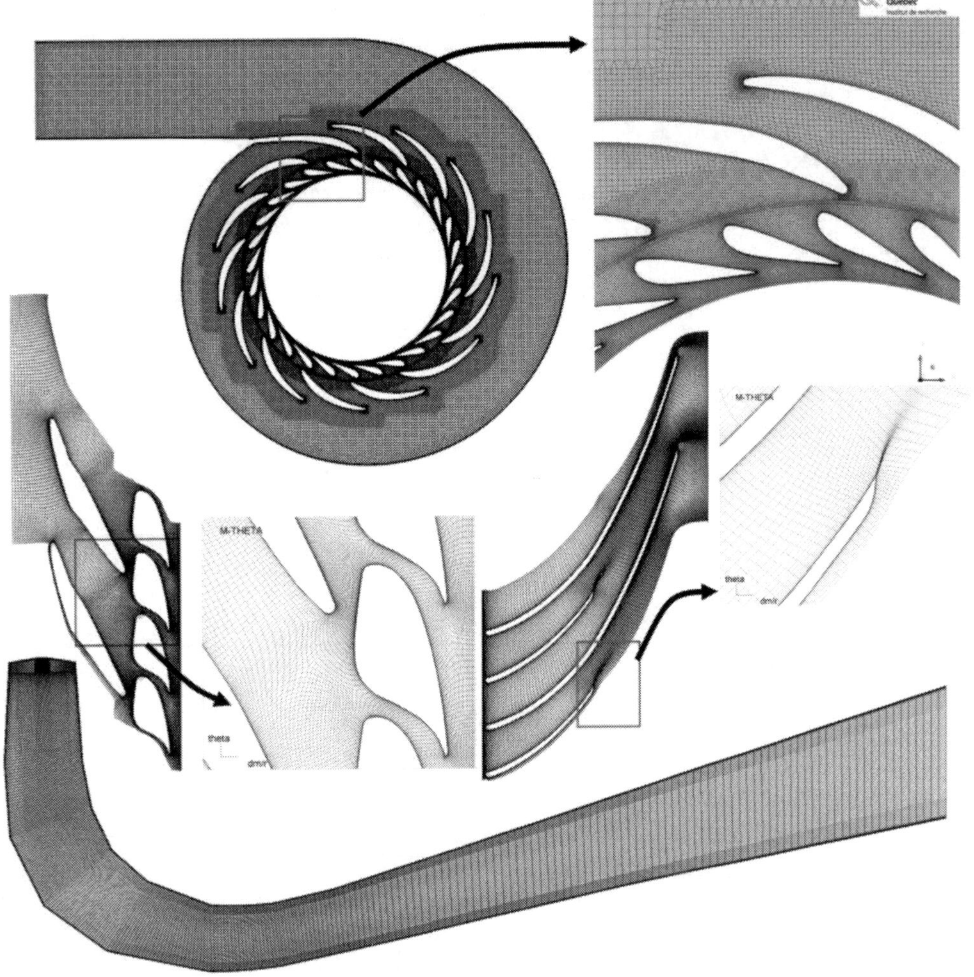

Figure 2. View of the spiral casing, distributor, runner and draft tube meshes.

For the spiral casing, Numeca Hexpress/Hydrid was used to produce a cartesian hexahedral dominant unstructured conformal mesh (i.e. without hanging node). The casing mesh extended to the stay vanes and was made to be used with a smaller structured guide vane domain. Autogrid5 was used to generate the meshes for the distributor and the runner. Beside the mesh including only a guide vane, a single sector mesh including a stay vane and two guide vanes was also generated for the distributor in order to allow the use of simplified models. Even with the unusual shape of the guide vane near the hub and the shroud with an eccentric rotation axis, the software was able to produce a high quality grid without introducing an interface between the blades. In the runner, a sharp topology was used at the trailing edge to keep a good resolution in the blade wakes. ANSYS ICEM CFD was used for the draft tube. Care was taken to ensure the resolution was similar on both sides of the various interfaces between the stationary and rotating parts. A view of the different meshes is provided in figure 2 and some quality metrics can be found in table 4.

3.2. Mesh size

First, an overall mesh independence study was performed with steady state simulations at the BEP. ANSYS CFX 15 was used for the simulations. These calculations were performed with the k-epsilon model with only a sector of the distributor (1/14 periodicity with one stay vane and two guide vanes) and of the runner (1/15 periodicity with a main blade and a splitter). The height of the first element off the wall was set to 0.2 mm as in [2]. The rotating and stationary domains were connected with mixing plane (stage) interfaces. Node-to-node matching was achieved for the periodic boundaries. Four different mesh densities were used. The measured flow rate was specified at the inlet while an opening was set at the outlet. Convergence was achieved when maximum residuals throughout the solution were lower than 1E-5. Spatial discretization of the momentum equations used the "high resolution" scheme which blends first and second orders in a way to balance stability and accuracy. As can be seen from table 2, the mesh resolution did not change the results much for efficiency or power. All these models overestimated the experimental power (around 8%) and efficiency (around 2%).

Table 2. Mesh resolution sensitivity.

Mesh size (k elements)					
Distributor	Runner	Draft Tube	Total	Efficiency [%]	Power [kW]
227	209	747	1116	94.67	23.78
451	277	1506	2234	94.82	23.72
913	567	3176	4658	94.77	23.72
1920	1150	6543	9613	94.70	23.72
EXP				92.61	21.97

3.3. Wall refinement

A second round of numerical sensitivity tests was performed to evaluate the effect of the near wall refinement. The first node off the wall distance was gradually lowered. In order to keep the element aspect ratio reasonable, the number of elements was increased accordingly. As shown in table 3, with y+ according to (2), the k-epsilon model was quite insensitive to this refinement. However, further inspection of the solution showed that the near wall limiter of the scalable wall function implemented in the k-epsilon model was already activated on most of the runner blades at 200 μm, which explains why the global values did not change. This limiter prevents modeling of the laminar sublayer and can be detected by monitoring the solverYplus value (3) which will then take a fixed value of 11.06. The percentage in parentheses in table 3 gives the area of the blade where the limiter is active.

$$y^+ = \frac{\sqrt{\tau_w / \rho} \cdot \Delta y}{\nu} \quad (2)$$

$$solverY^+ = \max(11.06, \frac{C_u^{1/4} k^{1/2} \cdot \Delta y}{\nu}) \quad (3)$$

The SST model was used to see if a different turbulence model would yield different results. The model had a completely different behavior than the k-epsilon and was very sensitive to the wall distance. While the power output was very large with the 200 μm mesh, it decreased rapidly along with the wall distance and stabilized when low Reynolds meshes (y+ < 5) were used. For this model, a wall function is used only when the first node does not allow to model the laminar sublayer [3]. Switching between the low Reynolds and wall function is transparent to the user.

Table 3. Wall refinement sensitivity.

Wall distance (μm)	Size (M elements)	Y+ Guide Vane (*)	Y+ Runner (*)	Power (kW)
k-epsilon				
200	2.23	70.0 (20)	22.8 (96)	23.72
100	2.27	34.6 (47)	11.6 (99)	23.52
50	2.31	17.3 (100)	5.9 (100)	23.55
20	2.31	6.90 (100)	2.5 (100)	23.64
SST				
200	2.23	66	24	24.68
100	2.27	34	12	24.08
50	2.51	17	6	23.45
20	2.82	7	4	23.11
5	3.61	2	1	22.60
2	3.61	0.7	0.4	22.60

* Shows the area percentage where the k-epsilon scalable wall function limiter was active (solverY+ = 11.06)

Further analysis of the total pressure drop and velocity angle at the runner inlet showed that most of the difference originated from the distributor. In figure 3, visual inspection clearly shows that the guide vane wakes are more intense with the SST model. This, in turn, modifies the inlet angle for the runner and, thus, the resulting torque.

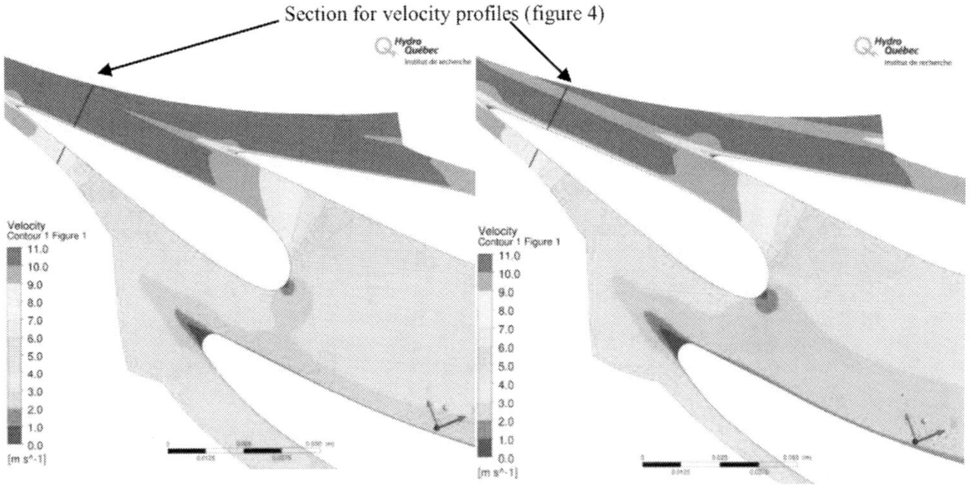

Figure 3. Velocity contours in the distributor for the k-epsilon and SST turbulence models.

These differences led us to investigate more closely the boundary layer evolution on the guide vanes. Velocity profiles were extracted along the black line visible near the trailing edge of one guide vane in figure 3. Numerical velocities were normalized by the local wall shear stress and dynamic viscosity and plotted against the wall distance. The logarithmic behavior of the boundary layer appears clearly in figure 4. The wake of the adjacent guide vane is also visible on the suction side. On the pressure side, the boundary layer is very thin (in the order of y+ 200 or 0.2 mm), due to local acceleration. On the suction side, the boundary layer extends to 0.6 mm. Because it seems to provide a better power output along with a good resolution in the thin boundary layer, the SST model with the low Re meshes of 5E-6 m was chosen to continue the study.

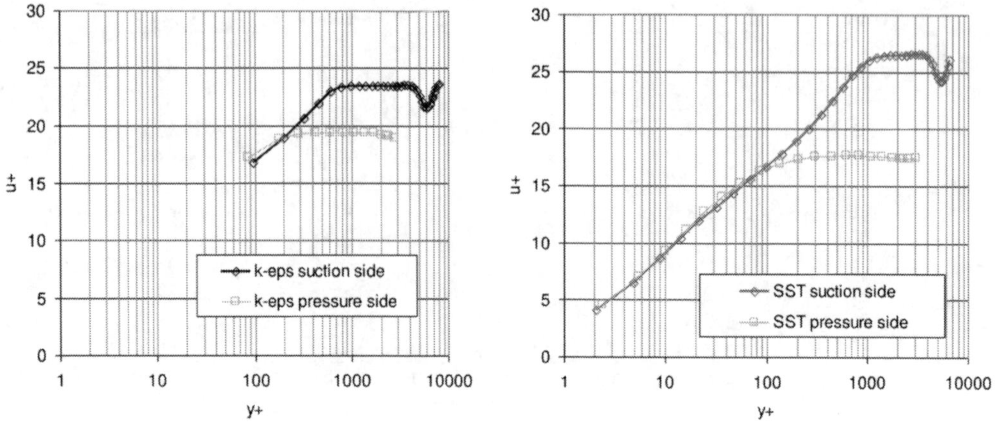

Figure 4. Boundary layer evolution on the guide vane for the k-epsilon and SST model, x-axis (y+) uses logarithmic scale to see the boundary layer extent.

Table 4 presents some of the final mesh characteristics. The minimal angle and maximum expansion factor are usually within an acceptable range for the structured meshes while the aspect ratio is high, given the need to resolve the boundary layer up to the wall. The unstructured spiral casing mesh shows some low angles and high expansion factors. The complete turbine simulation included 32M elements while the simplified models included between 4 and 6M elements.

Table 4. Final Mesh Characteristics

	Size (M elements)	Min. angle	Max. Exp. Factor	Aspect Ratio
Spiral Casing	7.28 (73 % Hex)	4	63	114
Distributor				
complete	7.4	24.1	5	1120
1 sector	0.98	23.4	6	1658
Runner				
Complete	15.5	29.1	5	1383
1 sector	1.04	29.1	5	1383
Draft Tube	2.04	33.7	6	2882

4. Numerical setup

Different strategies were used to get the dynamic pressure fluctuations (figure 5). The full machine simulation from the spiral casing inlet to the draft tube outlet is the most complete and is therefore considered as the reference. In order to limit the computational cost, two other setups were tested to reproduce the RSI fluctuations. These do not include the draft tube and spiral casing and were only performed at BEP. They use the profile transform (PT) and the Fourier transform (FT) approaches.

The profile transform, as its name states, involves scaling or stretching at the RS interface. This means that frequencies are expected to be slightly off, since there are 28 guide vanes and 30 runner blades. This gives a pitch ratio of 0.933, a bit far from the usual recommended ratio for this approach, which is between 0.98 and 1.02. Fitting this ratio would require modeling 22 guide vanes and 24 runner blades, which is almost a full machine simulation.

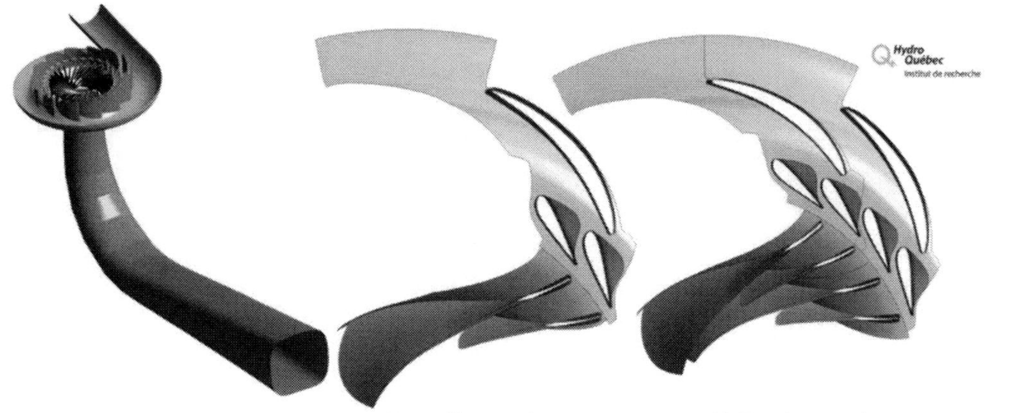

(a) Full turbine (b) Profile transform (c) Fourier transform

Figure 5. Tested configurations.

The Fourier transform is a beta feature of the transient blade row module (TBR) of CFX designed to capture RSI at a lower numerical cost than a complete simulation over 360° when there is unequal pitch between the stator and the rotor. Schematic representation of the FT is presented in figure 6. The basic principle of phased-shifted boundary conditions is that pitch-wise boundaries are periodic to each other but at different moments in time. The FT requires two passages for both the distributor and the runner. Flow variables are sampled in the central plane in order to evaluate the Fourier coefficients describing the signal according to (4). Since fluctuations are reproduced with coefficients, storing the full signal is not necessary. Fluctuations on the periodic boundaries are reconstructed with (5) using a phase lag (or time shift) based on the pitch differences and runner velocity (6). Conversion of the signal into coefficients means that fluctuations that do not match harmonic frequencies will be filtered in the process. Following the same line, the RS interface is further decomposed into theta harmonics resulting in a double (t-θ) Fourier series. More details about the method can be found in [3]. So far, to the authors' knowledge, only cases dealing with compressible flows have been published with the FT modeling approach [4][5].

$$\phi_{sampling}(t) = \sum_{m=-M}^{M} A_m e^{-j(\omega m t)} \quad (4)$$

$$\phi_{per1}(t) = \phi_{sampling}(t - \Delta t);\ \phi_{per2}(t) = \phi_{sampling}(t + \Delta t) \quad (5)$$

$$\Delta t = \frac{P_R - P_s}{\omega R} \quad (6)$$

On the numerical side, the FT method was found to be quite sensitive to the location of the periodic interfaces. In order to improve the numerical stability, the runner mesh was modified to move the periodic boundary closer to the center of the blade channel. This mesh manipulation involved rotating some elements and merging nodes but did not require any mesh regeneration. It can be seen in figures 2 and 5 that the original periodic boundary of the runner used in the full turbine and PT is very close to the splitter while for the FT it is in the middle of the blade channel. No such change was required in the distributor.

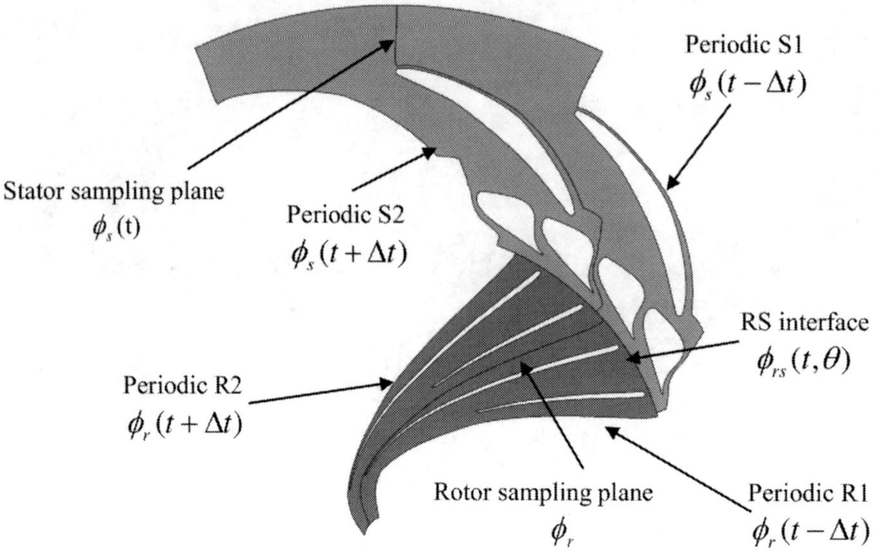

Figure 6. Fourier Transform of ANSYS CFX.

For all setups, the measured flowrate was specified at the inlet while a static pressure was prescribed at the draft tube outlet. A high resolution advection scheme was used for the momentum and turbulent quantities. The FT approach requires that an integer number of iterations is used to describe the RSI in both domains. The number of iterations per revolution therefore had to be a multiple of 210 (14 stators * 15 runners). Also, a minimum of 20 points was deemed necessary to capture the RSI. A combination of these criteria led to a timestep representing 840 iterations per runner revolution. This corresponds to a rotation of the runner of about 0.4 degrees/timestep. Convergence of the inner transient loop required maximal residuals of the momentum equations below 1E-4 before moving to the next iteration. Based on the flow rate and runner diameter, it was estimated that the flow took about half a revolution to travel through the runner. In order to get rid of initial effects, the time required by the flow to pass the runner was double so all the calculations were initialized with a complete runner revolution prior to recording the dynamic content. The acquisition period for the results was made on the second complete revolution. Since the simulation time was obviously too short to capture draft tube vortex ropes and similar low frequency phenomena, a simulation with only the draft tube was performed at PL. For this simulation, a 2D velocity profile coming from the full turbine simulation was imposed at the draft tube inlet. Since no fluctuation was obtained with the SST model, the SAS-SST turbulence model was used in order to allow coherent structures to develop in the draft tube. Based on previous experiences, a larger time step, about one order of magnitude greater than in the runner and corresponding to 3.2 degrees runner revolution, was used and more than 120 runner rotations were simulated.

5. Results

The average pressure at the different sensor locations is presented in figure 7 and is generally in good agreement with the experiments. Pressure is slightly overpredicted on the runner blades and underpredicted in the draft tube, for the three conditions tested. At BEP, the PT and FT curves superpose each other.

Journal of Physics: Conference Series **579** (2015) 012001 doi:10.1088/1742-6596/579/1/012001

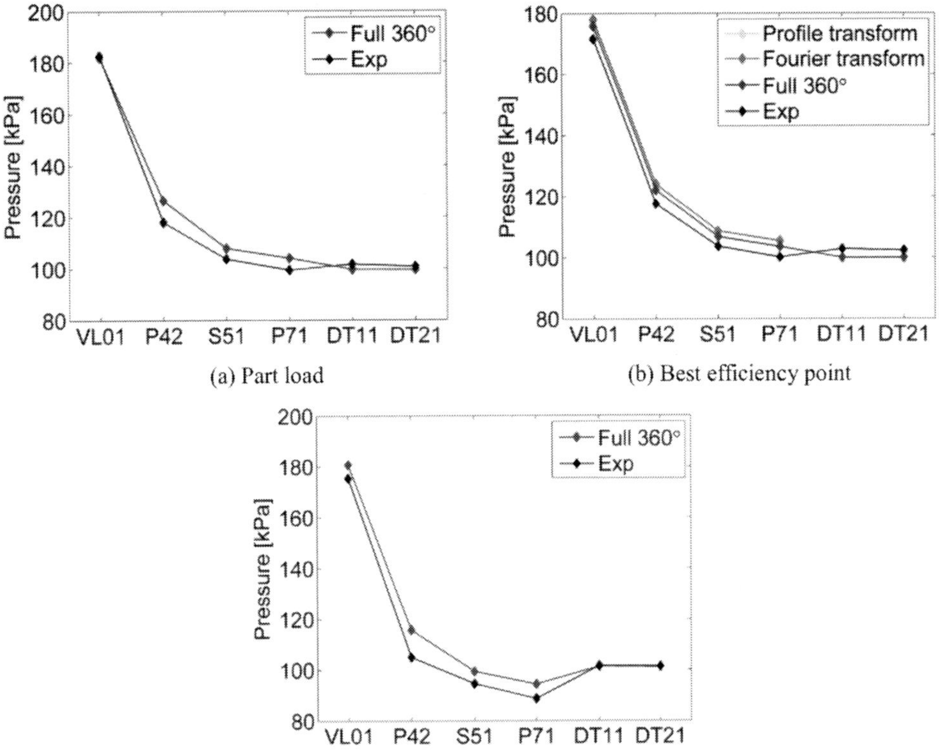

(a) Part load

(b) Best efficiency point

(c) Full load

Figure 7. Average pressure at different locations.

Pressure fluctuations at BEP are presented in figure 8 for the three different strategies. The full 360° FFT signal (figure 8a) shows good agreement with the experimental amplitudes for the main fluctuations. However, some differences can still be seen elsewhere in the spectrum. The first harmonic of the guide vane passing frequency, at twice the fundamental frequency (i.e. f/f_n =56 in the rotating part), is already pretty much diffused in the simulation whereas it is distinguishable in the experimental signal (figure 8d). Also for VL01, at f/f_n =15, the experimental signal in the distributor is able to perceive some difference between the main runner and splitter blades while there is no such distinction in the full 360° simulation.

The profile transform approach, shown in figure 8b, produced fluctuations only at f/f_n = 30 in both the rotating and stationary domains. As mentioned previously, this frequency shift was expected due to the interface scaling. While fluctuations are slightly overestimated in the stator, they are underestimated in the rotating part. Again, this might be due to the scaling. The distributor sees the equivalent of larger blades passing by due to the stretching in the circumferential direction while the runner sees thinner wakes coming in. A pitch ratio closer to unity could probably prevent this.

For the FT case, illustrated in figure 8c, the signal of the pressure sensors was reconstructed from the Fourier coefficients obtained after the second runner revolution. The right frequencies are predicted in the both domains. The fluctuation amplitudes are comparable to those obtained previously by the profile transform approach. They are overestimated for VL01 and underestimated in the rotating part. Also noticeable is the harmonic content present on the signal which comes from the Fourier decomposition.

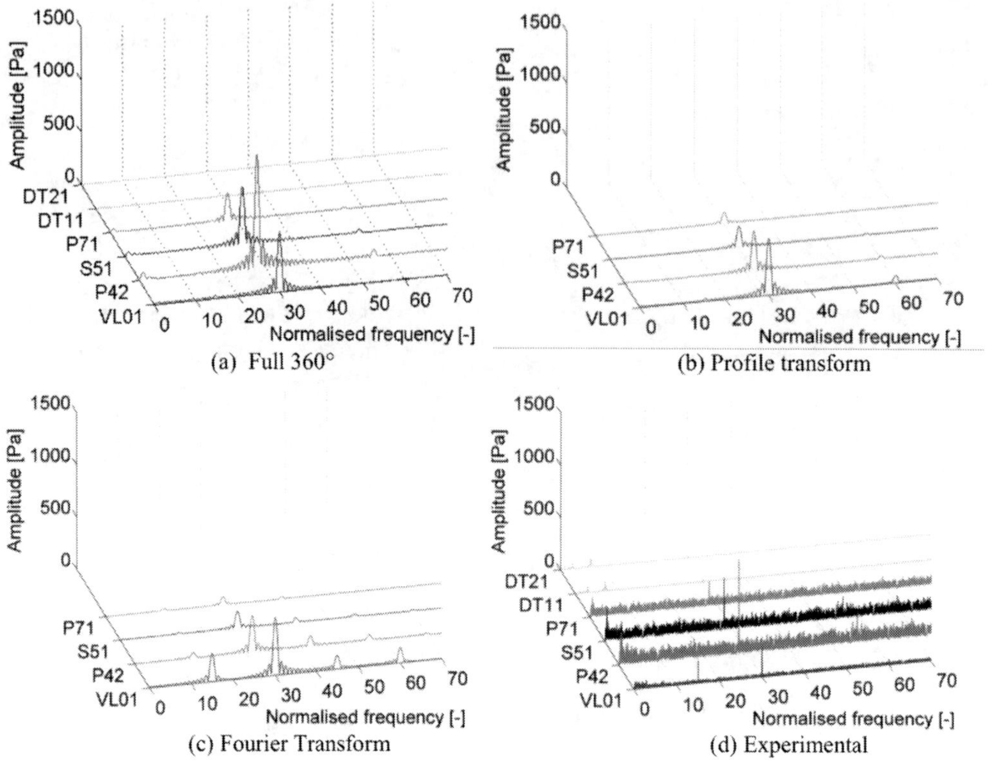

(a) Full 360°
(b) Profile transform
(c) Fourier Transform
(d) Experimental

Figure 8. FFT of the pressure signals at best efficiency point.

Because the same meshes and schemes were used for the guide vane and the runner in all the previous cases, the observed differences in the fluctuations could only be a consequence of the different boundary conditions or of the methods themselves. However, pressure fluctuations at the sensor locations in the runner only tell part of the story regarding blade loading and life expectancy. Since they are located halfway and downstream in the runner channel, wakes have already been mixed and diffused. To get another view on how well the RSI was actually captured, it is interesting to look at a global quantity like the blade torque developed on the runner and the splitter blades. A few RSI periods are presented in figure 9 at BEP. As shown in this figure, the predicted torque differs quite a lot depending on the model used. The largest fluctuations are obtained for the full 360° simulation. The FT approach captures some phenomena at $f/f_n = 14$ on both the runner and the splitter blade. This corresponds to the stay vane passing frequency but also to the fundamental frequency of the Fourier series. Thus, it is not yet clear if those fluctuations are numerical or real. The fact that both the stator (stay and guide vanes) and the rotor (blade and splitter) are made of tandem blades might complicate the use of this approach. The profile transform captures even fewer fluctuations than the FT approach. The frequency shift in the rotating frame mentioned earlier is also clearly seen.

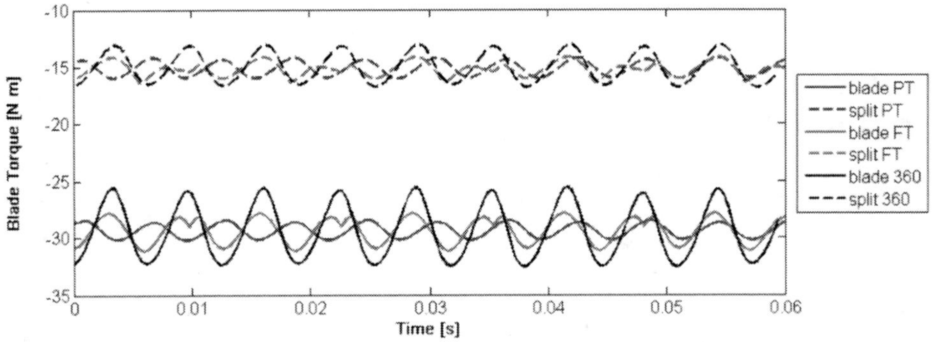

Figure 9. Torque developed on the runner main and splitter blades at BEP.

Averaged velocity profiles on lines 1 and 2 in the draft tube are presented in figure 10. In order to mimic the way the experimental data were captured, CFD quantities were extracted and averaged from a single line over a full runner revolution. The axial-velocity component is globally well reproduced by the simulation whereas the tangential component is overestimated in the center of the flow, indicating the presence of a swirl.

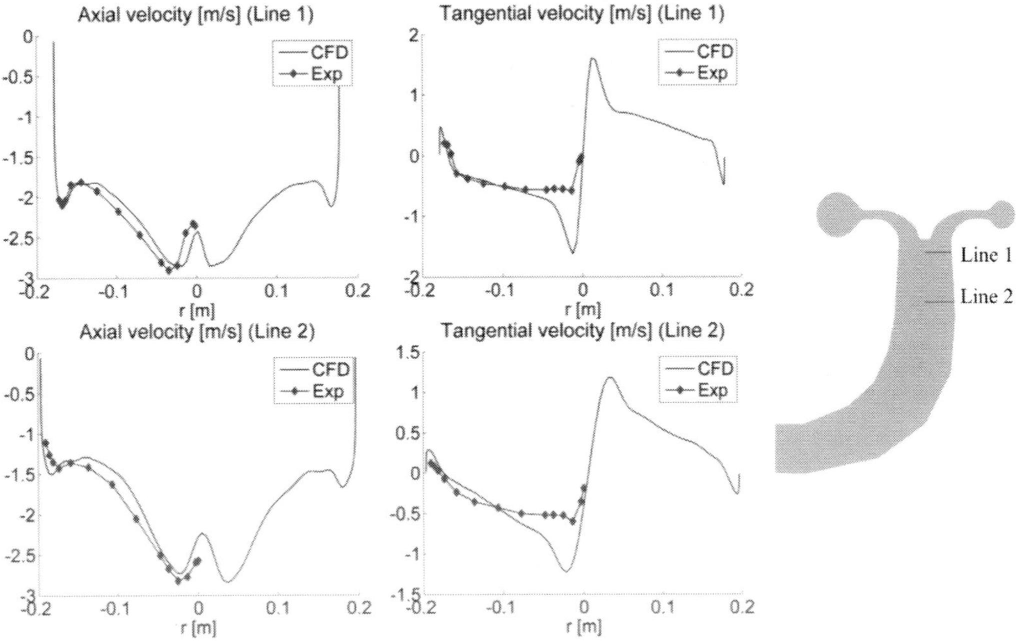

Figure 10. Averaged velocity profiles on lines 1 and 2 at best efficiency point.

Experimental phase resolved velocities on planes 1 and 2 are displayed in figure 11 and compared to instantaneous velocities obtained at the same elevation. At BEP, the axial velocity matches reasonably well with experimental values. Wakes due to the 15 long runner blades are still present on plane 1 in the experiment whereas they are hardly noticeable in the simulations. On plane 2, wakes have also disappeared from the experimental signal. Simulated tangential velocity is over predicted in the center of the flow on both planes as too much swirl seems to be predicted.

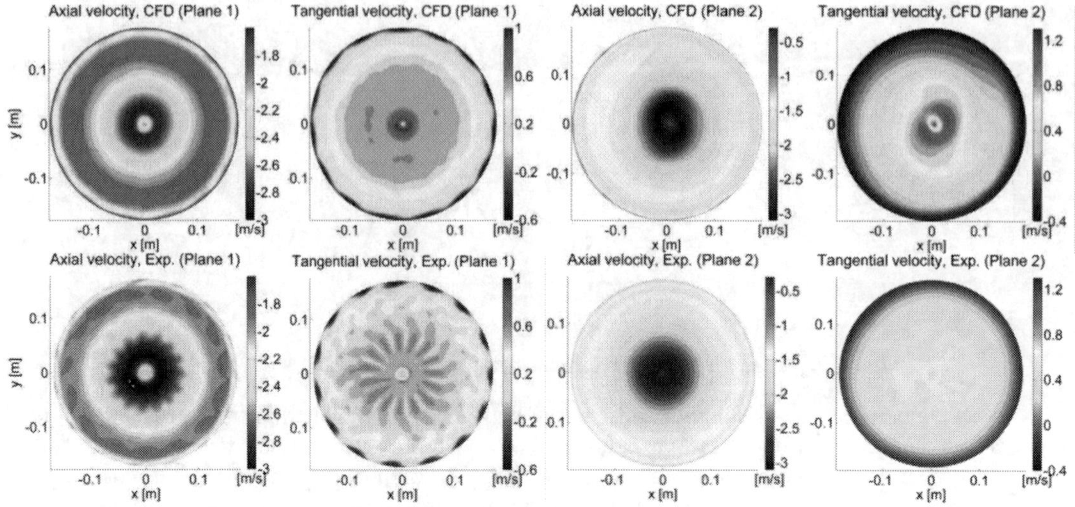

Figure 11. Velocity profiles (CFD on top, experimental on bottom) at BEP.

Looking at the FFT at PL and HL in figure 12, amplitudes of the RSI fluctuations are generally captured very well for both operating points. In the simulations, fluctuations are slightly stronger at HL than at PL in the runner due to stronger wakes coming in from the guide vane at higher discharge. One noticeable difference at HL between the simulation and experiment is the absence of energy content at f/f_n =15 for VL01. Also at HL, a significant low frequency can be seen at P42 in the simulation. No physical explanation has been found yet for this.

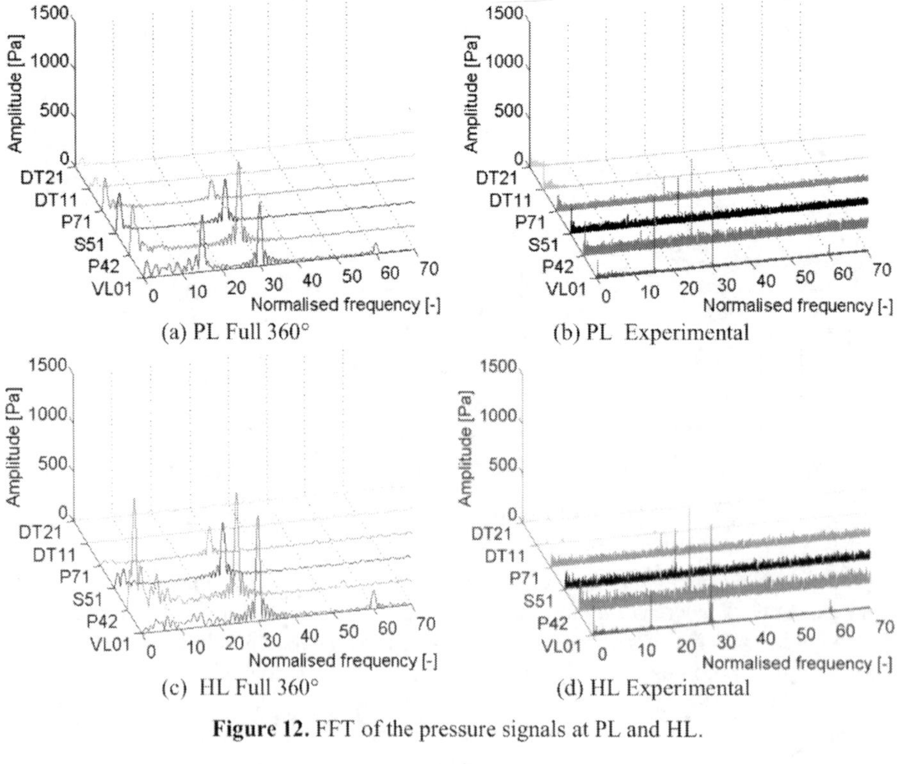

(a) PL Full 360° (b) PL Experimental

(c) HL Full 360° (d) HL Experimental

Figure 12. FFT of the pressure signals at PL and HL.

The average velocity profiles for the two measurement sections at PL are presented in figure 13. Axial and tangential components are generally well captured. Corresponding velocity contours confirm that in figure 14. Velocity profiles and contours at HL are displayed in figures 15 and 16. In order to improve the comparison at this point, the CFD velocity contours are phase averaged.

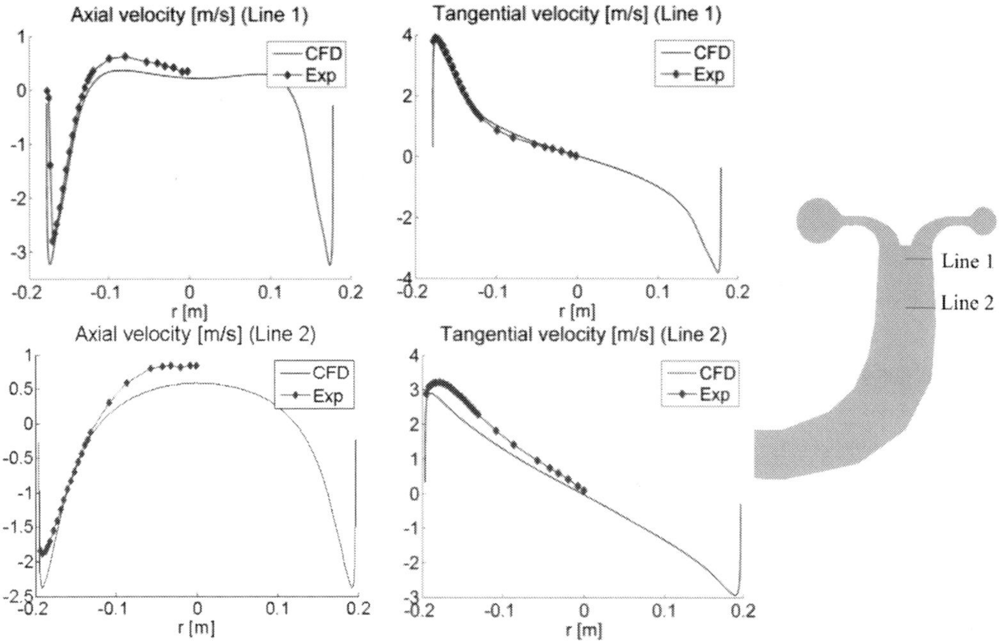

Figure 13. Averaged velocity profiles on lines 1 and 2 at PL.

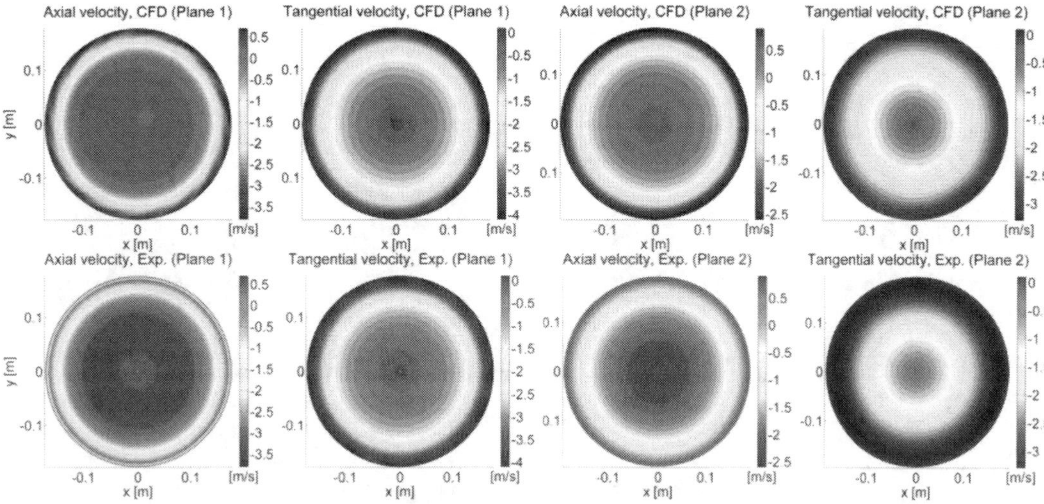

Figure 14. Velocity profiles (CFD on top, experimental on bottom) at PL.

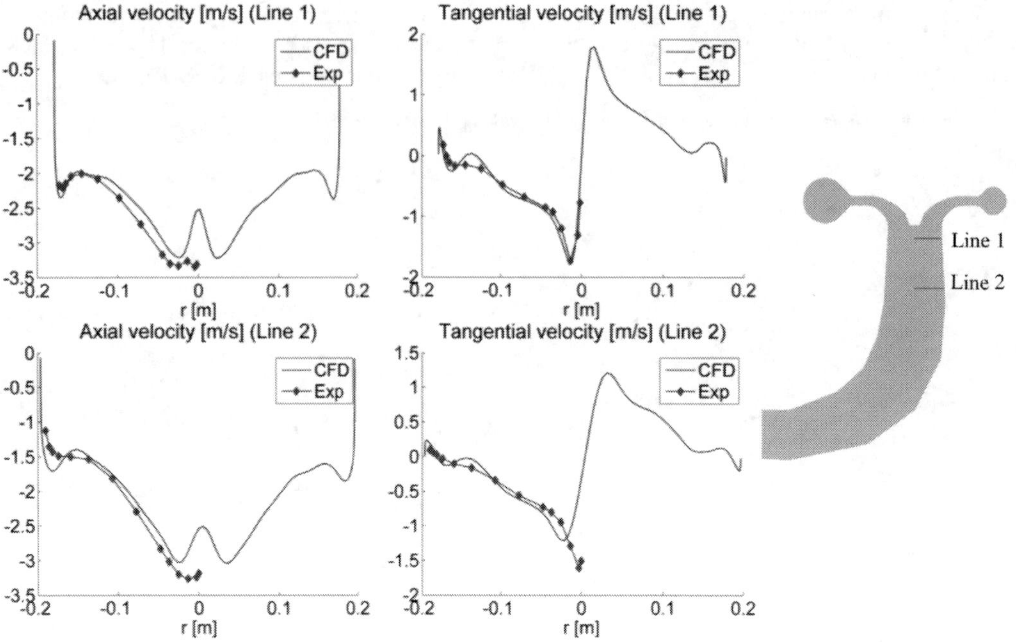

Figure 15. Averaged velocity profiles on lines 1 and 2 at HL.

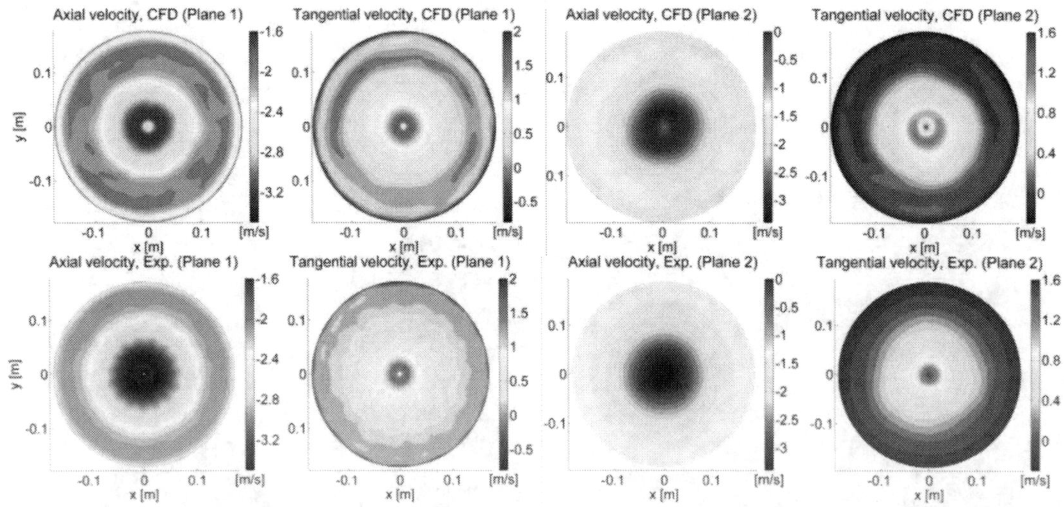

Figure 16. Velocity profiles (CFD on top, experimental on bottom) at HL.

In order to capture the draft tube fluctuations associated with the vortex rope, an additional simulation was performed at PL. As shown in the left part of figure 17, the SST model failed to reproduce the expected pulsating behavior of the vortex rope whereas the SAS-SST was able to reproduce some of the dynamic content. This model is well known to allow unsteadiness to develop [6]. The right part of figure 17 presents a comparison between the experimental and numerical FFT analysis. The amplitude and frequency of the draft tube sensors seem well reproduced. However, the amplitude of the experimental signal is larger on DT11 than on DT21 while it is the opposite in the numerical signal.

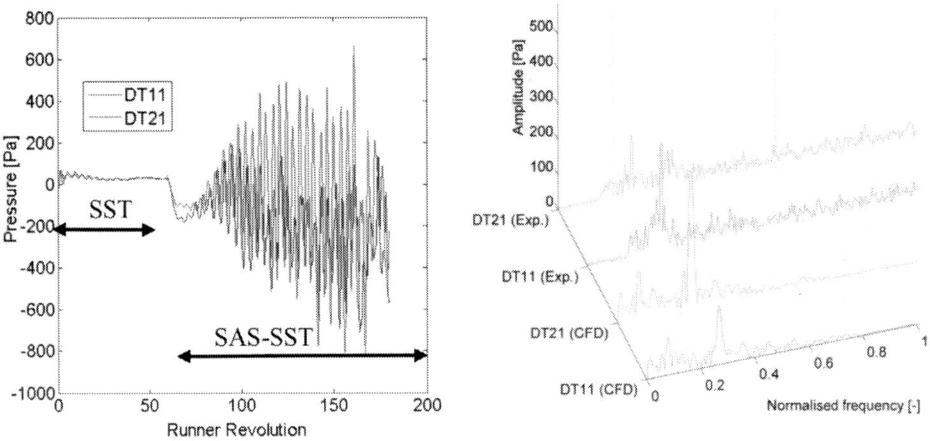

Figure 17. Left: pressure in the draft tube for the SST and SAS-SST model. Right: FFT of the signal.

Further analysis of the draft tube pressure signals in the draft tube is possible according to [7]. The synchronous component of the fluctuations (DT_{sync}) can be obtained by averaging DT11 and DT21. The rotating component (DT_{rot}) is obtained later by subtracting the synchronous component from the signal. Since there are only two points, the rotating component has the same magnitude on both sensors but is opposite in sign. The FFT (figure 18, left) of those signals shows that the synchronous and rotating components have the same amplitude. On the right of figure 18 is an illustration of the flow structures at part load highlighted with the Q-criterion for the SAS-SST simulation. Below the runner, a discernable vortex rope (shown in blue with iso-pressure) is present; however, after the flow reorientation in the elbow, this structure breaks down creating many smaller vortices.

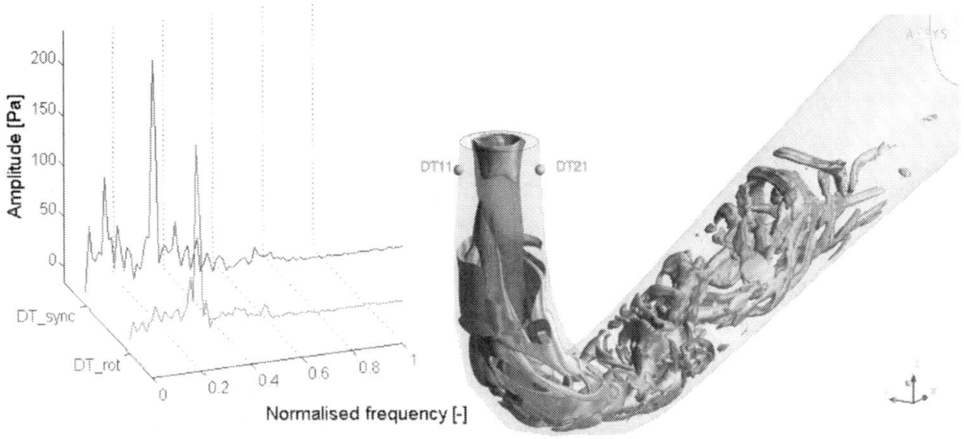

Figure 18. Left: FFT of the synchronous and rotating pressure fluctuations for the CFD signal. Right: Vortex structures identified by Q-criterion at part PL for the SAS-SST model.

6. Conclusion

In this paper, the challenges of obtaining the right pressure fluctuations were investigated, with potential impact on the turbine's life expectancy in mind. One of the inherent difficulties in reproducing the complete dynamic behavior of a hydraulic turbine is the numerical cost of capturing fluctuations happening on different timescales.

It is always good practice to perform some steady simulations before using heavier unsteady models. Those simulations showed that the power and efficiency of the machine were consistently over predicted, indicating that something was probably missing in the simulation. Sensitivity to near-wall refinement led to the use of low Reynolds meshes together with the SST turbulence model.

While expensive in terms of numerical cost, unsteady simulations of the complete turbine did reproduce the fluctuations in the runner and the distributor well. A separate model of the draft tube along with the SAS-SST results was used to complement the results at PL. The frequency of the vortex rope was captured.

The profile transform and Fourier transform approaches were tentatively used as a quick way to obtain RSI fluctuations. However, both the blade torque fluctuations and the dynamic content of the pressure according to FFT analysis were underestimated in the runner for the simplified models. Numerical diffusion due to the meshes or schemes does not seem responsible for this under prediction. Indeed, the same meshes and schemes were used in the full turbine simulation and provided more accurate results. Further investigations about the setup and models themselves are therefore necessary to explain the discrepancy and better define their useful range. Possible improvements to the FT method are also possible since it is still a beta feature.

Nomenclature and abbreviations

PL	Part load
BEP	Best efficiency point
HL	High load
CFD	Computational fluid dynamic
RSI	Rotor/stator interaction
SST	Shear Stress Transport
SAS	Scale Adaptive Simulation
FFT	Fast Fourier Transform
LDV	Laser Doppler velocimetry
TBR	Transient blade row
PT	Profile transform
FT	Fourier transform
BPF	Blade passing frequency
BL	Boundary layer
RS	Rotor/stator
DT	Draft tube

U	Velocity [m/s]
H	Head [m]
τ_w	Wall shear stress [N/m^2]
r	Density [kg/m^3]
C_u	k-Epsilon Turbulence model constant
k	Turbulent Kinetic energy [m^2/s^2]
Δy	Wall distance [m]
ν	Dynamic viscosity [m^2/s^3]
t	Time [s]
θ	Angular position [rad]
ω	Angular velocity [rad/s]
R	Radius
y+	Normalized wall distance [-]
n_{ed}	Normalized rotation speed [-]
Q_{ed}	Normalized flowrate [-]
T	Period [s]
ϕ	Flow variable
Pr	Angular span of the rotor [rad]
Ps	Angular span of the stator [rad]
A_m	Fourier coefficients

Journal of Physics: Conference Series **579** (2015) 012001 doi:10.1088/1742-6596/579/1/012001

References

[1] Francis-99 workshop internet site, 2014 [online]. Available at http://www.ltu.se/research/subjects/Stromningslara/Konferenser/Francis-99

[2] Chirag T., Cervantes M.J., Gandhi B.K. and Dahlhaug O.G., 2013. Experimental and numerical studies for a high head Francis turbine at several operating points. J of Fluid Engineering, November 2013, Vol. **135**

[3] Ansys CFX 15.0 Theory Guide, 2013

[4] Connel S., Hutchinson B., Galpin P., Campregher R. and Godin P., 2012, The efficient Computation of Transient Flow in Turbine Blade Rows Using Transformation Methods. ASME Turbo Expo 2012, doi:10.1115/GT2012-69019

[5] Sharma G., Zori L., Connell S. and Godin P., 2013, Efficient computation of large pitch ratio transonic flow in a fan with inlet distortion, ASME Turbo Expo 2013, doi:10.1115/GT2013-95059

[6] D'agostini Neto A., Jester-Zuerker R., Jung A. and Maiwald M., 2012, Evaluation of a Francis turbine draft tube flow at part load using hybrid RANS-LES turbulence modelling, IOP Conf. Series Earth and Environmental Science **15**, doi: 10.1088/1755-1315/15/6/062010

[7] Doerfler P. K. and Ruchonnet N., 2012, A statistical method for draft tube pressure pulsation analysis, IOP Conf. Series Earth and Environmental Science **15**, doi:10.1088/1755-1315/15/6/062002

Proper Orthogonal Decomposition of Pressure Fields in a Draft Tube Cone of the Francis (Tokke) Turbine Model

D Stefan[1] **and P Rudolf**

Brno University of Technology, Faculty of Mechanical Engineering, V. Kaplan Department of Fluid Engineering, Technická 2896/2, 616 69 Brno.

E-mail: david.steffan@gmail.com, rudolf@fme.vutbr.cz

Abstract. The simulations of high head Francis turbine model (Tokke) are performed for three operating conditions - Part Load, Best Efficiency Point (BEP) and Full Load using software Ansys Fluent R15 and alternatively OpenFOAM 2.2.2. For both solvers the simulations employ Realizable k-ε turbulence model. The unsteady pressure pulsations of pressure signal from two monitoring points situated in the draft tube cone and one behind the guide vanes are evaluated for all three operating conditions in order to compare frequencies and amplitudes with the experimental results. The computed velocity fields are compared with the experimental ones using LDA measurements in two locations situated in the draft tube cone. The proper orthogonal decomposition (POD) is applied on a longitudinal slice through the draft tube cone. The unsteady static pressure fields are decomposed and a spatio-temporal behavior of modes is correlated with amplitude-frequency results obtained from the pressure signal in monitoring points. The main application of POD is to describe which modes are related to an interaction between rotor (turbine runner) and stator (spiral casing and guide vanes) and cause dynamic flow behavior in the draft tube. The numerically computed efficiency is correlated with the experimental one in order to verify the simulation accuracy.

1. Introduction

The numerical simulations of hydraulic turbines are recently common for industrial companies. The main goal is to predict hydraulic efficiency of the machine for an entire range of operating conditions and thus to verify turbine hydraulic design. Nevertheless in the industry there are not usually enough computational resources and time to deal with complex calculations to verify turbulence models, discretisation schemes and perform time-consuming post-processing on large computational grids. Therefore the Francis-99 test case is a good option to deal with the post-processing methods e.g. proper orthogonal decomposition (POD) and compare it with the experimental results of the model Francis turbine [1]. POD offers complete spatio-temporal description of unsteady flows and it is especially suited to flows with strong coherent structures. The model turbine of specific speed $n_s = 3.65nQ^{0.5}H^{0.75} = 86.03$ with the runner composes of 15 splitters and 15 full length blades ($D_{runner} = 0.349$ m, H = 12 m) is a scaled down (1:5.1) 110 MW prototype turbine ($D_{runner} = 1.779$ m, H = 377 m) operated at Tokke power plant in Telemark region of south Norway. The scheme of experimental test rig is shown in figure 1.

[1] Corresponding author

The numerical calculations are carried out using both commercial software Ansys Fluent R15 and open source software OpenFOAM 2.2.2 for three operating points - Part Laod (Q = 0.071 m^3s^{-1}), BEP (Q = 0.203 m^3s^{-1}) and High Load (Q = 0.221 m^3s^{-1}).

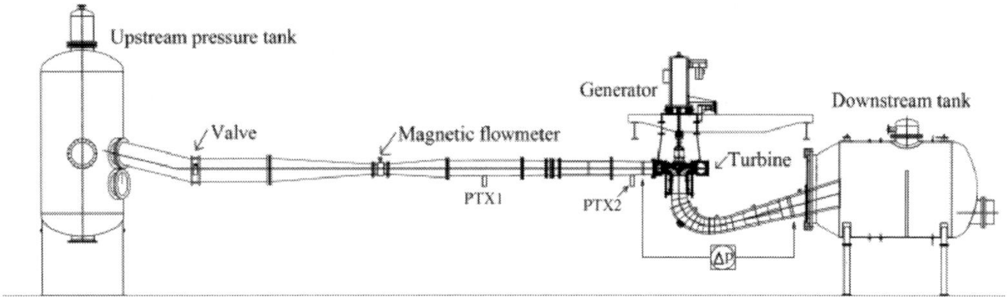

Figure 1. Experimental test rig of Tokke Franciss turbine model [2]

2. Computational study

In order to perform numerical calculations, both commercial and open source software were chosen with opportunity to test code performance. The turbine geometry and meshes are prepared in ICEM software and provided by the workshop organizers through the website [2]. Only small necessary corrections were made to enhance quality of the interfaces between the spiral case and turbine runner. The meshes contain two stationary (spiral casing and draft tube) and one rotational (runner) parts. The mesh for each operating point includes approximately 12.5 million of elements. All computations ran employing linux cluster partitioned into 16 cores in case of Ansys Fluent and into 64 cores in case of OpenFOAM software. The unsteady solutions were performed for all three operating points until the periodic flow state was achieved, then the time averaged flow fields and pressure monitors were recorded. The geometrical view of computational domain is presented in figure 2.

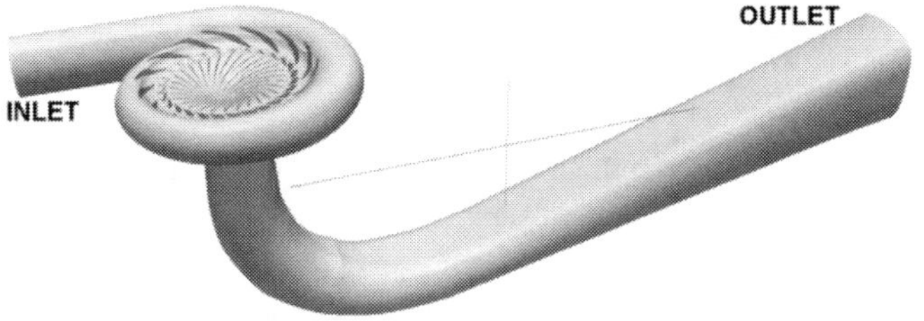

Figure 2. Computational domain geometry

2.1. Computation using Ansys Fluent R15

Using commercial software brings relatively easy case set-up due to proved setting options and graphic user interface. On the other hand user is restricted by a number of available licences and inaccessible code modifications. The Ansys Fluent was chosen because it is not as widely spread in turbo-machinery computations as for example CFX and also due to authors' previous experiences.

The case set-up of Ansys Fluent calculation presented in table 1 is identical for all three operating regimes. The material properties and operating conditions (density, kinematic viscosity, runner speed and flow rate) were defined according to the measurements [2]. The time-step size was chosen to correspond with 1° of the runner rotation.

Journal of Physics: Conference Series **579** (2015) 012002 doi:10.1088/1742-6596/579/1/012002

Table 1. Ansys Fluent case set-up

Turbulence Model	Realizable k-ε [3]
Near-Wall Treatment	Non-Equilibrium Wall Function
Pressure-Velocity Coupling	SIMPLE/PISO
Spatial Discretization	
Gradient	Least Square Cell Based
Pressure	Second Order Upwind/PRESTO!
Momentum	Second Order Upwind
Turbulent Kinetic Energy	Second Order Upwind
Turbulent Dissipation Rate	Second Order Upwind
Transient Formulation	Second Order Implicit
Boundary conditions	
Inlet	Velocity inlet – 5% of turbulence intensity
Outlet	Pressure outlet – 10% of turbulence intensity The value of pressure computed from measured inlet pressure and differential pressure between inlet and outlet
Time-step Size	
Part Load	4e-4 s
BEP	5e-4 s
High Load	4.5e-4 s

2.2. Computation using OpenFOAM

The open source software OpenFOAM has become arguably the best option for the license-free calculations. We have used OpenFOAM in version 2.2.2 and incompressible pimpleDyMFoam solver with implemented AMI interfaces between the mesh parts. The calculations are carried out using identical meshes as in case of Ansys Fluent which were transformed to the OpenFOAM format via fluent3DMeshToFoam utility.

The case set-up for calculations using OpenFOAM is similar with Fluent except particular differences in the spatio-temporal discretization, see table 2. At the outlet of the domain the value of pressure boundary condition was set to zero considering that the OpenFOAM use kinematic pressure in form $^{p}/_{\rho}$.

Table 2. OpenFOAM case set-up

Turbulence Model	Realizable k-ε [3]
Near-Wall Treatment	kqRWallFunction/epsilonWallFunction/nutkWallFunctions
Pressure-Velocity Coupling	PIMPLE
fvSchemes	
ddtSchemes	Euler
gradSchemes	Gauss linear
divSchemes	Gauss upwind (U,k,epsilon), Gauss linear
laplacianSchemes	Gauss linear corrected
interpolationSchemes	linear
snGradSchemes	limited corrected 0.333
Boundary conditions	
Inlet	Velocity – fixedValue, value corresponding to flow rate Pressure – zeroGradient Turbulent Kinetic Energy – fixedValue, value corresponding to 5% of turbulence intensity Turbulent Dissipation Rate – fixedValue, value corresponding to 5% of turbulence intensity
Outlet	Velocity – zeroGradient Pressure – FixedValue, Value 0 Turbulent Kinetic Energy – zeroGradient Turbulent Dissipation Rate – zeroGradient
Time-step Size	1.5e-5 s (for Co = 1)

3. Velocity profiles in the draft tube

The experimental velocity profiles measured in two lines situated downstream in the draft tube cone were provided in order to verify numerically computed flow exiting the turbine runner [2]. The locations of LDA measuring lines are shown in figure 3. The locations are 64mm (LDA-1) and 382mm (LDA-2) below the draft tube inlet.

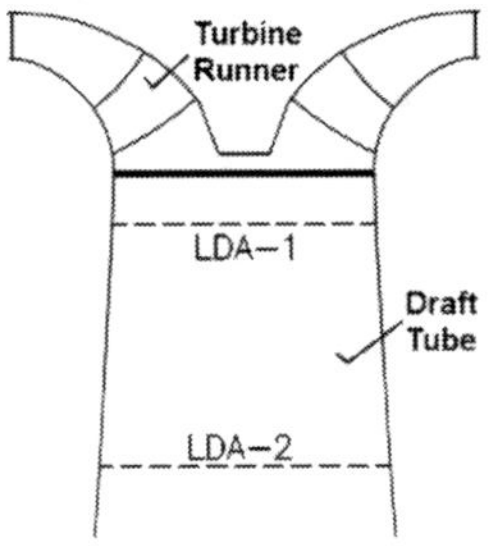

Figure 3. Locations of LDA measuring lines

In figures 4-6 the numerically computed axial and tangential profiles are compared with the experimental measurements. At Part Load the OpenFOAM fails in prediction of tangential velocity and large discrepancy in the axial velocity can be observed in the wall region, see figure 4. The tangential velocity computed using Fluent is in better agreement but magnitude is underestimated, even more in the LDA-2 location, see figure 4b. At BEP and High Load the largest discrepancies can be observed close to the draft tube axis, where especially in LDA-2 location the magnitude of numerically computed axial velocity is significantly underestimated. The velocity profiles at BEP and High Load computed using OpenFOAM show very good agreement for axial velocity apart the area close to the centre of the draft tube cone where the computed results underestimate experimental ones. On the other hand tangential velocity profiles computed using OpenFOAM have better agreement with the experimental one than in case of Ansys Fluent calculation. Analysis shows that axial velocity profiles are much better predicted than the tangential ones for both Fluent and OpenFOAM. While in the wall region the inaccuracy between measurements and computations is probably caused by the near wall modelling, the significant variation in the central region (especially at BEP and FullLoad) is attributed to both inaccuracy of LDA measurements (see figure 6b, where the measured tangential velocity in the axis suppose to be zero) and inaccuracy in turbulence models (large discrepancy between Fluent and OpenFOAM).

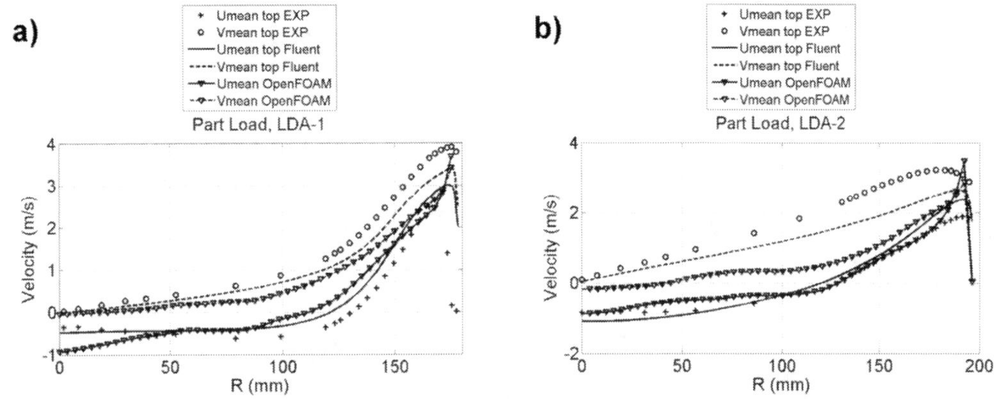

Figure 4. Profiles of axial (Umean) and tangential (Vmean) velocities in LDA-1 (a) and LDA-2 (b) locations at Part Load

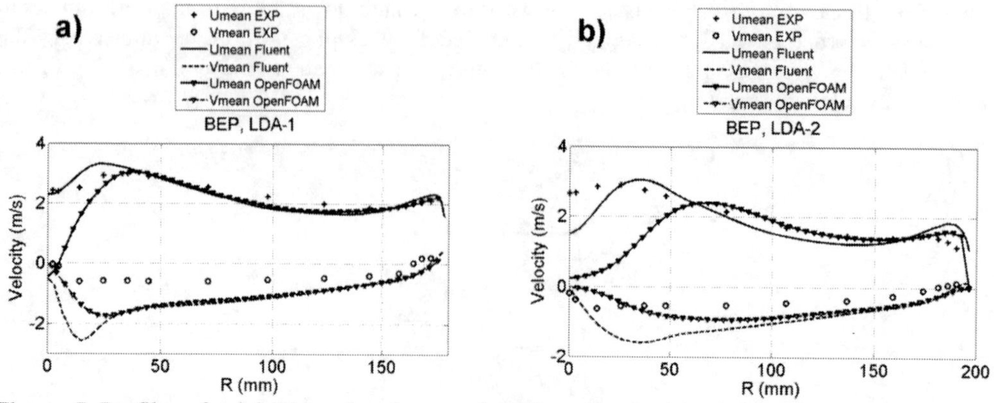

Figure 5. Profiles of axial (Umean) and tangential (Vmean) velocities in LDA-1 (a) and LDA-2 (b) locations at BEP

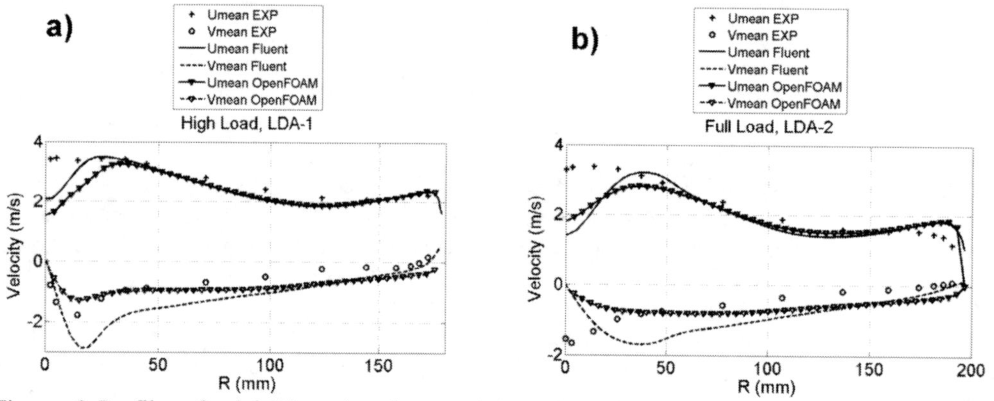

Figure 6. Profiles of axial (Umean) and tangential (Vmean) velocities in LDA-1 (a) and LDA-2 (b) locations at High Load

It has to be mentioned, that slight difference between measured and computed profiles at BEP and Full Load can arise due to different absolute values of the head, flow rate and runner speed used for the measurements in order to suppress vibrations induced in the test rig [2].

4. Pressure pulsations

The numerical unsteady pressure pulsations were evaluated in a three different probe locations. The first probe (VL01) is situated in the gap between the spiral distributor and the turbine runner, the second probe (DT11) and the third probe (DT21) are situated in the cone of the draft tube. The probe locations correspond with the locations of pressure transducers during the experiment, see figure 7.

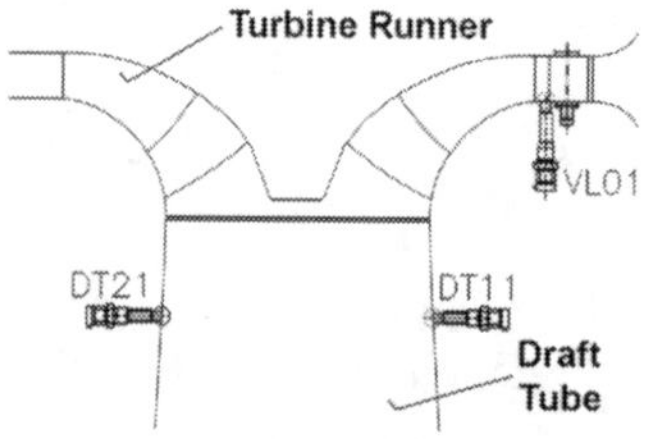

Figure 7. Location of pressure probes

First, the average values of computed pressure are estimated and compared with the experiment, see table 3. The numerical results are presented in a form of percentage deviation from the experimental measurements. The largest discrepancy between computed and measured pressure was found at High Load for DT11 probe where the value of numerical pressure underestimates the experimental one by 7.7 % (Fluent) and 8.3 % (OpneFOAM). The best agreement between the computed and measured pressure is at BEP for OpenFOAM where the numerically computed pressure in DT21 probe underestimate the experimental one just by 0.3%.

Table 3. Average pressure in DT11, DT21 and VL01 probes

Probe	Part Load			BEP			High Load		
	EXP (Pa)	Deviation Fluent (%)	Deviation OF* (%)	EXP (Pa)	Deviation Fluent (%)	Deviation OF* (%)	EXP (Pa)	Deviation Fluent (%)	Deviation OF* (%)
DT11	101603	-2.6	-2.2	102779	-2.5	-0.8	101651	-7.7	-8.3
DT21	101389	-2.3	-2.4	102284	-2.1	-0.3	100780	-6.7	-7.4
VL01	175408	4.6	5.6	171644	6.2	5.9	182636	-1.3	-4.9

*OpenFOAM

The amplitude-frequency spectra are carried out using general FFT function in MATLAB software for all three operating points. The frequencies are normalized assuming the runner angular speed, see table 4. The runner has 15 full length and 15 splitter blades, thus the normalized frequency, corresponding with the blade passing frequency is $f_n = 30$ with its higher harmonics $f_n = 60, 90, \ldots$ Due to influence of splitter blades the normalized frequency $f_n = 15$ also appears.

The frequency spectra of DT11 and VL01 probes are presented in figures 8, 10 and 12 and figures 9, 11 and 13 respectively. The spectra of measured signals can be compared with the spectra of computed signal of both Ansys Fluent and OpenFOAM (please note different scales of vertical axes in all plots). The best agreement for VL01 probe was found at High Load (figure 13) where the amplitude magnitudes and normalized frequencies of both numerical spectra well correspond with the measured one. On the other hand, the largest disagreements in amplitude magnitudes were found for DT11 probe (figures 8b, 10b and 12b) in case of numerical simulations using Ansys Fluent where the high frequency noise can be observed. It is clear that the source of this noise arises from the numerical calculation and we think that the reason can be influence of the mesh interfaces interaction produced downstream to the draft tube.

Furthermore, the particular discrepancies in the amplitude magnitudes can arise from the different lengths of pressure signal and different sampling frequencies used for the measurements and computations.

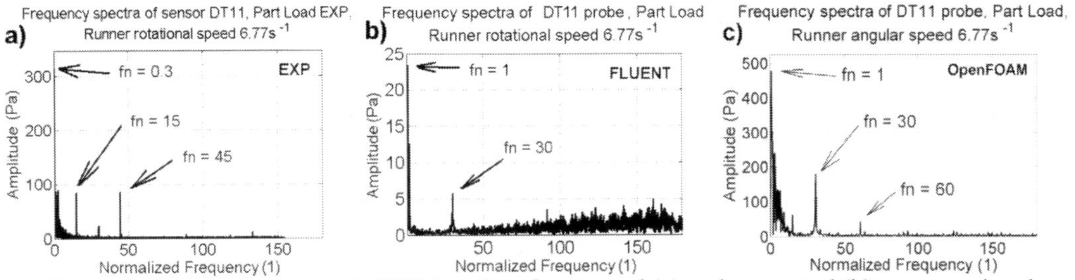

Figure 8. Frequency spectra in DT11 probe of measured (a) and computed (b) pressure signal, Part Load Q = 0.071m^{3}s^{-1}

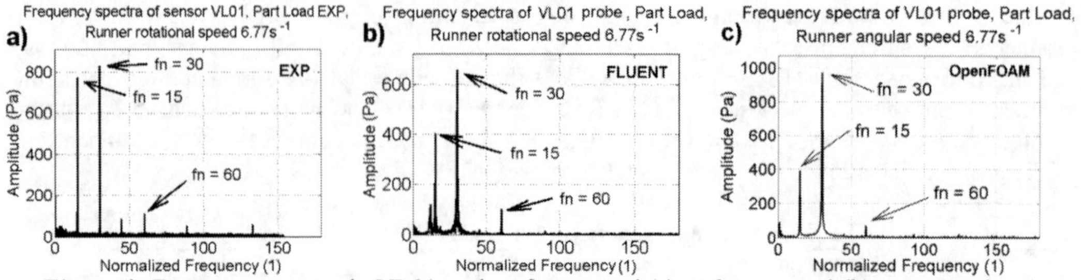

Figure 9. Frequency spectra in VL01 probe of measured (a) and computed (b) pressure signal, Part Load $Q = 0.071 m^3s^{-1}$

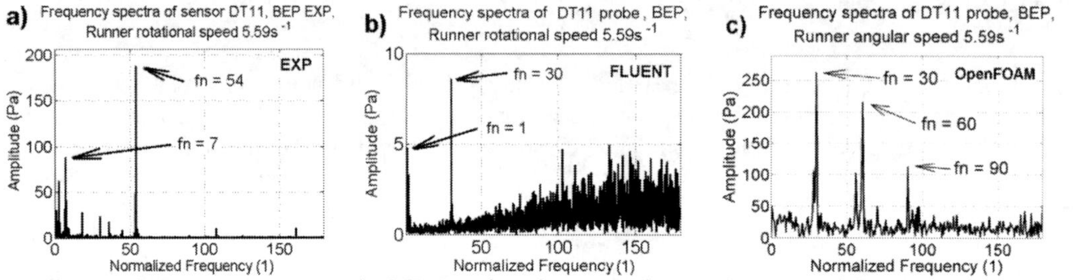

Figure 10. Frequency spectra in DT11 probe of measured (a) and computed (b) pressure signal, BEP $Q = 0.203 m^3s^{-1}$

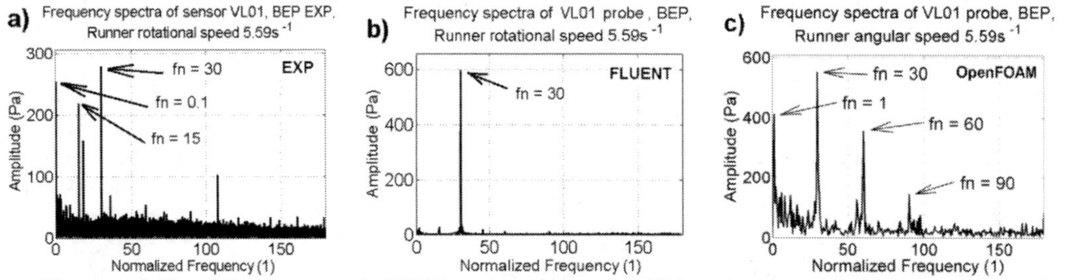

Figure 11. Frequency spectra in VL01 probe of measured (a) and computed (b) pressure signal, BEP $Q = 0.203 m^3s^{-1}$

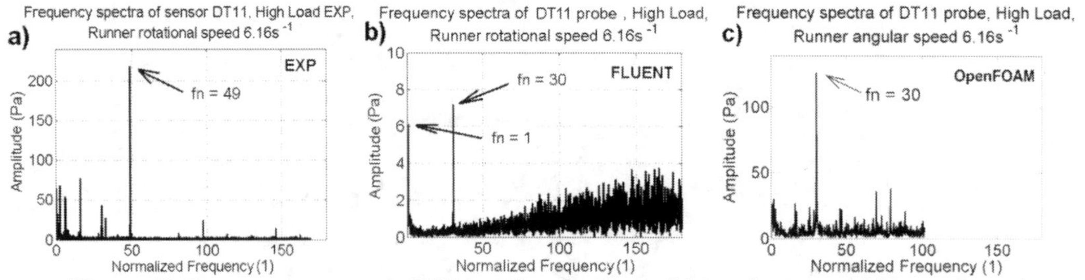

Figure 12. Frequency spectra in DT11 probe of measured (a) and computed Fluent (b) and OpenFOAM (c) pressure signals at High Load $Q = 0.221 m^3s^{-1}$

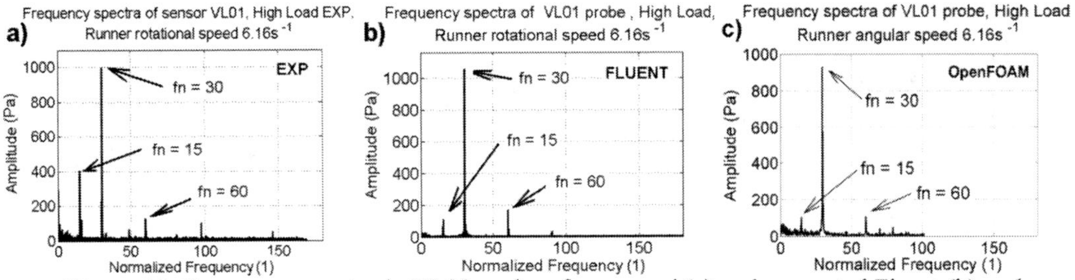

Figure 13. Frequency spectra in VL01 probe of measured (a) and computed Fluent (b) and OpenFOAM (c) pressure signals at High Load $Q = 0.221 m^3s^{-1}$

5. Integral energy characteristics

The calculated (Fluent/OpenFOAM) and measured (EXP) turbine parameters are summarized in table 4. Initially the Second Order Upwind discretization scheme for the pressure was chosen in Ansys Fluent. In order to compute hydraulic efficiency (1) the integral values of pressure and velocity are calculated using mass-weighted integrals (7) over the inlet and outlet of the domain.

$$\eta_h = \frac{T \cdot \omega}{\rho \cdot Q \cdot g \cdot H} = \frac{(T_p + T_\mu) \cdot 2 \cdot \pi \cdot n}{Q \cdot (\Delta p_s + \Delta p_d)} \quad (1)$$

The T_p and T_μ is pressure and friction torque respectively, n is the runner angular speed, Q is the flow rate and Δp_s and Δp_d is differential static pressure and differential dynamic pressure respectively. In case of OpenFOAM the integral values of pressure and velocity were calculated using patchAverage utility on corresponding patches.

Table 4. Computed and measured turbine parameters

	Part Load			BEP			High Load		
	EXP	Fluent	OF*	EXP	Fluent	OF*	EXP	Fluent	OF*
Net Head H (m)	12.29	14.16	15.6	11.91	14.04	14.6	11.84	12.37	12.17
Flow Rate Q (m^3s^{-1})	0.071	0.071	0.071	0.203	0.203	0.203	0.221	0.221	0.221
Runner Angular Speed n (s^{-1})	6.77	6.77	6.77	5.59	5.59	5.59	6.16	6.16	6.16
Torque T (Nm)	144.06	188.8	253.5	628.41	733	767.7	605.62	639.3	683.8
Differential Pressure Δp (kPa)	120.394	138.6	152.3	114.978	135.3	141.2	114.033	118.4	116.5
Hydraulic Efficiency η (%)	71.69	81.41	99.5	92.61	92.02	92.41	90.66	92.3	100.3

*OpenFOAM

The best agreement between computed and measured hydraulic efficiency is at BEP. While the Fluent calculated hydraulic efficiency with 0.64% error compare to the experiment the OpenFOAM was more accurate with error only 0.22%. However the significant discrepancies were obtained for off-design conditions at Part Load and High Load. At Part Load the Fluent calculated hydraulic efficiency with 11.87% error compare to the experiment and at High Load with 1.75% error. Moreover the hydraulic efficiency calculated using OpenFOAM shows irrational results of hydraulic efficiency $\eta = 99.5\%$ for Part Load and 100.3% for High Load. At High Load the head and differential pressure computed using OpenFOAM are in much better agreement with the experiment than the Fluent calculation. Nevertheless the runner torque is considerably overestimated which is here the main reason of nonsensical hydraulic efficiency. At Part Load the OpenFOAM wrongly computes both the runner torque and net head which is influenced by the inaccuracy of tangential velocity profile at the runner outlet, see figure 4. Considering the Euler equation ($gH = u_1 v_{u1} - u_2 v_{u2}$) the computed head respectively specific energy is higher than in case of measurements.

Due to discrepancy in results at Part Load and High Load the PRESTO! scheme for pressure discretization was tested for Ansys Fluent calculations. The PRESTO! scheme is generally recommended for highly swirling flows which are supposed to appear in the draft tube, when the Francis turbine is operated out of the best efficiency point. Nevertheless one can see that for both Part Load (12.66% error) and High Load (2.2% error) the PRESTO! scheme brings worse accuracy.

The comparison of computed and measured efficiency versus flow rate is presented in figure 20 for all calculations.

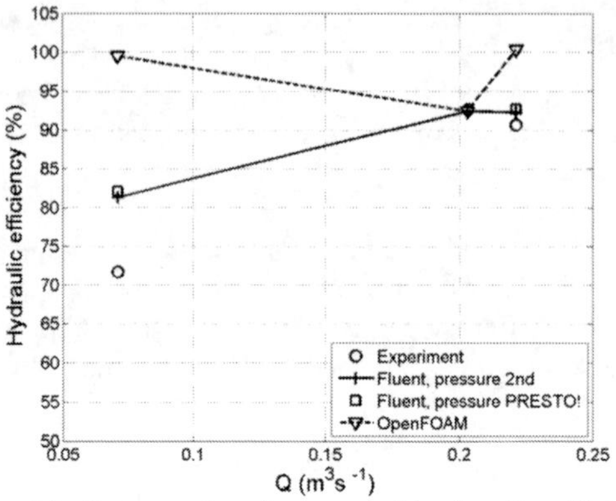

Figure 20. Computed and measured hydraulic efficiencies

6. Proper orthogonal decomposition

The proper orthogonal decomposition (POD) was introduced into the field of fluid mechanics by Lumley [4] as a technique which enables to bridge time and frequency domain and obtains spatio-temporal information about the flow in a form of spatio-temporal eigenfunctions. Comprehensive theoretical description of POD, including classical approach and its properties, can be found in [5]. The POD provides a basis for the modal decomposition of a set of functions and most efficient way of capturing the dominant components, i.e. coherent structures. The POD can be used to analyze experimental as well as numerical data being applied to scalar or vectorial functions. The scalar function $p(\boldsymbol{x};t)$, e.g. pressure field, is considered in following theoretical explanation. For instance, a set of data $p(\boldsymbol{x},t)$ as a function of physical space $\boldsymbol{x}$ and time t, can be expressed using POD as a set of orthogonal spatial basis functions $\phi_i^k(\boldsymbol{x})$ (i.e. spatial modes), and temporal functions $a_k(t)$ (i.e. temporal modes), respectively. Where $i = 1,2,\ldots,N$, N is the number of grid points and $k = 1,2,\ldots,M$, M the number of snapshots. Accordingly, the approximation of the data set onto the first k snapshots can be written in terms of the spatial and temporal functions as follow

$$\hat{u}_i(\boldsymbol{x},t) = \sum_{k=1}^{K} a_k(t)\,\phi_i^k(\boldsymbol{x}) \quad (1)$$

where $K < M$ has the largest mean square projection [6].

In the method of snapshots suggested by Sirovich [7], the general N x N eigenvalue problem is reduced to M x M eigenvalue problem. Especially, this solution brings a substantial reduction of computational effort if the number of grid points N significantly exceeds number of data set snapshots. Numerical computation is an appropriate example to use snapshot method. The inner product between every pair of pressure fields, i.e. snapshots, is the temporal correlation matrix $C(t,t')$ used as the kernel and defined by eq.

$$C(t,t') = \frac{1}{M}\int_{\Omega} p(\boldsymbol{x},t), p(\boldsymbol{x},t')\,d\boldsymbol{x} \quad (2)$$

Temporal modes $a_k(t)$ are the eigenvectors of matrix $C(t,t')$ being obtained by solving the eigenvalue problem of the form (3):

$$\int_T C(t,t')\,a_m(t')dt' = \lambda_m a_m(t) \quad (3)$$

The spatial modes $\phi_i^k(x)$ are calculated based on orthogonal property:

$$\phi_i^k(\boldsymbol{x}) = \sum_{i=1}^{M} a_k(t) p_i(\boldsymbol{x}, t) \tag{4}$$

Particularly, the flow structure is fully described using spatio-temporal representation of modes $a_k(t)$, $\phi_i^k(\boldsymbol{x})$. The correlation matrix $C(t,t')$ is symmetric and positive definite. Thus the eigenvalues are real and define the contribution of the coherent structure into complex flow [8]. Correspondingly, the eigenvalue magnitude of one single mode represents an amount of this contribution from the total sum $\int_\Omega \langle p_i(x) p_i(x) \rangle dx$ [7]. Finally, the computed modes are sorted out taking into account the largest value of the eigenvalue magnitude in order to identify the most significant structures. As a result, the temporal modes $a_k(t)$ are calculated by projecting the data function $p(\boldsymbol{x};t)$ on the eigenfunction $\phi_i^k(\boldsymbol{x})$.

$$a_k(t) = \sum_{i=1}^{M} p_i\,(\boldsymbol{x}, t)\,\phi_i^k\,(\boldsymbol{x}) \tag{5}$$

Throughout, the decomposed function $p(\boldsymbol{x};t)$ is considered in dimensionless form in order to compare the values associated to different investigated cases. The dimensionless form of the pressure field is done as follows

$$p(\boldsymbol{x};t) \equiv \frac{\Delta p(x,t)}{\frac{1}{2}\rho v_{ref}^2} \tag{6}$$

where $\Delta p(\boldsymbol{x},t) = p(\boldsymbol{x},t) - \overline{p_{ref}}$, $p(\boldsymbol{x},t)$ is the static pressure field obtained in numerical simulation at each time step *t*, $\overline{p_{ref}}$ static pressure mean value on the outlet surface of the domain (draft tube outlet) computed as a mass-weighted surface integral in form:

$$\overline{p_{ref}} = \frac{1}{Q_m} \int_0^{Q_m} p\, dQ_m \tag{7}$$

and v_{ref} corresponds to bulk velocity at the draft tube inlet.

7. POD of pressure fields

The successful application of the POD for investigation of the vortex rope dynamic in the draft tube of the hydraulic turbine can be found in [9] and [10]. The previous investigations were carried out for a constrained domain of the simplified draft tube. In this paper the entire turbine domain ensure full dynamical influences, primarily the rotor-stator interaction. The POD analysis is applied on the numerical data from the Ansys Fluent calculations for all three operating points. The node values of the static pressure are extracted in the longitudinal slice along the draft tube cone, starts180 mm and ends 638 mm below the dividing plane of the spiral casing. The pressure data are stored with the specified number of constant time-step intervals. Then the POD is executed using script written in Matlab software where the eigevalues and eigenvectors are calculated employing the *eig* function. The resulting spatio-temporal representations of the static pressure modes can be correlated with the dynamic behaviour extracted from the pressure monitoring points. Thus we are able to decide which mode is consequence of the rotor-stator dynamic and which is caused by the self-induced instability in the draft tube. It is necessary to say that when the significant coherent structure (e.g. vortex rope) does not appear in the draft tube (as is case of these operating points), the resulting modes have very small eigenvalue magnitude compare to the mean flow, see table 5.

Table 5. Eigenvalue magnitudes compare to the mean flow

	mean flow	1st mode	2nd mode	3rd mode	4th mode	5th mode
Part Load	99.9691%	0.0182%	0.0040%	0.0032%	0.0014%	0.0012%
BEP	99.9882%	0.0062%	0.0021%	0.0019%	0.0005%	0.0004%
High Load	99.9881%	0.0041%	0.0017%	0.0016%	0.0009%	0.0008%

In the following subsections the spatio-temporal representations of the first five modes for each operating point are presented and their relation to the turbine regime is discussed. The value of static pressure is normalized according to the equation (6). For all three operating points no significant vortex rope leading to the strong modes of self-induced instability appears [9]. Therefore in the present analysis we will distinguish two kinds of the static pressure modes. The modes related to the forced instability with the frequency at ratio of the runner rotational frequency and modes related to the rotor-stator interaction at the blade passing frequency.

7.1. Part Load Q=0.071 m^3s^{-1}

In figure 14a the normalized mean static pressure p* at Part Load operating point is shown. At this operating point the main flow is realized close to the cone wall with the strong swirl exiting turbine runner. The first mode of static pressure at Part Load (figure 14b) has very low normalized frequency $f_n = 0.1$, see figure 15. This mode together with fourth mode (figure 14e) $f_n = 1$ is consequence of forced instability. The second, third and fifth modes (figure 14c, d, f) are related to the rotor-stator interaction with normalized frequencies correspond to the blade passing frequency $f_n = 30$ and its higher harmonics $f_n = 60$. It is important to notice, that the third mode with $f_n = 15$ is a consequence of the splitter blades.

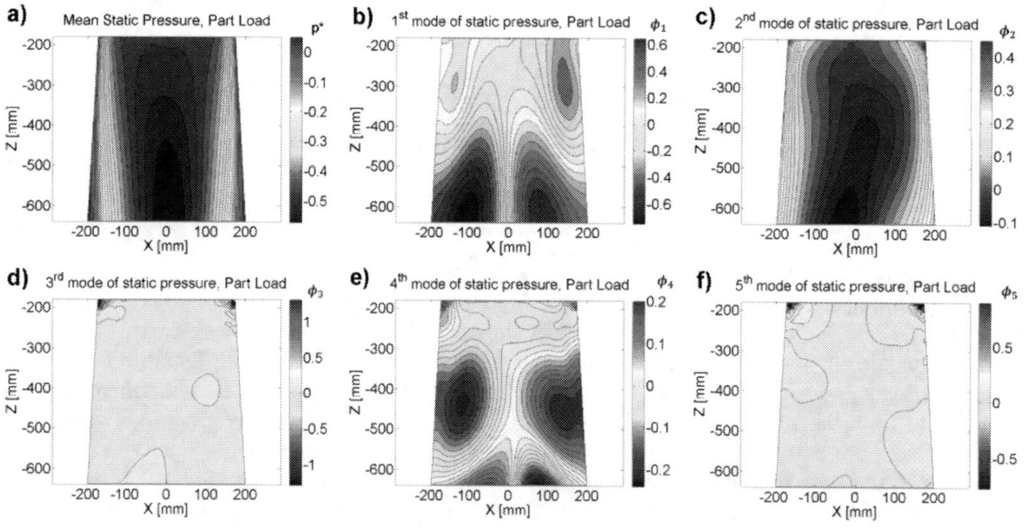

Figure 14. Mean static pressure (a) and spatial representation of the first five modes (b-f)

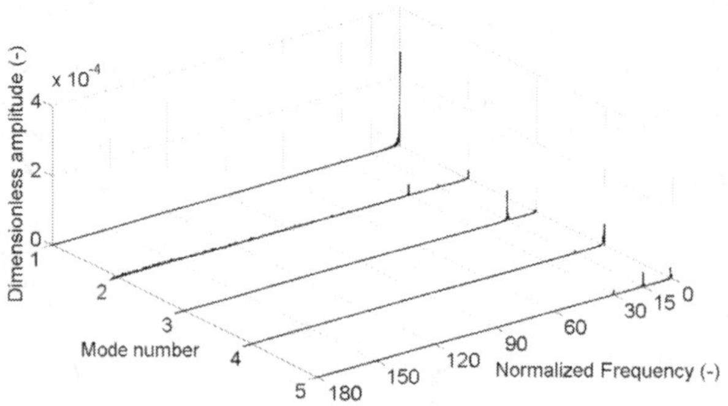

Figure 15. Amplitude-frequency spectra of the first five temporal modes, Part Load

7.2. BEP Q=0.203 m³s⁻¹

In the figure 16a the contours of normalized mean static pressure p* at BEP operating point are presented. Similarly as at Part Load two main kinds of the modes appear. The modes related with the rotor-stator interaction (figure 16b, c, d) have main normalized frequency $f_n = 30$ (the first mode) and $f_n = 15/30$ (the second and the third mode), see figure 17a. The fourth and the fifth mode (figure 16e, f) represent forced instability with normalized frequency $f_n = 1$ caused by the swirl at the runner rotational speed. The second and third modes have $f_n = 30$ and $f_n = 15$ with the difference that in case of the second mode the higher normalized frequency is $f_n = 15$ and in case of the third mode it is $f_n = 30$. It is clear that the pressure field is influenced differently whether the guide vanes interact with the full length blade or the splitter blade. The fourth and the fifth mode caused by the forced instability have same order of magnitude but they are phase shifted by 0.27 radians, see figure 17b.

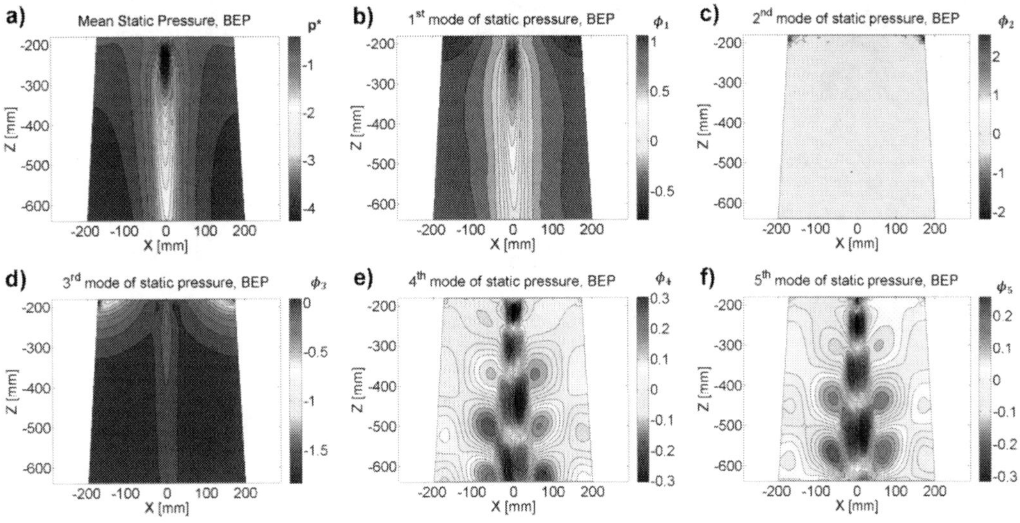

Figure 16. Mean static pressure (a) and spatial representation of the first five modes (b-f)

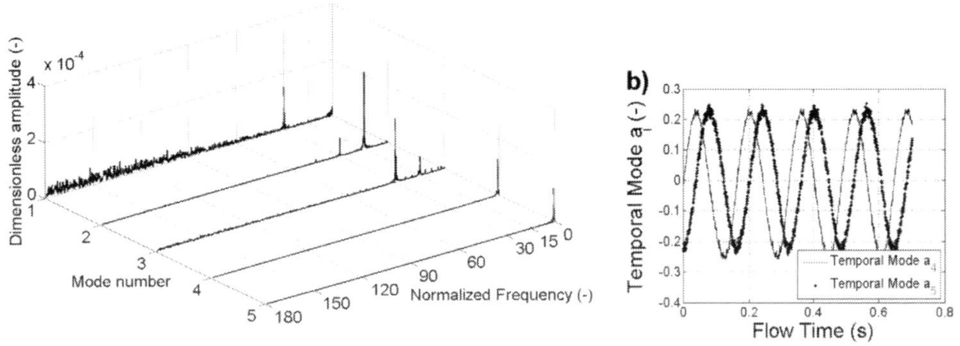

Figure 17. Amplitude-frequency temporal modes (a) and phase shift between the fourth and the fifth temporal modes (b)

7.3. High Load Q=0.221 m³s⁻¹

The contours of normalized mean static pressure p* at High Load (figure 18a) are very similar to ones at BEP operating point. In case of High Load the first mode (figure 18b) has two dominant frequencies, see figure 19. It is clear that the normalized frequency $f_n = 30$ is response of the rotor-stator interaction and $f_n = 2$ corresponds to the forced instability. The second and the third mode (figure 18c, d) with normalized frequency $f_n = 2$ and the fifth mode (figure 18f) with $f_n = 0.2$ are consequence of the forced instability forming unstable swirl rotating in the draft tube cone two times faster (the second and third mode) respectively five times slower (the fifth mode) than the turbine runner. The fourth mode (figure 18e) has dominant normalized frequencies $f_n = 15$ and $f_n = 30$ and spatial representation shows correlation to the rotor-stator interaction combining full length blades with the splitter blades.

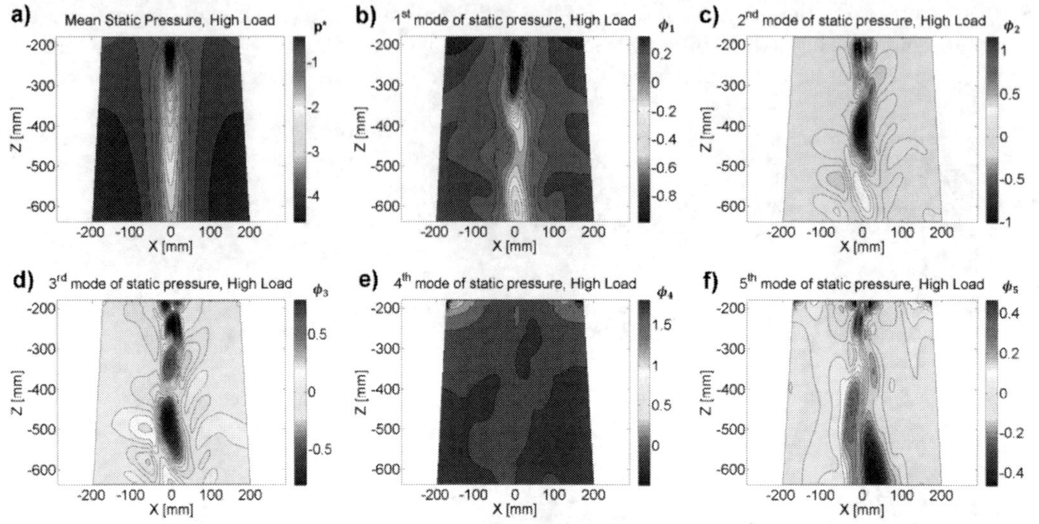

Figure 18. Mean static pressure (a) and spatial representation of the first five modes (b-f)

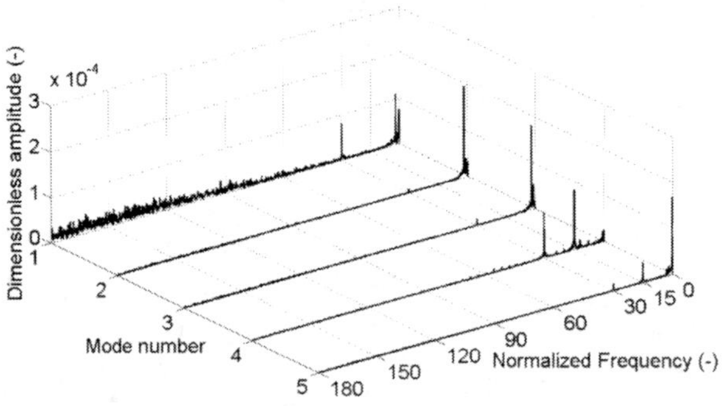

Figure 19. Amplitude-frequency spectra of the first five temporal modes, High Load

8. Result discussion

At the BEP operating point one can see that the computed head using both Ansys Fluent and OpenFOAM is considerably overestimated. The cause lays in a wrongly computed tangential velocity profiles at the turbine outlet where the velocity magnitudes overestimate experimental measurements. Nevertheless the computed runner torques are also overestimated, thus the resulting numerically computed hydraulic efficiencies adequately correspond with the experimental ones.

Different situation is for off-design conditions. While the Fluent is able to compute hydraulic efficiency with a relatively small error at High Load and with larger error at Part Load, the results obtained using the OpenFOAM are not credible. The main reason of OpenFOAM failure was found in a computation of the runner torque which is for both Part Load and High Load significantly overestimated. This supposes to be further investigation for the future reliable usage of the

OpenFOAM solver. The main difference between Fluent and OpenFOAM can be found in the different numerical schemes used in each solver and in the estimation of the runner torque.

The unsteady pressure analysis shows large disagreement between computed and measured amplitudes in a case of DT11 probes. The spectra of numerical signal from Ansys Fluent include the high frequency noise which is not presented in the experimental results. We suppose that this noise is a main cause of the amplitudes disagreement. Nevertheless [2] informs about vibrations induced in the test rig at BEP and High Load, which interfered with the LDA measurements at these conditions. Regarding to the pressure measurements we think that the high amplitudes at the measured frequency $f_n = 54$ (BEP) and $f_n = 49$ (High Load) in DT11 spectra could be potentially cause of these vibrations as well.

From the POD analysis at Part Load and High Load operating points it was found that some temporal modes related to the forced instability have more than only one dominant frequency of the vortex rotation, the normalized frequencies related to the rotor-stator interaction are also visible. That means the rotor-stator interaction is significant source of the pressure pulsations in the draft tube. Moreover in a case of the rotor-stator interaction some modes with normalized frequency $f_n = 15$ appear. These modes are influence of fifteen splitter blades. From the eigenvalue magnitude presented in table 4 it is clear that the Part Load is the most dynamic operating regime. Nevertheless when the strong coherent structure, in a form of the vortex rope, does not appears as it is case of these investigated turbine regimes, the dynamic behaviour of the pressure field in the draft tube is very low as can be seen in the percentage amount of the eigenvalue presented in table 5. The most dominant pressure modes were found modes related to the rotor-stator interaction and the modes caused by the forced instability in a relation to the runner rotation.

Due to the large discrepancy in the OpenFOAM results at Part Load and High Load, different numerical schemes were tested for the disretization of the time (backward) and the vector fields (Guass limitedLinear). It was found that the numerical schemes have large influence on the results accuracy especially on the runner torque calculation. Nonetheless using these schemes did not lead to more realistic value of the hydraulic efficiency at Part Load and High Load respectively.

Acknowledgement
This work is an output of cooperation between projects of Technology Agency of the Czech Republic No. TE02000232 "Special rotary machine engineering centre" and NETME Centre, regional R&D centre built with the financial support from the Operational Programme Research and Development for Innovations within the project NETME Centre (New Technologies for Mechanical Engineering), Reg. No. CZ.1.05/2.1.00/01.0002 and, in the follow-up sustainability stage, supported through NETME CENTRE PLUS (LO1202) by financial means from the Ministry of Education, Youth and Sports under the „National Sustainability Programme I"

References

[1] Chirag T, Cervantes M J, Gandhi B K and Dahlhaug O G 2013 *J. Fluids Eng.* **135** 111102-17
[2] http://www.ltu.se/research/subjects/Stromningslara/Konferenser/Francis-99
[3] Shih T-H, Liou W W, Shabbir A, Yang Z and Zhu J 1995 *Computers Fluids* **24** 3 227-238
[4] Berkooz G, Holmes P and Lumley J L 1993 *Annual Review of Fluid Mechanics* **25** 539-575
[5] Holmes P, Lumley J L, Berkooz G and Rowley C W 2012 *Turbulence, Coherent Structures, Dynamical Systems and Symmetry*, Second Edition (Cambridge University Press) p 402
[6] Grinberg L, Yakhot A and Karniadakis G E 2009 *Annals of Biomedical Engineering* **37** 17
[7] Sirovich L 1987 *Quarterly of Applied Mathematics* **XLV** 561-571
[8] Lumley J L 1970 *Stochastic Tools in Turbulence*, Academic Press.
[9] Rudolf P and Štefan D 2012 *IOP Conf. Series: Earth and Environ. Sci.* **15** 062008 1-8
[10] Rudolf P and Jízdný M 2011 *Proc. 4th IAHR International Meeting on Cavitation and Dynamic Problems in Hydraulic Machinery and Systems*, (Belgrade, Serbia) 131-141

Journal of Physics: Conference Series **579** (2015) 012003 doi:10.1088/1742-6596/579/1/012003

Numerical Study of a High Head Francis Turbine with Measurements from the Francis-99 Project

H. Wallimann and R. Neubauer

Andritz Hydro AG, Hardstrasse 319, 8021 Zürich, Switzerland

E-mail: Henrik.Wallimann@andritz.com

Abstract. For the Francis-99 project initiated by the Norwegian University of Science and Technology (NTNU, Norway) and the Luleå University of Technology (LTU, Sweden) numerical flow simulation has been performed and the results compared to experimentally obtained data. The full machine including spiral casing, stay vanes, guide vanes, runner and draft tube was simulated transient for three operating points defined by the Francis-99 organisers. Two sets of results were created with differing time steps. Additionally, a reduced domain was simulated in a stationary manner to create a complete cut along constant prototype head and constant prototype discharge. The efficiency values and shape of the curves have been investigated and compared to the experimental data. Special attention has been given to rotor stator interaction (RSI). Signals from several probes and their counterpart in the simulation have been processed to evaluate the pressure fluctuations occurring due to the RSI. The direct comparison of the hydraulic efficiency obtained by the full machine simulation compared to the experimental data showed no improvement when using a 1° time step compared to a coarser 2° time step. At the BEP the 2° time step even showed a slightly better result with an absolute deviation 1.08% compared with 1.24% for the 1° time step. At the other two operating points the simulation results were practically identical but fell short of predicting the measured values. The RSI evaluation was done using the results of the 2° time step simulation, which proved to be an adequate setting to reproduce pressure signals with peaks at the correct frequencies. The simulation results showed the highest amplitudes in the vaneless space at the BEP operating point at a location different from the probe measurements available. This implies that not only the radial distance, but the shape of the vaneless space influences the RSI.

1. Introduction

Computational fluid dynamics (CFD) is an established tool in industry for the design of Francis and other types of turbines. The history of numerical flow simulation is illustrated in [1]. Continuous effort is needed to improve and validate the CFD tools. The principle task of the CFD tools is to predict the hydraulic efficiency and pressure distribution as accurately as possible and capture flow phenomena of importance. The interaction between stationary and rotary parts is an important phenomena of the hydraulic system. It is a source of vibration that can lead to serious damage to the machine [2]. Predictions for low part load and high load away from the best efficiency points gain in importance due to the increasing demand to operate a turbine with flexible power output.

The Francis-99 project offers an opportunity to test the state of the art tools available today. It also allows assessing CFD setups that are not yet integrated in the day to day business as computational time and resources are often prohibitive, but might become part of the standard design procedure in the near future. The measurements provided by the organizers of the Francis-99 project exceed the data produced during a model test and allow a more detailed comparison between experimental and numerical values.

Similar studies have been performed in the past for pump turbines in turbine mode, for example in [3] and [4]. In these studies, the SST turbulence model was found to cope best with the pressure fluctuations due to RSI and was consequently used again for this study.

In this paper the results of CFD simulations of the full machine and a reduced domain are presented. The hydraulic efficiency is compared qualitatively and quantitatively. The results of the full machine simulations are compared with data from several pressure probes installed at the measured model and the pressure fluctuations due to RSI are discussed.

2. Experimental Model

The experimental values used for comparison and validation in this paper were obtained at the test rig at the Waterpower laboratory, NTNU in Norway. For the discussed operation points, the test rig was operated in an open loop manner with the draft tube connected to a downstream tank open to air. A more detailed description of the test rig and instrumentation can be found in [5]. The tested model was designed based on the prototype values of the Tokke power plant in Norway. The corresponding prototype values are H_P=377m, P_P=110MW, D_P=1.779m and Q_P=31m3/s.

The diameter of the model is D_M=0.349m (scale = 1:5.1). The model runner consists of 15 full length and 15 splitter blades. The number of guide vanes is 28. Experimentally the best efficiency point was found at a H_M=11.91, Q_M=0.203m^3/s and a runner speed of n_M=5.59rev/s. It amounts to η_M=92.61%. A list of the provided operating point data is shown in Table 1.

	flow rate [m^3/s]	net head [m]	guide vane angle [°]	Runner angular speed [rev/s]	Efficiency [%]
part load	0.071	12.29	3.91	6.77	71.69
BEP	0.203	11.91	9.84	5.59	92.61
high load	0.221	11.84	12.44	6.16	90.66

Table 1. Experimental operating point data for part load, BEP (best efficiency point) and full load

3. Numerical Model:

3.1. Full Machine

To numerically simulate the experimentally investigated model the commercial software Ansys CFX, solver version 14.0, was used. The numerical grid was created with ICEM CFD using the ICEM files offered by the Francis-99 organizers. The complete machine consisted of a grid combining spiral casing, stay vanes and guide vanes (different for each operating point), a second grid representing the runner and a third grid containing the draft tube. The number of nodes was 3.6mio, 5.9mio and 3.6mio respectively. This added up to a total of 13.1mio nodes. A grid sensitivity test performed in [5] for the BEP showed a similar number of nodes to be a reasonable resolution. Except for merging two separately provided runner parts (runner with blades and part just below the hub) and improvement of the geometric match of the interface between the guide vanes and the runner, no changes were made to the provided grids. Figure 1 shows a picture of the complete domain.

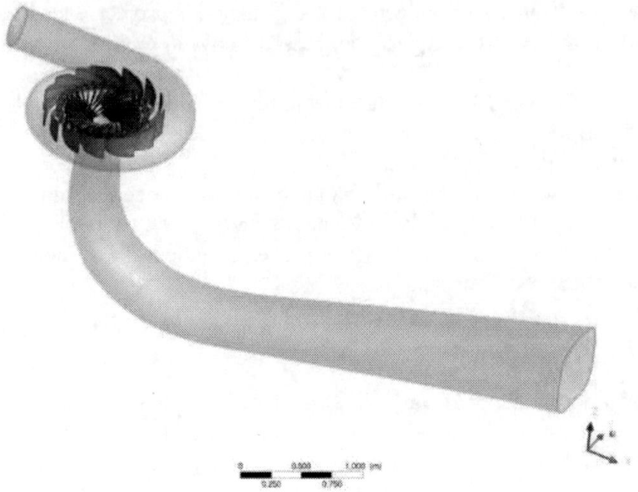

Figure 1. Computational domain of full machine including spiral casing, stay vanes, guide vanes, runner and draft tube

The shear stress transport (SST) model was chosen for turbulence modelling. The high resolution scheme was used as advection scheme and for the turbulence numerics first order was chosen. The transient scheme was set to second order backward Euler.

At the spiral casing inlet the discharge of the operating point was imposed as boundary condition. The draft tube outlet was modelled as outlet with an average pressure of 1bar, representing the open loop setup of the model test rig. The interfaces between guide vanes and runner as well as runner and draft tube were modelled as transient interfaces covering the full 360°. All walls were modelled as hydraulically smooth. A first set of three simulations was started without initial solution and run for five revolutions. The time step was set for each operating point individually, depending on the runner speed, to result in a step of 2° in respect to the runner rotation. A second set of simulations was run with a time step corresponding to a 1° angular step using the 2° results as initial condition and run for two more revolutions.

Monitor points were set to represent the pressure probes used during the experimental investigation. A Fortran routine was integrated to conduct a discrete Fourier transformation on the last revolution and store the amplitudes of the first three harmonics. For the rotating runner domain the first harmonic was chosen according to the guide vane passing frequency (28 per revolution) and for the stationary domains the first harmonic was chosen based on the number of blades with full length (15). Other numerical results as e.g. efficiency were recorded and averaged over the last revolution of the simulation.

Simulations were performed for the three given operation points; one in part load operation, one at the experimentally found best efficiency point and one at high load operation according to the provided grids.

3.2. Reduced Domain

A second numerical setup was used to evaluate the qualitative prediction of the hydraulic efficiency. The domain was reduced to the runner, simulating a single channel including one full length blade and one splitter blade as well as periodic surfaces by adjusting the provided ICEM CFD files. A discharge boundary condition was used and the operating point fully defined by setting an inflow angle representing the flow coming from the guide vanes. As for the full domain the SST model was used for turbulence modelling as well as first order for turbulence numerics. For the advection scheme a blend factor of 1.0 was used, enforcing second order accuracy.

Journal of Physics: Conference Series **579** (2015) 012003 doi:10.1088/1742-6596/579/1/012003

Steady state simulations were performed for the 3 given operating points as well as for a constant prototype H cut near H_P=386.9m and a constant Q cut Q_P=30.06m which correspond to the found BEP by the model test and transposed to prototype values by hydraulic similitude according to IEC 60193 [6].

4. Results

4.1. Hydraulic Efficiencies: Full Machine

Figure 2 shows the experimentally obtained hydraulic efficiency compared to the efficiencies calculated from the CFD simulation of the full machine. Although the values are plotted against model discharge, the three operating points have to be regarded as individual points on the hill chart and do not represent a cut along a constant head. The CFD value is the average taken over the last revolution. Generally the CFD predictions should be slightly too high as no leakage losses through the labyrinth gaps can occur and no disc friction is present. As can be seen from the figure, the BEP of the simulation with a 2° step matches best with a deviation of 1.08%. The 1° time step simulation results in a slightly higher efficiency, with a deviation of 1.24%. The high load point at a model discharge of 0.221m^3/s is similarly over predicted by both settings (+3.6%) as well as the efficiency at the part load operating point (+11.0%).

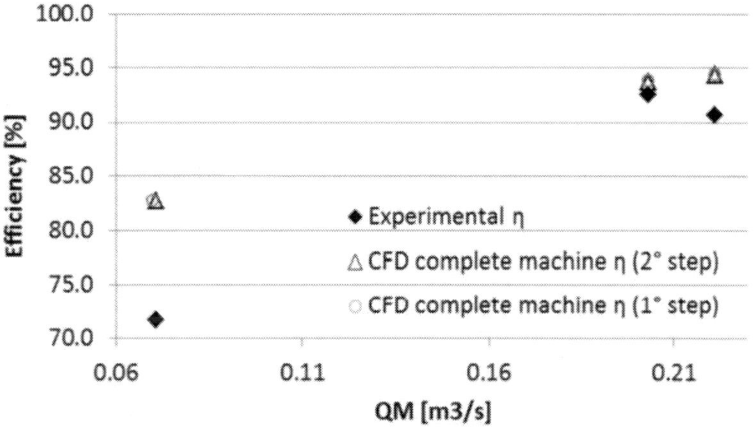

Figure 2. Comparison of experimental and CFD predicted efficiency simulating the complete machine

4.2. Hydraulic Efficiencies: Reduced Domain

Figure 3 shows the comparison of the three operating points plotted over the discharge in model scale. Due to the reduced domain the absolute values cannot be compared but only the qualitative distribution. Although it looks like the BEP was correctly predicted, a look at Figure 4 showing the circumferential velocity reveals that the simulated operating point is a high load operating point. The negative circumferential values indicate swirl in counter rotating direction. The CFD BEP is located at lower discharges (dashed curve representing the CFD BEP of Figure 5), where the swirl tends to zero.

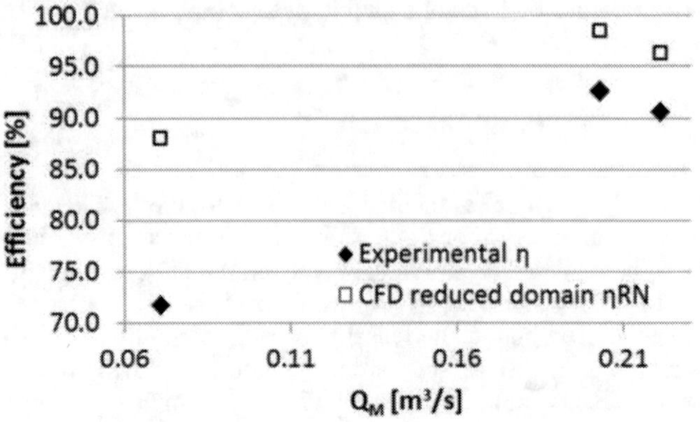

Figure 3. Qualitative comparison of experimental and CFD predicted efficiency simulating a reduced domain

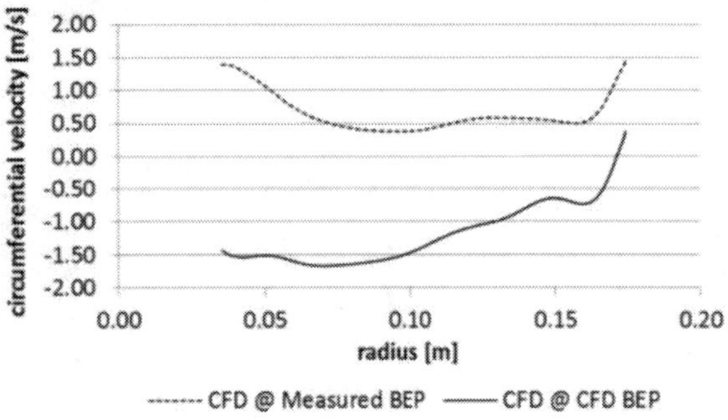

Figure 4. Circumferential velocities at runner outlet of CFD reduced domain simulation

The following figures (Figure 5 and Figure 6) show a cut at constant prototype head or prototype discharge. For a given prototype diameter of 1.779m and a rotational speed of 375rpm the prototype BEP derived from the measured BEP would be at a net head of 386.9m and a discharge of 30.1m3/s. Therefore the cuts were made at these values. The peak of the efficiency curve is shifted to lower discharge. This is often observed for low specific speed runners as the guide vane losses are not included and don't increase linear with increasing discharge. The curve over H is very flat and the measured and transposed BEP is located at the plateau.

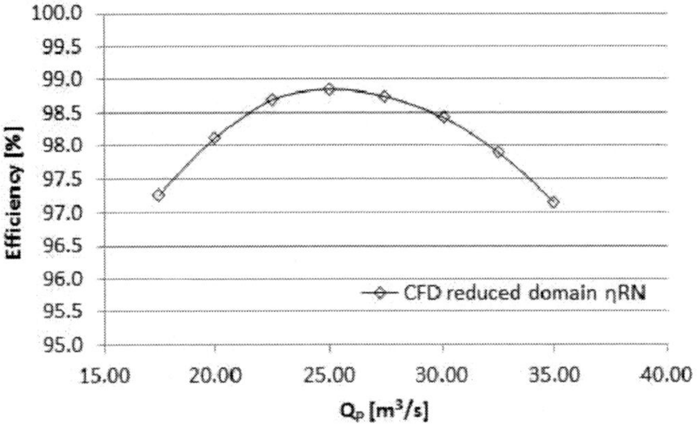

Figure 5. CFD reduced domain simulation predicted runner efficiency at constant prototype net head of 386.9m

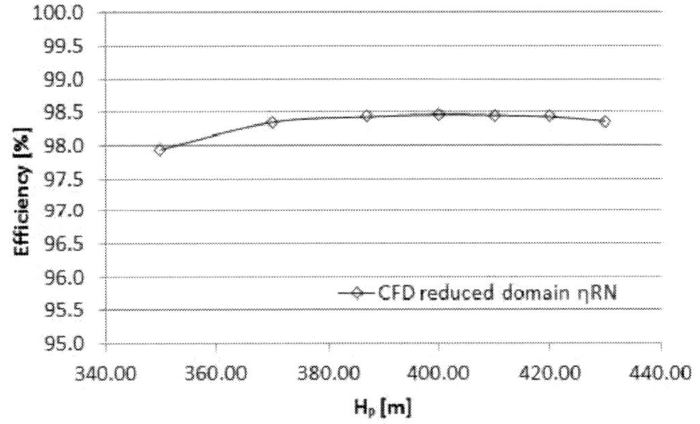

Figure 6. CFD reduced domain predicted runner efficiency at constant prototype discharge of $30.1 m^3/s$

4.3. Rotor Stator Interaction

The three operating points (part load, BEP and high load) simulated with a 2° step have been evaluated in respect to pressure amplitudes originating from the interaction between the rotary and stationary components. Table 2 shows an overview of the results. The top three rows indicate the geometric parameters influencing the RSI. Changing the guide vane angle leads to a different extension of the vaneless space from the trailing edge of the guide vanes (TE) to the leading edge of the runner blades (LE). The minimum radial distance is reached at high load.

Figure 7 shows the top view of the 3D geometry. A circular plane of evaluation has been created to evaluate the area average of the pressure amplitudes in the vaneless space. The radius of the area has been chosen so that the surface includes the location of pressure probe VL01. The probe position is indicated by a pink sphere. All frequencies have been normalized by the runner frequency f_{RN}.

location	variable	normalized frequency	unit	part load	BEP	high load
guide vanes	guide vane angle	-	[°]	3.91	9.84	12.44
vaneless space	radial distance between guide vane TE and runner blade LE	-	[mm]	24	19	17
			[%D_M]	6.9	5.5	4.9
vaneless space	area averaged amplitude	$f/f_{RN} = 30$	[kPa]	0.859	1.555	1.312
stay vane surface	maximum amplitude	$f/f_{RN} = 30$	[kPa]	0.364	0.408	1.141
guide vane surface	maximum amplitude	$f/f_{RN} = 30$	[kPa]	0.999	2.377	1.829
runner blades surface	maximum amplitude	$f/f_{RN} = 28$	[kPa]	2.161	5.448	3.834
runner blade 4	maximum amplitude	$f/f_{RN} = 28$	[kPa]	2.137	5.218	3.757
draft tube surface	maximum amplitude	$f/f_{RN} = 15$	[kPa]	0.744	0.278	0.234
draft tube surface	maximum amplitude	$f/f_{RN} = 30$	[kPa]	0.470	0.219	0.198

Table 2. RSI Evaluation

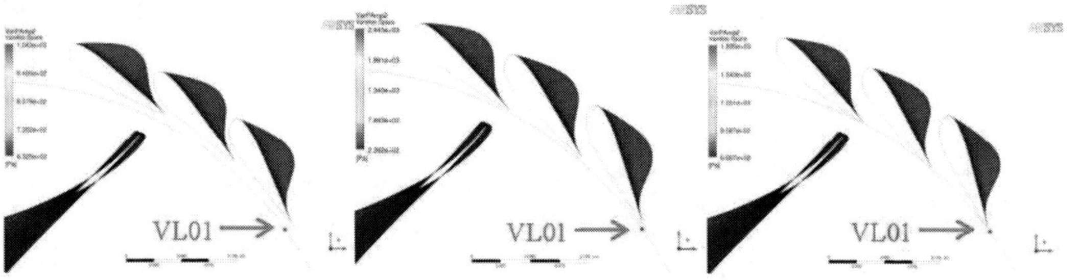

Figure 7. Top view of vaneless space for part, BEP and high load (left to right)

The distribution of pressure amplitudes for the different operating points on the circular evaluation surface is shown by Figure 8, Figure 9 and Figure 10. The area averaged pressure amplitudes are 0.859kPa, 1.555kPa and 1.312kPa for part, BEP and high load. These values show that the strongest interaction is to be expected at the BEP. It implies that not only the radial distance from the trailing edge of the guide vane to the leading edge of the passing runner blade, but the shape of the vaneless space is of importance. The larger the guide vane angle the smaller the radial distance, but at the same time the distance between guide vane trailing edge and the surface of the adjacent guide vane is increased. The pressure amplitude distribution shows a circumferential pattern. Part load operation and BEP operation show a similar pattern with the high amplitude patches moving with the rotational movement of the guide vanes. High load operation shows high amplitude patches that are split up.

The overall maximum amplitude value on the runner blades surface is found at the BEP and amounts to 5.448kPa. The maximum amplitude for the guide vanes is also found at the BEP. The stay vanes experience amplitudes reduced up to 83% compared to the guide vane values. The maximum value appears at the high load operating point. For the draft tube the maximum values for two harmonics are listed in Table 2 as the splitter blades are shortened and the first harmonic of 15 (normalized frequency) shows slightly higher amplitudes than the second harmonic of 30. The highest amplitude of 0.744kPa is found at the part load operating point for the first harmonic. High load shows the lowest amplitudes in the draft tube cone.

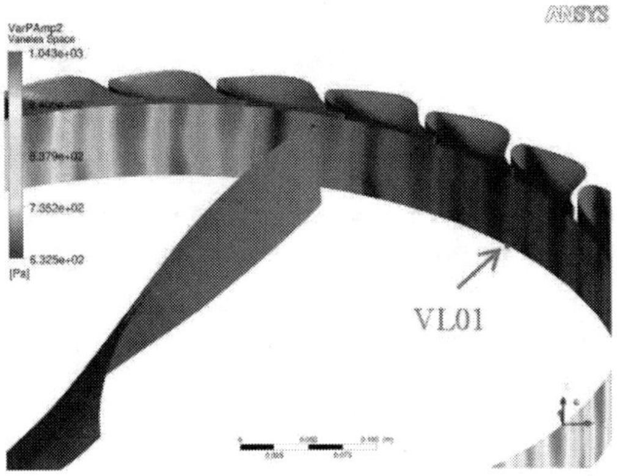

Figure 8. Distribution of pressure amplitudes in vaneless space at part load

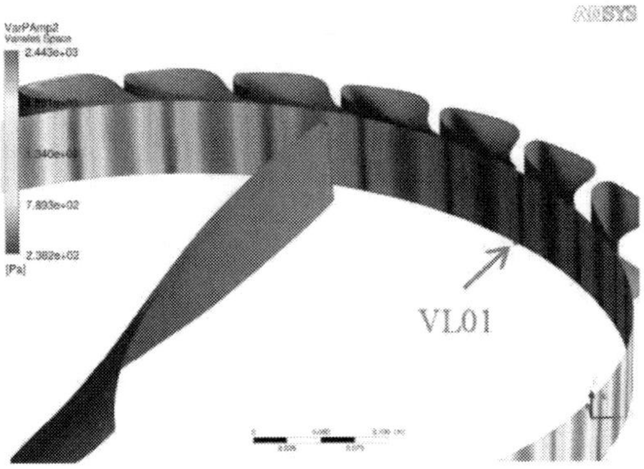

Figure 9. Distribution of pressure amplitudes in vaneless space at BEP

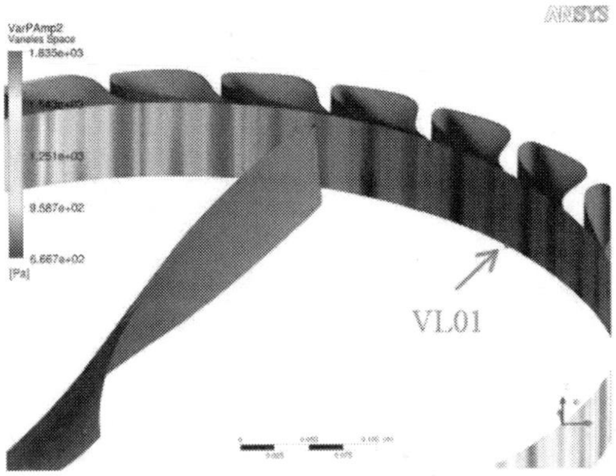

Figure 10. Distribution of pressure amplitudes in vaneless space at high load

To compare CFD and experimental values the CFD pressure signal at the location of pressure probe VL01 has been extracted. Both signals were processed by a discrete Fourier transformation. The complete experimental signal was used as input whereas of the CFD data the last revolution was used.

Below, in Figure 11, the pressure amplitudes are plotted versus the frequency normalized by the runner frequency. For better visibility the amplitude spectra have been plotted separately. As can be seen both, experimental and CFD signals, show peaks at 15 and 30 times the runner frequency which corresponds to the first and second harmonic of the runner blade passing frequency. In part and high load a small peak at 60 times the runner frequency is indicated as well. The peak of the experimental signal at the BEP near 17.9 and at high load near 16.2 times the runner frequency (100Hz) was identified as system excitation frequency in [5] and therefore not present in the CFD signal.

Quantitatively the pressure probe signal shows the maximum amplitude in the vaneless space at the high load operating point. Figure 8 to Figure 10 indicate the location of pressure probe VL01 as pink sphere and reveal that the amplitude is dependent on the spatial position and the overall maximum amplitude in the vaneless space is at BEP operating conditions.

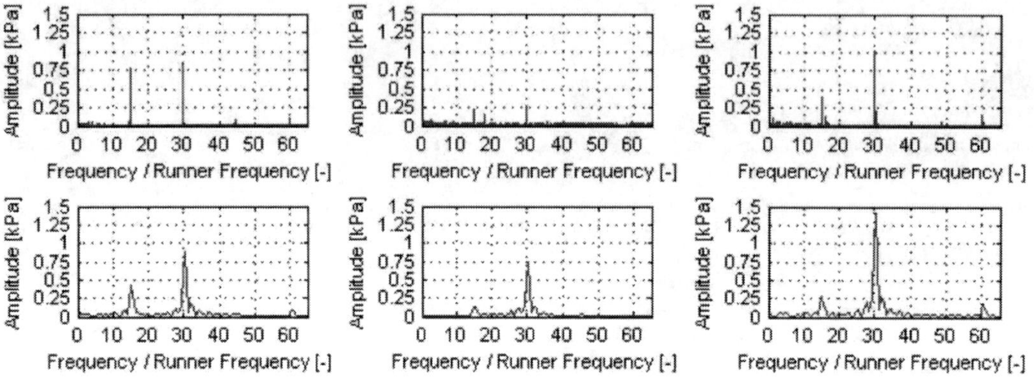

Figure 11. Comparison of experimental (top) and CFD (bottom) amplitudes spectrum of sensor VL01 for part, BEP and high load (from left to right)

Figure 12 shows the pressure side of runner blade 4, where the pressure probes P42 and P71 (near trailing edge) are mounted. The pressure amplitudes are shown for the BEP and correspond to the guide vane passing frequency of 28 per runner revolution. The amplitude values for P42 and P71 are 0.648kPa and 0.309kPa respectively. Figure 13 shows the suction side of the same blade. The maximum pressure amplitude for the chosen blade is 5.218kPa and occurs at the leading edge surface coming closest to the guide vane trailing edges. On the pressure side the amplitudes first decrease and then increase slightly again when moving downstream along the blade. On the suction side the amplitudes gradually become smaller towards the trailing edge.

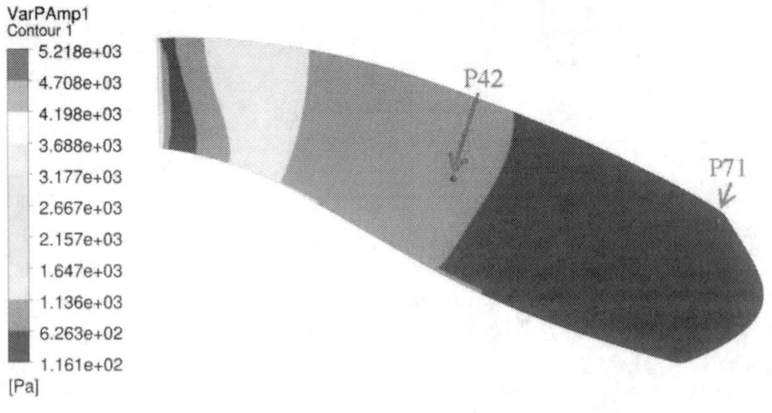

Figure 12. Pressure amplitude distribution on runner blade 4 pressure side at BEP

Figure 13. Pressure amplitude distribution on runner blade 4 suction side at BEP

The following figures (Figure 14 to Figure 16) show the pressure amplitudes of experimental and CFD signals of all the blade mounted probes: P42 and P71 on the pressure side and S51 on the suction side. As for the VL01 sensor amplitudes the experimental data shows several peaks not observed in the CFD results. There are 4 peaks in each graph that change position depending on the runner frequency. The grid frequency is 50Hz and appears between a normalized frequency of 7.4 to 8.9. The three large peaks are observed system excitation frequencies 100Hz, 200Hz and 300Hz [5].

The normalized frequency of interest for comparison is the guide vane passing frequency which is 28 and captured by all signals. As already seen from the amplitude distribution across the blade, P71 shows much lower amplitudes then P42 located more upstream. Note, that the y-axis is adjusted for each sensor. The CFD signals show lower amplitudes at part load for all sensors compared to BEP and high load, which are close to equal. The experimental signals do not show a variation across the operating points, suggesting that amplitude of the oscillation downstream the blade surface is not influenced anymore due to the guide vane position, but only the oscillation frequency is triggered by the interaction.

Experimental and CFD value agree on the tendency of the value when comparing each sensor for a given operating point. For the BEP operating point P42 experiences the largest oscillation with an amplitude of 0.904kPa, P71 an amplitude of 0.236kPa and S51 an amplitude of 0.494kPa. The corresponding CFD values are 0.648kPa, 0.309kPa and 0.469kPa respectively.

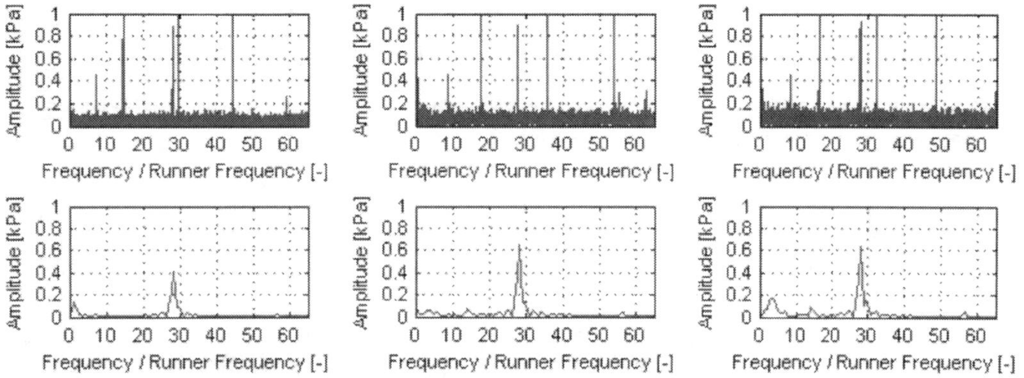

Figure 14. Comparison of experimental (top) and CFD (bottom) amplitudes spectrum of sensor P42 for part, BEP and high load (from left to right)

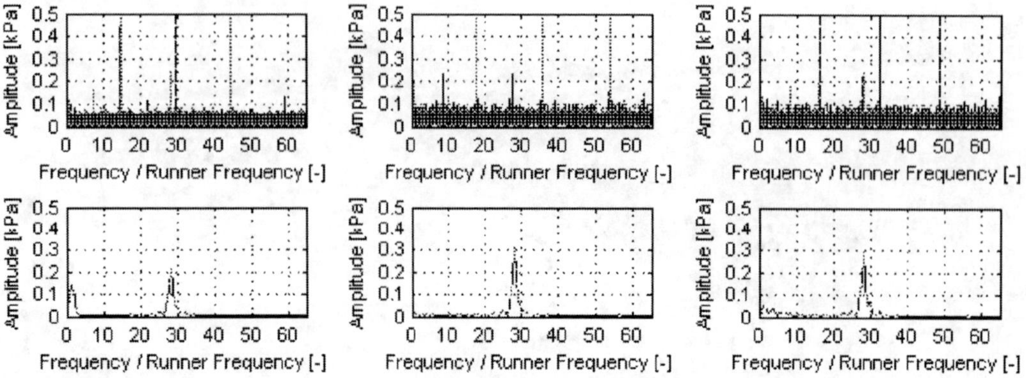

Figure 15. Comparison of experimental (top) and CFD (bottom) amplitudes spectrum of sensor P71 for part, BEP and high load (from left to right)

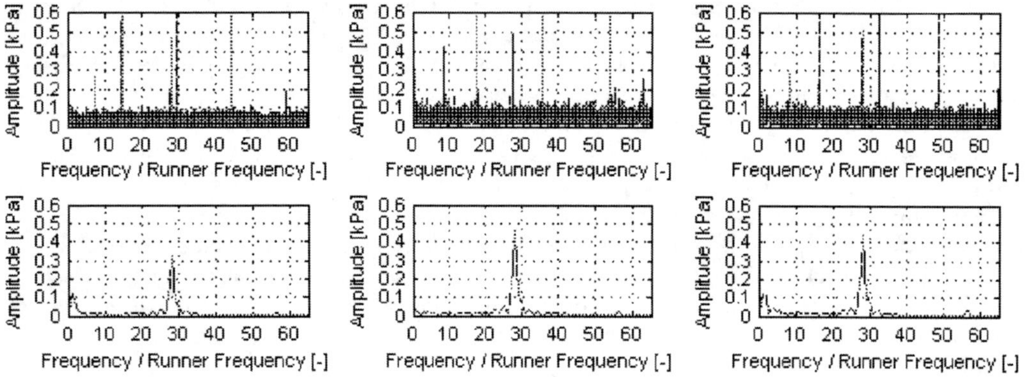

Figure 16. Comparison of experimental (top) and CFD (bottom) amplitudes spectrum of sensor S51 for part, BEP and high load (from left to right)

Figure 17 and Figure 18 compare the amplitude spectra of the sensors installed in the draft tube cone. Both sensors are at the same height, but they are rotated by 30° around the rotational axis of the turbine so that DT11 is closer to the outer and DT21 closer to the inner bend of the draft tube. Therefore an asymmetry in the signals is expected. The single peak between 10 and 30 in all six experimental signals is again the above mentioned 100Hz excitation frequency.

At the BEP, the amplitude at a normalized frequency of 30 matches very well for DT11 and DT21. The CFD shows no low frequencies as no vortex is present. For part load, low frequencies originating from a part load vortex appear in both experimental and CFD signals. The CFD amplitudes over predict the effect of the vortex. The runner passing frequency is barely present anymore. At high load, contrary to the CFD amplitudes for the runner passing frequency the experimental values show an increase compared to the BEP.

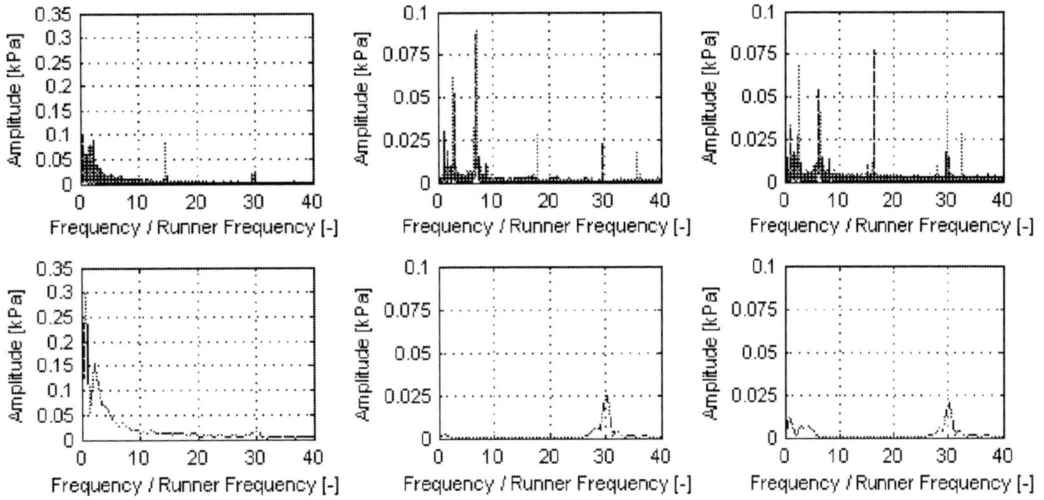

Figure 17. Comparison of experimental (top) and CFD (bottom) amplitudes spectrum of sensor DT11 for part, BEP and high load (from left to right)

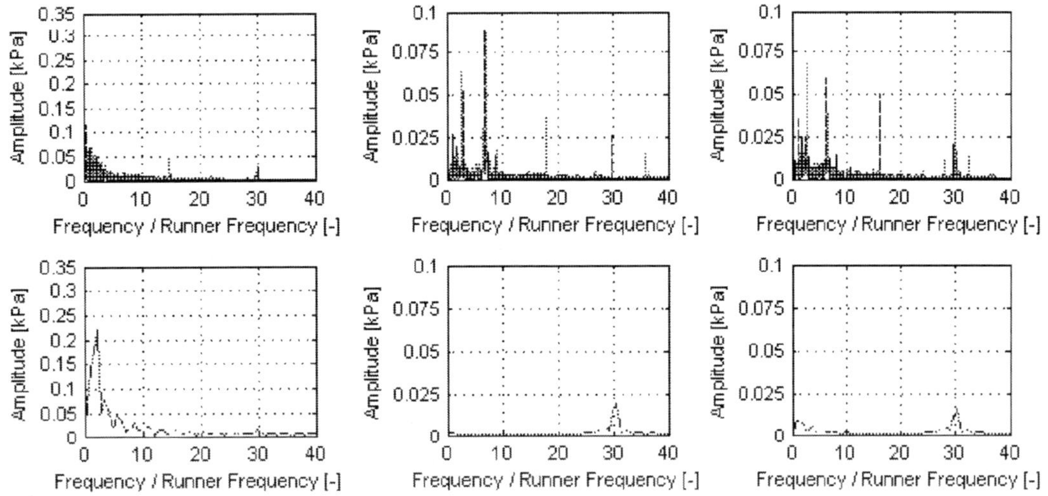

Figure 18. Comparison of experimental (top) and CFD (bottom) amplitudes spectrum of sensor DT21 for part, BEP and high load (from left to right)

5. Conclusion

Transient simulations have been conducted to compare with experimental data provided by the organisers of the Francis-99 project. The best match of hydraulic efficiencies was found at the BEP with a +1.08% deviation from the experimental value. Larger deviations were found at part and high load. No improvement was observed when using a 1° time step instead of a 2° time step. The 1° time step showed a worse match at the BEP (+1.24%) and identical values for the part load (+11.0%) and high load operating point (+3.6%).

The efficiency results shows that CFD is not yet suitable to predict absolute efficiency values away from the best efficiency point and further improvements have to be made. Possible errors are the accuracy of the geometry, grid dependent effects as well as modelling. It would be interesting to compare the discharge boundary condition used with a total pressure boundary condition. Further, it was shown that with a reduced domain simulation it is possible to predict the shape of the efficiency curve except for a shift to lower discharges.

Detailed analysis of the pressure amplitudes in the vaneless space showed a maximum average amplitude of 1.555kPa at the BEP. The pressure probe installed in the vaneless space showed the maximum amplitude for the high load operating point. Due to the spatial distribution of the pressure amplitudes this was found to be only a local maximum. The result suggests that the shape of the vaneless space and not only the minimal distance between guide vane trailing edge and runner blade leading edge influence the pressure amplitudes occurring due to RSI.

Comparison of the amplitude spectrum between measured and simulated pressure signals showed that CFD captures the relevant frequencies. The tendency of the amplitude values was correct for the probe in the vaneless space and when comparing the sensors on the blade for a given operating point. The comparison of the signals in the draft tube cone compared well for the runner passing frequency but fell short of the experimental values for low frequencies associated with a part or full load vortex. To improve the quality of the CFD results concerning the absolute amplitude values, further effort is needed.

6. Acknowledgements

The authors would like to thank the organisers of the Francis-99 project for the opportunity to participate in this project and test the state of the art CFD tools.

References

[1] Keck H, Sick M 2008 *Thirty years of numerical flow simulation in hydraulic turbomachines* Acta Mech **211** pp 211-229

[2] Egusquiza E, Mateos B, Escaler X 2002 *Analysis of Runner-Stator Interaction in Operating Pump-Turbines* Proceedings of the 21st IAHR Symposium on Hydraulic Machinery and Systems, Lausanne, Switzerland

[3] Keller M, Doerfler P, Sallaberger M, Sick M 2005 *Rotor-Stator Interaction in Pump Turbines – Numerical Prediction and Experiment* Proceedings of Hydro2005 Villach Austria

[4] Liang Q W, Keller M, Sick M 2009 *Rotor Stator Interaction during no Load Operation of Pump Turbines* Proceedings of Hydro2009 Lyon France

[5] Trivedi C, Cervantes J M, Gandhi B K, Dahlhaug O G 2013 *Experimental and Numerical Studies for a High Head Francis Turbine at Several Operating Points* ASME J. of Fluids Engineering **135** 111102

[6] IEC 60193 1999-11 IEC *Hydraulic Turbines, Storage Pumps and Pump-Turbines – Model acceptance tests* International Electrotechnical Commission

Journal of Physics: Conference Series **579** (2015) 012004 doi:10.1088/1742-6596/579/1/012004

Numerical modeling of flow in the Francis-99 turbine with Reynolds stress model and detached eddy simulation method

A V Minakov[1,2], A V Sentyabov[1,2], D V Platonov[1], A A Dekterev[1,2] and A A Gavrilov[1,2]

[1]Siberian Federal University, Krasnoyarsk, Svobodniy 79, 660079 Russia
[2]Institute of Thermophysics SB RAS, Novosibirsk, Lavrentyeva 1, 630090 Russia
[3]Novosibirsk State University, Novosibirsk, Pirogova 2, 630090 Russia

E-mail: tov-andrey@yandex.ru, sentyabov_a_v@mail.ru, platonov-08@yandex.ru, dekterev@mail.ru

Abstract. The paper presents numerical simulation of flow in Francis-99 water turbine under three operation modes: part load, best efficiency point and high load. Calculations were performed by means of Reynolds stress model and detached eddy simulation based on k-omega SST model. The paper focuses on the flows in the draft tube. The calculated mean velocity components in the draft tube are in close agreement with experimental results. Calculated r.m.s velocity components agree with experimental pulsations qualitatively.

1. Introduction

Increased stability, increased efficiency of hydraulic machines is impossible without studying physical mechanisms of hydrodynamic processes, among which an important role is played by transient phenomena associated with generation of large-scale vortex structures. Flow inside hydraulic turbines is connected with a variety of transient phenomena [1–5]. However, the greatest problems are caused by low-frequency pulsations in draft tube. There are different mechanisms of the excitation, part of which is associated with the precession of the vortex core (PVC). Precession of the vortex core occurs under party load (at a discharge of 0.5 – 0.85 of the optimal), where the flow has considerable residual swirl. Vortex breakdown in such a flow leads to formation of the recirculation zone at the flow axis and the rotation of the spiral vortex rope around it. Precession of the vortex rope is a serious danger for the hydraulic turbine equipment. Powerful flow-induced pulsations can lead to strong vibrations of hydraulic turbine construction and in the case of resonance can lead to rupture.

The development of computer technology has made it possible to use modern methods of computational fluid dynamics to description of turbulent flows in geometrically complex spatial objects, such as hydraulic turbines. Recently there appeared a large number of papers in which three-dimensional formulation in the calculations of various processes in hydraulic turbines were made. From recent studies one can mention references [6–10], which are devoted to the study of the PVC in various applications. Among researchers one can mentions papers made by Doerfler P. [1, 3], Nicolet C. [4, 5], Ruprecht A. [7], Avellan F. [9] who made significant contribution to the development of computational modeling techniques applied to hydraulic turbines. Thus, it becomes clear that the use of numerical modeling can serve not only as an alternative approach to experiment with which it is possible to calculate such turbine parameters as discharge, torque and efficiency, but also provides with fundamentally new possibilities. Numerical modeling allows analyzing the local structure of the flow and on this basis predicting the behavior of the pulsation characteristics. Of course, there is always a question of adequacy of the results of such simulations. Therefore, the main objective of this work was testing methods of the numerical modeling of water turbine flow.

2. Mathematical model and numerical method

Describing the flows in water turbines one has to face several problems. The first challenge is related to modeling turbulence in channels of complex geometric shapes and the second one is strong swirl flows modeling. At the same time engineering calculations requires turbulence models, quite accurate describing not only average fields, but also large-scale flow pulsations. Widely spread in engineering calculations k-ε and k-ω turbulence models poorly describe such flows. To improve the modeling of turbulent swirling flows scientists are trying either to modify existing RANS turbulence models or use techniques which resolve large-scale turbulent eddies (LES, DES).

Swirling flow in hydraulic turbines can be accompanied by the precession of the vortex core. For modeling such phenomenon it is necessary to use non-stationary simulation methods such as the method of large-eddy simulation (Large Eddy Simulation – LES). However, this method requires a very fine grid, especially near the walls. At the same time RANS models are quite economical and have a good description of the boundary layers. To combine the advantages of these approaches the method of detached eddy simulation (DES) was proposed [11].

In the simulation of hydraulic turbines it is necessary to consider the rotation of the runner. There are several approaches for modeling flows with rotating bodies, they include: dynamic grid, sliding mesh, moving mesh and method based on a moving reference frame. The most common and simple way to model the runner rotation is to use a rotating reference frame. Rotating reference frame allows simulating flow in an approximation in which the runner is fixed and swirl flows pass through it. In this paper, modeling of the runner rotation was performed with rotated reference frame for runner zone. Numerous test calculations show the correctness of this approach, both in the description of mean turbine parameters and flow pulsations [12–14].

Below, the basic equations of the mathematical models expressing the conservation laws in the rotating reference frame are presented. Einstein summation convention is used.

The continuity equation (conservation of mass):

$$\frac{\partial u_i}{\partial x_i} = 0$$

Momentum equations (conservation of momentum) in a rotating reference frame for absolute velocities:

$$\frac{\partial \rho u_i}{\partial t} + \frac{\partial}{\partial x_j}\left(\rho u_j^r u_i\right) = -\frac{\partial p}{\partial x_i} + \frac{\partial}{\partial x_j}(\tau_{ij}^{\mathrm{m}} + \tau_{ij}^{\mathrm{t}}) - \rho \varepsilon_{ijk} \Omega_j u_k$$

Where: u_i – absolute velocity components, u_j^r – relative velocity components, τ_{ij}^{m} – viscous stress tensor, τ_{ij}^{t} – turbulent stress tensor, Ω – angular velocity of runner rotation, p – static pressure, ρ – density, ε_{ijk} – Levi-Civita symbol.

Components of the viscous stress tensor are defined as:

$$\tau_{ij}^{\mathrm{m}} = \mu\left(\frac{\partial u_i}{\partial x_j} + \frac{\partial u_j}{\partial x_i}\right)$$

Where μ is dynamic molecular viscosity.

Two turbulence modeling approach were used. The first one is Reynolds stress model (RSM) and the second one is detached eddy simulation (DES).

Reynolds stress $\rho\overline{u_i'u_j'}$ transport equations are solved in RSM model:

$$\frac{\partial}{\partial t}\left(\rho\overline{u_i'u_j'}\right)+\frac{\partial}{\partial x_l}\left(\rho u_l\overline{u_i'u_j'}\right)=\rho\left(P_{ij}+Diff_{ij}-\varepsilon_{ij}+\phi_{ij}\right).$$

Where P_{ij} is production term, $Diff_{ij}$ is diffusion term, ε_{ij} is dissipation term, ϕ_{ij} is pressure strain term. Production term does not require any modeling:

$$P_{ij}=-\left(\overline{u_i'u_l'}\frac{\partial u_j}{\partial x_l}+\overline{u_j'u_l'}\frac{\partial u_i}{\partial x_l}\right).$$

Diffusion term is modeled by means of turbulent viscosity:

$$Diff_{ij}=\frac{\partial}{\partial x_l}\left(\left(\mu+\frac{\mu_t}{\sigma_k}\right)\frac{\partial\overline{u_i'u_j'}}{\partial x_l}\right),\ \sigma_k=0.82.$$

Where turbulent viscosity:

$$\mu_t=\rho C_\mu\frac{k^2}{\varepsilon},\ k=\frac{1}{2}\overline{u_i'u_i'},\ C_\mu=0.09.$$

Dissipation tensor ε_{ij} is modeled by isotropic tensor:

$$\varepsilon_{ij}=\frac{2}{3}\delta_{ij}\varepsilon,$$

Where scalar dissipation rate ε is computed with a transport equation:

$$\frac{\partial\rho\varepsilon}{\partial t}+\frac{\partial}{\partial x_j}\left(\rho u_j\varepsilon\right)=\frac{\partial}{\partial x_j}\left(\left(\mu+\frac{\mu_t}{\sigma_k}\right)\frac{\partial\varepsilon}{\partial x_j}\right)+\rho\frac{C_{\varepsilon1}P_k-C_{\varepsilon2}\varepsilon}{T},$$

$$T=\frac{k}{\varepsilon},\ \sigma_\varepsilon=1.0,\ C_{\varepsilon1}=1.44,\ C_{\varepsilon2}=1.92$$

Pressure strain term is modeled by LRR-IP strain model:

$$\phi_{ij}=\phi_{ij,1}+\phi_{ij,2},$$

$$\phi_{ij,1}=-c_1\varepsilon a_{ij},\ c_1=1.8,$$

$$\phi_{ij,2}=-c_2\left(P_{ij}-\frac{2}{3}P_k\delta_{ij}\right),\ c_2=0.6,$$

where

$$P_k=\frac{1}{2}P_{ii},\ a_{ij}=\frac{\overline{u_i'u_j'}}{k}-\frac{2}{3}\delta_{ij},$$

In constructing two-equation models of turbulence for defining the components of Reynolds' stress tensor τ_{ij}^t, Boussinesq's hypothesis of isotropic turbulent viscosity is used:

$$\tau_{ij}^{t} = \mu_t \left[\left(\frac{\partial u_i}{\partial x_j} + \frac{\partial u_j}{\partial x_i} \right) - \frac{2}{3} \delta_{ij} \rho k \right]$$

In this paper we use the DES approach based on k-ω SST model [15]. The shear-stress transport (SST) k-ω model was developed by Menter to effectively blend the robust and accurate formulation of the k-ω model in the near-wall region with the k–ε model in the free stream. The standard equations of k-ω SST model [15, 16]:

$$\frac{\partial(\rho k)}{\partial t} + \frac{\partial}{\partial x_j}(\rho u_j k) = \frac{\partial}{\partial x_j}\left(\Gamma_k \frac{\partial k}{\partial x_j} \right) + G_k - Y_k,$$

$$\frac{\partial(\rho \omega)}{\partial t} + \frac{\partial}{\partial x_j}(\rho u_j \omega) = \frac{\partial}{\partial x_j}\left(\Gamma_\omega \frac{\partial \omega}{\partial x_j} \right) + G_\omega - Y_\omega + D_\omega$$

Here, k – turbulence kinetic energy, ω – the specific dissipation rate, the term G_k represents the production of turbulence kinetic energy, G_ω represents the generation of ω, Γ_k and Γ_ω represent the effective diffusivity of k and, respectively, Y_k and Y_ω represent the dissipation of k and ω, D_ω represents the cross-diffusion term.

The dissipation term of the turbulent kinetic energy is modified for the DES turbulence model as described in [17]. In this model dissipative term in k-equation is modified by means of switcher F_{DES}:

$$\frac{\partial(\rho k)}{\partial t} + \frac{\partial}{\partial x_j}(\rho u_j k) = \frac{\partial}{\partial x_j}\left(\Gamma_k \frac{\partial k}{\partial x_j} \right) + G_k - Y_k \cdot F_{\mathrm{DES}},$$

$$F_{\mathrm{DES}} = \max\left(\frac{l_t}{C_{\mathrm{DES}}\Delta}, 1 \right), \qquad l_t = \frac{k^{1/2}}{\beta^* \omega}, \qquad C_{\mathrm{DES}} = 0.61$$

Where: l_t – turbulent length scale, C_{DES} – empirical constant, and Δ is defined as the maximum of three sizes of the control volume $\Delta_x, \Delta_y, \Delta_z$.

Earlier test calculations have shown that this technique can reliably consider large-scale turbulent fluctuations in water turbines and estimate their frequency [12–14].

One should says a few words about the numerical approach. Discretization of transport equations was carried out by the control volume method on unstructured grids. Coupling of the velocity and pressure fields for incompressible flow was realized using the SIMPLE-like procedure. For the approximation of the convective terms in the equations for momentum components second order scheme was used. For the approximation of the convective terms in the equations for turbulent characteristics up-wind scheme was used.

3. Testing of the numerical algorithm

As a test the analysis of low-frequency pressure pulsations for hydraulic turbines RO120 and RO230 was made. These turbines were manufactured at the OJSC "Power Machines" (LMZ), St. Petersburg, Russia. Main characteristics of the turbines are shown in table 1. 3D geometry was used. The geometry represents an exact copy of the hydraulic turbine. The computational domain includes all the basic components of the turbine (spiral case, guide vanes, runner, and draft tube) except inlet penstock.

Unstructured grids were built for each of the geometry of the hydraulic units. Computational grids include about 4–5 million cells. The grid at the runner blades and the guide vanes was fined. Calculations of unsteady flow were performed for the wicket gate opening values in the range from

A_0 = 100 mm to 550 mm in case of turbine RO120 and in the range from A_0= 206 mm to 550 mm in case of turbine RO230. These ranges of the opening correspond to regimes with maximum pressure pulsations. In all cases, the calculations were performed on single-phase formulation. Cavitation and air admission in the calculations were not considered.

For the analysis of flow-induced pulsations of great interest is the behavior of the flow structure in a turbine under different operating conditions. Figure 1 shows typical vortex structure of the flow behind the runner for several modes of the hydraulic turbine. The flow structure is shown using instant isosurface of static pressure. Runner in the figures is colored with static pressures.

Table 1. Turbines basic parameters.

	First turbine	Second turbine
Turbine type	Francis	Francis
Runner type	RO120	RO230
Head	113 m	210 m
Maximum power	340 MW	640 MW
Runner rotational speed	125 r.p.m.	143 r.p.m.
Diameter of runner	6.38 m	6.77 m
Number of blades of runner	14	16
Number of vanes of wicket gate	20	20

Since precession vortex rope is a serious danger for hydraulic turbines, especially for high-head water power plants, great interest is the frequency and intensity of pressure fluctuations in a turbine. Figure 2 shows the comparison of the calculated and experimental spectra of pressure fluctuations at point D for regimes with a strong vortex rope. Comparison is carried out for a point D, lying on the wall of the draft tube cone at a distance of about 3m from the lower edge of the runner. The graph shows the dimensionless peak-peak pressure pulsations on the wall related to full head. Frequency is dimensionless on the runner rotation frequency. Experimental spectra for the pressure pulsations were obtained on the model stands in the hydraulic laboratory of OJSC "Power Machines" (LMP), St. Petersburg. Stands models are reduced by about 15 times of considered hydro turbine. The scheme of the experimental setup is shown in figure 3. Nominal runner diameter model is equal to 460 mm.

As it can be seen (figure 2) for modes with pronounced vortex precession the frequency is dominant in the spectrum of pressure pulsation. Value of the frequency for both turbines is between $0.2f_n$ to $0.3f_n$, where f_n is the runner rotation frequency. Calculated values of the vortex precession frequency with an error of 5–10% reproduce the experimental data for both PVC frequencies of the water turbines. As it can be seen, the levels of pressure pulsations of the experiment and calculations agree well with each other.

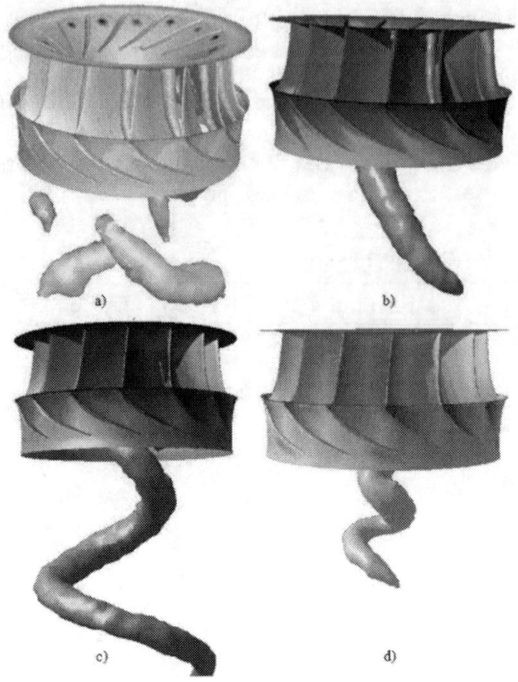

Figure 1. Vortex flow structure in the hydraulic turbine RO120 for different openings of the guide vanes: a) A_0 = 150 mm, b) A_0 = = 260 mm, c) A_0 = 300 mm, d) A_0 = 390 mm.

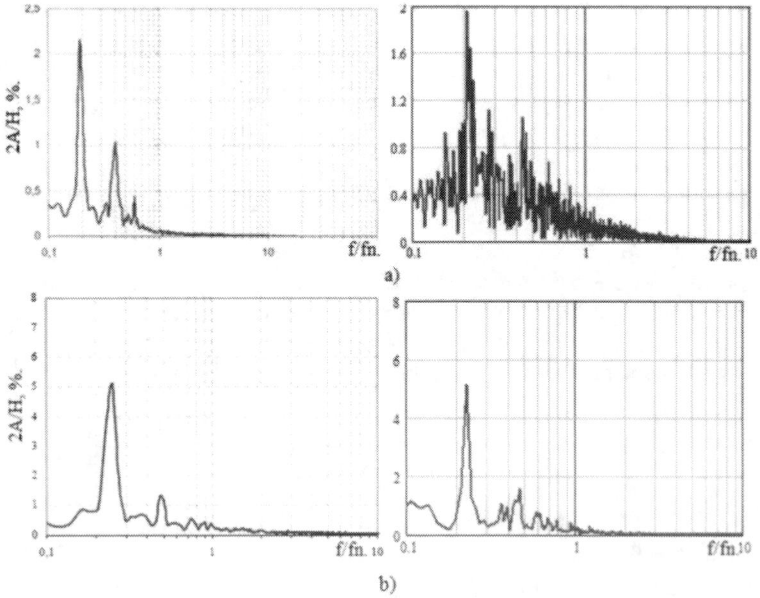

Figure 2. Experimental (left) and calculated (right) spectra of the pressure fluctuations for hydro turbine RO230 (a) at the A_0 = 350 mm, and for turbine RO120 (b) at the A_0 = 275 mm.

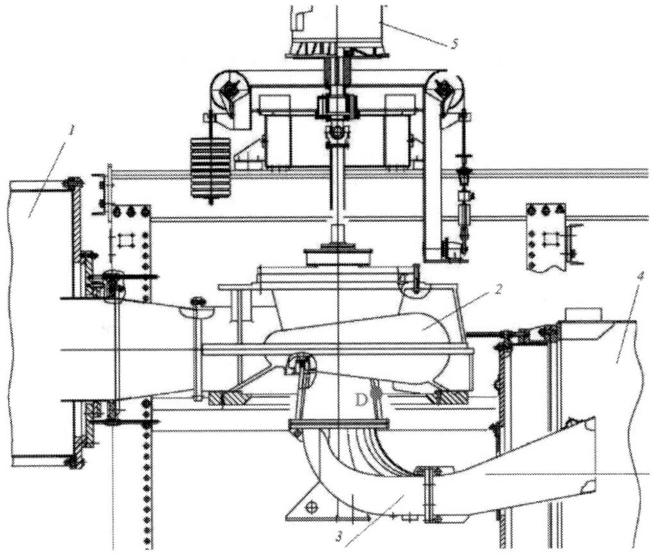

Figure 3. Scheme of the experimental stand: 1 – vacuum tank, 2 – spiral case, 3 – draft tube 4 – pressure vessel tank, 5 – brake generator.

Dependences of the pressure intensity pulsations from the wicket gate opening and the power are shown in figure 4. Squares show the calculated values of the pressure pulsations in point D of the draft tube diffuser for the considered values of the wicket gate openings. Calculated values of the pressure fluctuations intensity for extreme pressures (peak-peak) in the monitoring point were provided. Turbine RO120-situ measurements of pressure fluctuations are given for comparison with the results of the calculations. Experimental data for turbine RO230 were obtained on the model stand OJSC "Power Machines" (LMP), St. Petersburg (figure 3).

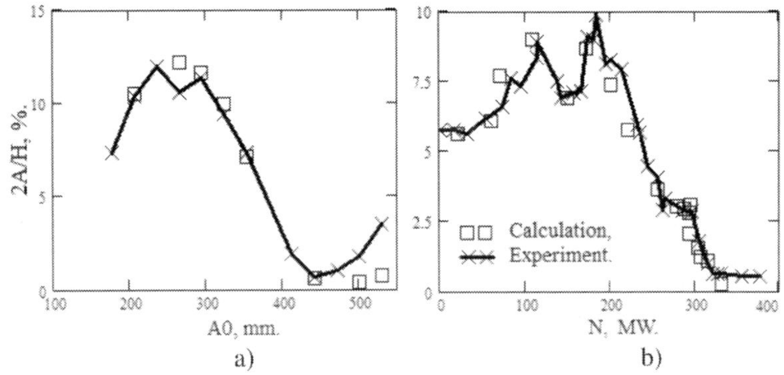

Figure 4. The pressure pulsations in the draft tube cone at the point D, depending on the wicket gate opening for turbine RO230 (a) and depending on the power of the hydraulic turbine RO120 (b).

Comparison of in-situ measurements with the calculated values shows, that the calculated data satisfactorily describe the behavior of pressure fluctuations in the draft tube in a wide range of wicket gate openings. The differences between calculations and experiment observed at very large openings of turbine RO230 takes place likely due to the presence of cavitation processes occurring in the experiment, whereas they are not considered in the calculations. Nevertheless, if we consider pulsations representing the greatest danger, here the calculated values of pressure pulsations are consistent with in situ measurements with accuracy of about 10%. It shows that the numerical simulation is possible to be used not only for the calculation of the turbine parameters, but also for an adequate assessment of the intensity of time-dependent processes in hydraulic turbines. In calculations the behavior of the turbine parameters (efficiency, power and discharge) was also analyzed. For example, in figure 5 the results of comparison of the calculated and experimental values of the water discharge through the turbines at various wicket gate openings A_0 are shown. Calculated values of the turbine parameters were obtained by averaging the non-stationary values. Experimental data were obtained on the hydraulic modeling stand LMP (figure 3) and converted to full-scale settings. As it can be seen from figure 5, close agreement between the calculated and experimental data is shown. In general, the analysis of the accuracy of calculations of the turbine parameters showed that for these turbines discrepancy of the calculation and measurement results does not exceed 0.5% of discharge, 0.5% of power and 1% of efficiency over the entire range of the wicket gate openings. Thus, modern methods of computational modeling can be successfully used in the design and optimization of hydraulic turbines.

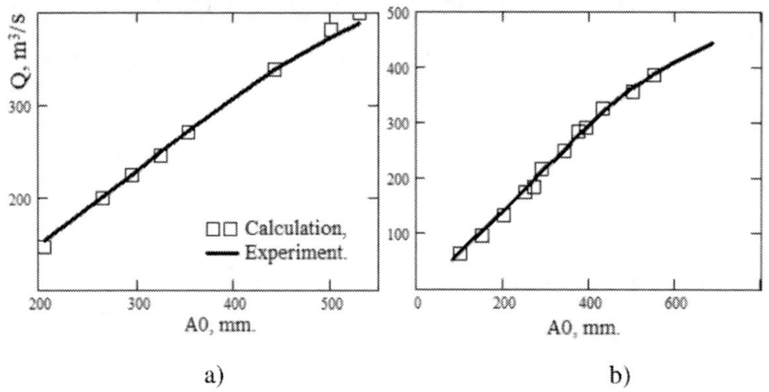

Figure 5. Dependence of the water discharge through the turbine from the wicket gate opening for turbine RO230 (a) and RO120 (b).

4. Results of Francis-99 simulations

The section presents results of simulation of the flow in the turbine Francis-99. Francis-99 is test case in the same spirit as Turbine-99 workshops [18]. In contrast Turbine-99, Francis-99 considers scaled down model of high head Francis turbine. Three operation point proposed for the Francis-99 workshop were calculated. They are party load mode (PL), best efficiency point (BEP) and high load mode (HL). Geometry data provided by the workshop committee were used. Computational domain includes the runner and the draft tube (figure 6, *a*). The runner diameter was 0.349 m, and the net head was approximately 12 m. Experimental results such as net head, discharge and torque were presented by the organizer (tables 2 – 4). Velocity distributions along two lines in the draft tube cone and pressure pulsations were also available.

The mesh consists of 6 million control volumes (figure 6, *b*). Constant radial and tangential velocity components were set at the runner inlet. Angle between velocity direction and the runner inlet corresponds to guide vane angle. Outflow boundary conditions were set at the draft tube outlet.

Three operation points were calculated with Reynolds stress model. Part load and high load operation points were also calculated with DES method. Figure 7 shows velocity magnitude in central cross-section for three operation points. Flow patterns for HL and BEP operating points are similar. Narrow concentrated vortex is formed under the runner tip. On the other hand, at the PL operation point the downward axial flow is close to the draft tube wall. At this operation point a wide recirculation zone is formed at the axis in the draft tube cone.

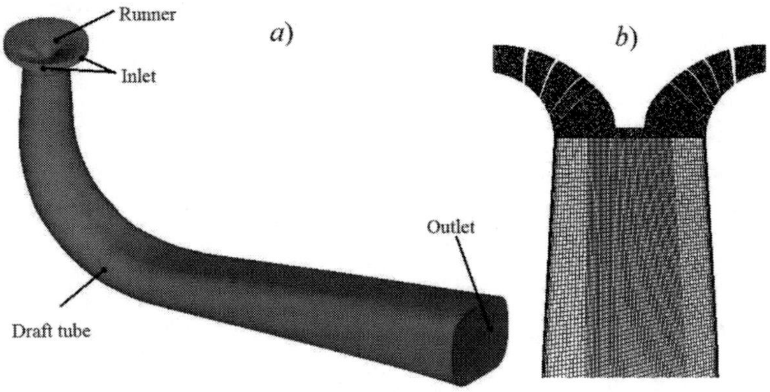

Figure 6. Francis-99 test case; *a*) computational domain; *b*) mesh in central cross-section

Tables 2 – 4 present parameters of the turbine for three operation points (net head in calculations was defined between runner inlet and the draft tube outlet). Computational results were obtained with steady Reynolds stress model. The results of calculations agree with experimental results qualitatively. The results of simulation for the part load operating point shows the largest discrepancy of efficiency with experimental data.

Table 2. Part load.

	RSM	Experiment
Net head [m]	11.30	12.29
Pressure torque [N m]	165.7	137.52
Friction torque [N m]	-3.5	-6.54
Total torque [N m]	162.2	
Efficiency [%]	80.60	71.69

Table 3. Best efficiency point.

	RSM	Experiment
Net head [m]	11.61	11.91
Pressure torque [N m]	650.8	619.56
Friction torque [N m]	-6.9	-8.85
Total torque [N m]	643.9	
Efficiency [%]	95.34	92.61

Table 4. High load.

	RSM	Experiment
Net head [m]	11.19	11.84
Pressure torque [N m]	608.1	597.99
Friction torque [N m]	-9.5	-7.63
Total torque [N m]	598.6	
Efficiency [%]	90.67	90.66

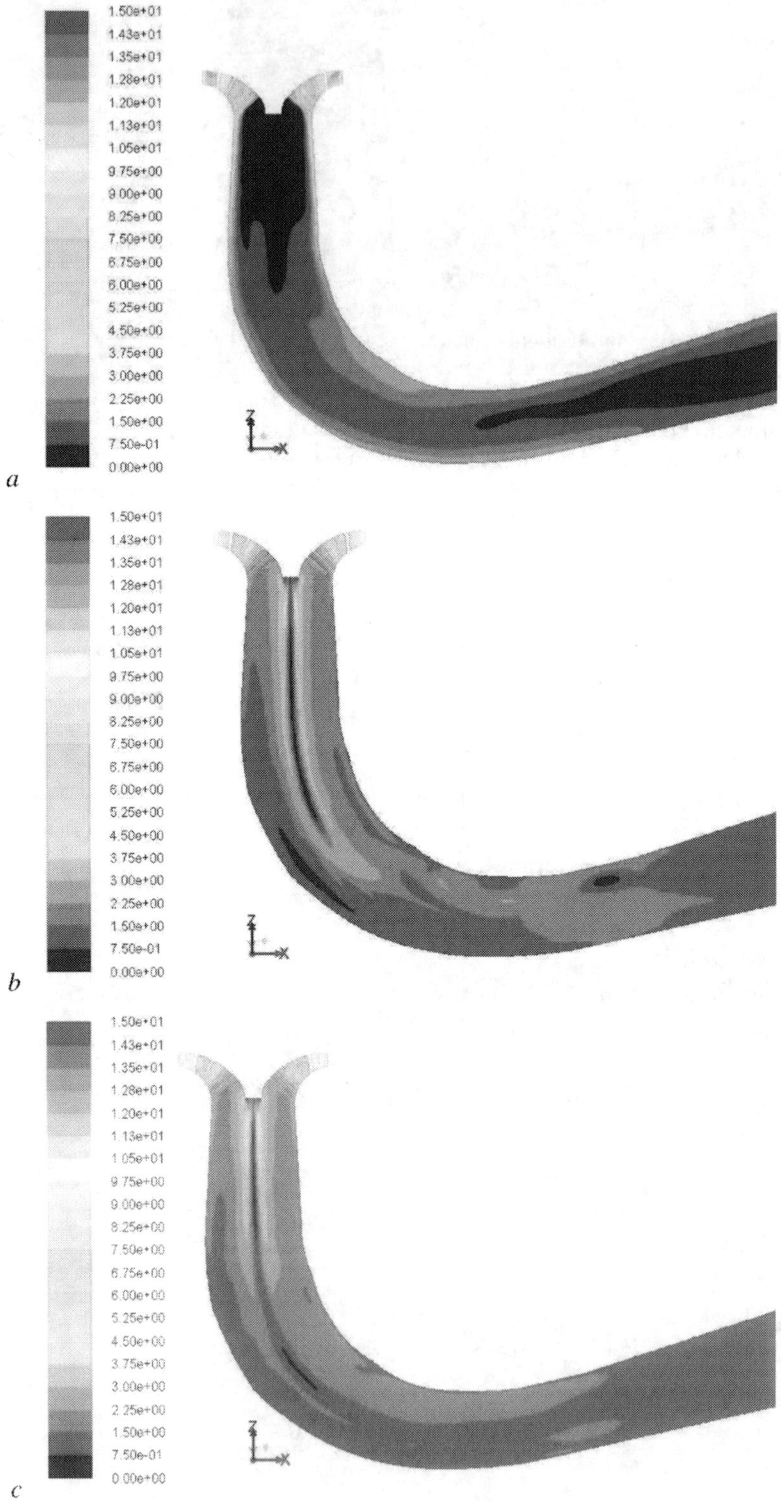

Figure 7. Velocity magnitude in central plane; *a*) part load, *b*) best efficiency point, *c*) high load.

Figures 8 – 10 show velocity components along the top and bottom lines in the draft tube for three operation points. The results of the calculations are in close agreement with experimental data both for Reynolds stress model and DES method. Part load operation point shows the small discrepancy. At the best efficiency point (BEP) swirl velocity is significantly overestimated near the axis. It seems that due to low swirl intensity at BEP the flow is sensitive to small variation of the inlet conditions. Results of BEP and HL simulation significantly underestimate axial velocity near the axis of the draft tube cone. This narrow stagnant zone corresponds to the concentrated vortex under the runner tip. Experimental results do not show this stagnant zone.

Figure 11 shows r.m.s. pulsations of velocity components at the part load operation point. Resolved velocity pulsations are presented for the detached eddy simulation. Simulation and experimental results agree well in the bottom cross-section of the draft tube. In the top cross-section simulation underestimates velocity pulsations, perhaps, due to insufficiently fine computational mesh of the runner.

Figure 12 shows the spectra of pressure at the point DT21. The spectra were obtained using DES method. It can be observed that at part load mode the intensity and frequency of the pressure fluctuations in the draft tube satisfactory agree with experiment. At high load mode calculated pulsations are several times lower than the experimental pulsations. Perhaps, rotor-stator interaction modeling is required at this operation mode. This assumption will be verified later.

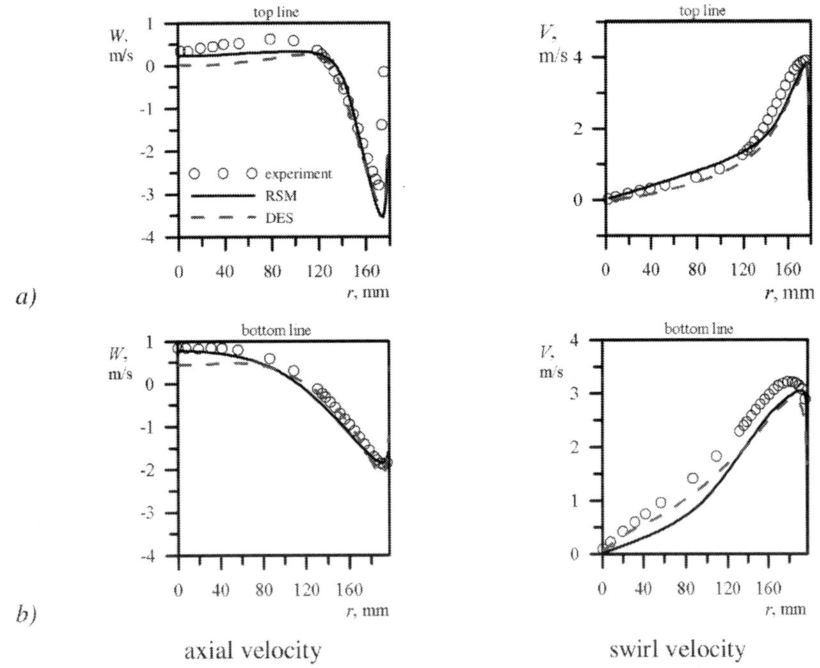

Figure 8. Velocity components in the draft tube, part load; *a*) top line, *b*) bottom line.

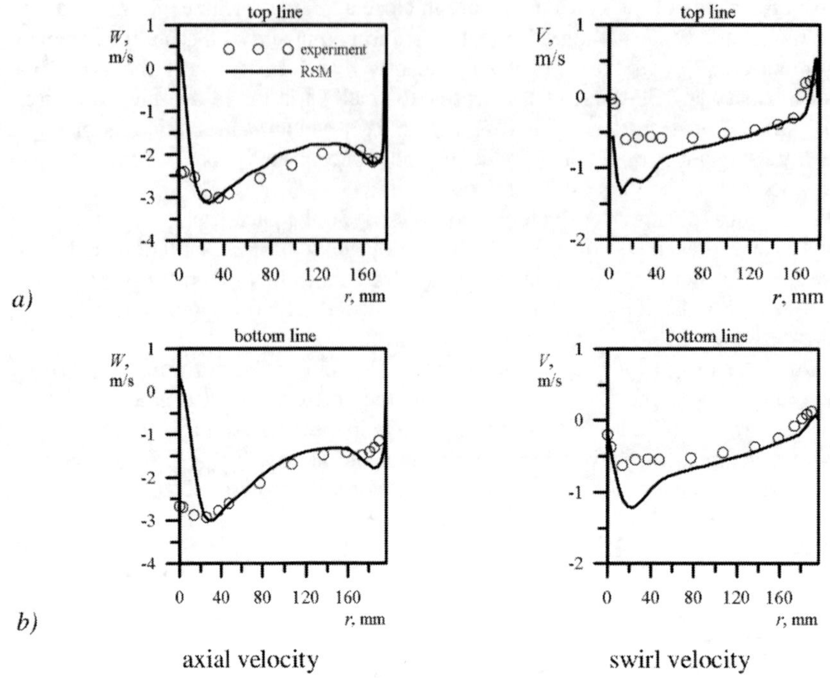

Figure 9. Velocity components in the draft tube, best efficiency point; *a*) top line, *b*) bottom line.

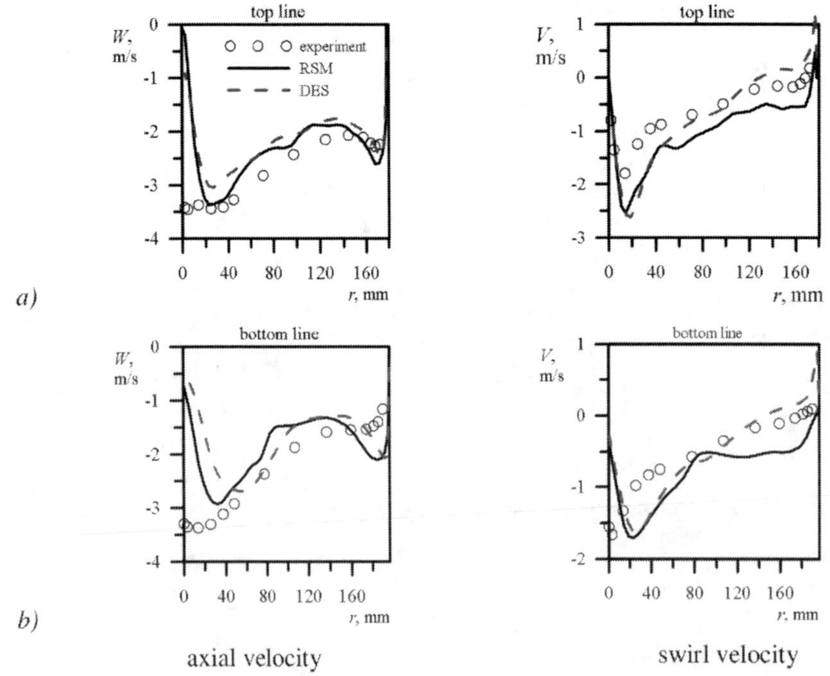

Figure 10. Velocity components in the draft tube, high load; *a*) top line, *b*) bottom line.

Journal of Physics: Conference Series **579** (2015) 012004 doi:10.1088/1742-6596/579/1/012004

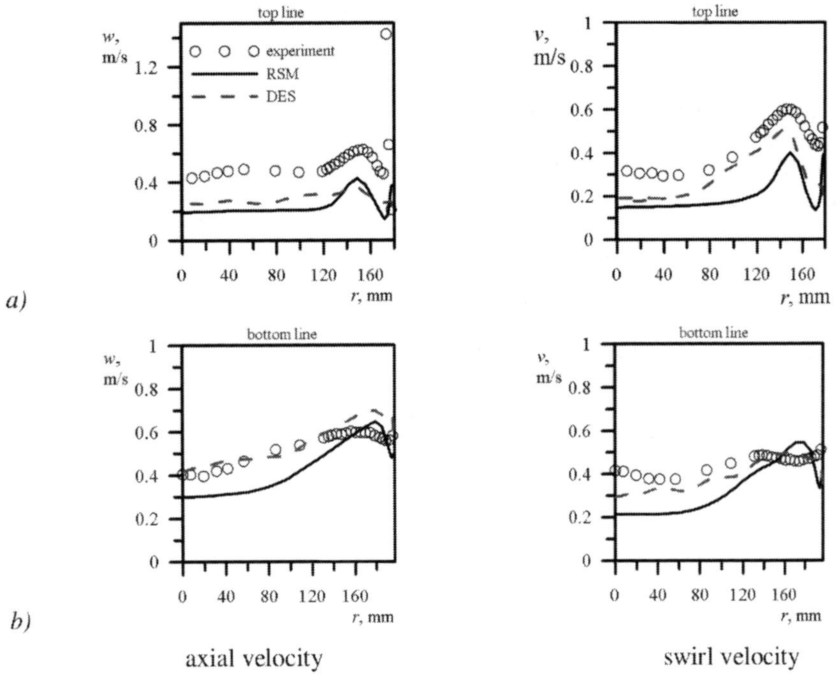

Figure 11. R.m.s. velocity components in the draft tube (DES – resolved, RSM – total), part load; *a*) top line, *b*) bottom line.

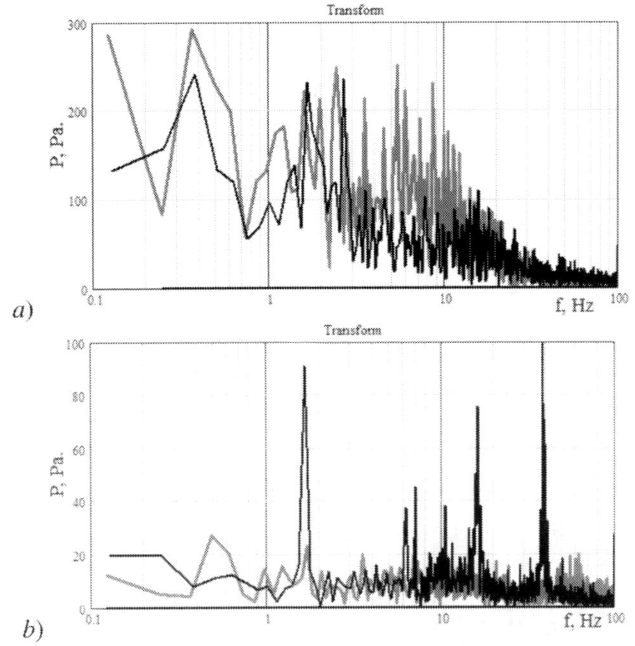

Figure 12. Experimental (black) and calculated (red) spectra of the pressure fluctuations at point DT21 for part load (a) and high load (b).

5. Conclusions

The paper presents the results of simulation of unsteady three-dimensional flow in the Francis-99 water turbine. Modeling technique is based on the DES method and RSM turbulence model and the approach of rotated reference frame for runner zone.

Three operation point proposed for the Francis-99 workshop were calculated. Narrow concentrated vortex is formed in the draft tube cone for HL and BEP operating points. For the PL operation point a wide recirculation zone is formed at the axis in the draft tube cone.

Calculations of turbine parameters agree with experimental results qualitatively. Part load operation point results show the largest discrepancy in efficiency.

Calculated flows in the draft tube cone are in close agreement with experimental data. Averaged velocity components are rather similar both for Reynolds stress model and DES method. At the same time, RSM model underestimated velocity pulsations in the draft tube for part load operating point. On the contrary, for the DES method r.m.s. velocity components in the bottom cross-section of the draft tube agree with experimental data. In the top cross-section both simulations underestimated velocity pulsations, perhaps, due to insufficiently fine computational mesh in the runner.

Acknowledgements

This work was supported in part by the Russian Foundation for Basic Research grants №13-08-98115 and №14-08-31586 and Russian government grant №11.G34.31.0046 designed to support scientific research under the guidance of leading scientists in Russian universities (lead Scientist K. Hanjalić).

References

[1] Doerfler P 1982 System dynamics of the Francis turbine half load surge *Proc. IAHR Symp. on Hydraulic Machinery and Systems* (Amsterdam, Netherlands)

[2] Alekseenko S V, Kuibin P A, Okulov V L and Shtork S I 1999 Helical vortices in swirl flow *J. of fluid mechanics.* **382**. 195–243

[3] Doerfler P, Sick M and Coutu A 2013 *Flow-Induced Pulsation and Vibration in Hydroelectric Machinery. Engineers Guidebook for Planning, Design and Troubleshooting* (London: Springer)

[4] Nicolet C 2007 Hydroacoustic modelling and numerical simulation of unsteady operation of hydroelectric systems *PhD Thesis* (EPFL № 3751, http://library.epfl.ch/theses/?nr=3751)

[5] Nicolet C, Ruchonnet N and Avellan F 2006 One-Dimensional Modeling of Rotor Stator Interaction in Francis Pump-Turbine *Proc. of ISROMAC-11, ASME: International Symp. on Transport Phenomena and Dynamics of Rotating Machinery*

[6] Cherny S G, Chirkov D V, Bannikov D V, Lapin V N, Skorospelov V A, Eshkunova I and Avdushenko A 2010 3D numerical simulation of transient processes in hydraulic turbines *Proc. IAHR Symp. on Hydraulic Machinery and Systems* (Timisoara, Romania)

[7] Ruprecht A 2006 Numerical prediction of vortex instabilities in turbomachinery *Notes on Numerical Fluid Mechanics and Multidisciplinary Design* **93** 211–24

[8] Flemming F, Foust J, Koutnik J and Fisher R K 2008 Overload surge investigation using CFD data *Proc. IAHR Symp. on Hydraulic Machinery and Systems* (Foz do Iguassu, Brazil)

[9] Alligne S, Maruzewski P, Dinh T, Wang B, Fedorov A, Iosfin J and Avellan F 2010 Prediction of a Francis turbine prototype full load instability from investigations on the reduced scale model *Proc. IAHR Symp. on Hydraulic Machinery and Systems* (Timisoara, Romania)

[10] Alekseenko S V, Dekterev A A, Litvinov I V, Minakov A V and Stork S I 2011 Numerical and experimental simulation of the flow in the water turbine draft tube *J. of Siberian Federal University. Engineering & Technologies* **5** 489–503

[11] Spalart P R, Jou W-H, Strelets M, Allmaras S R 1997 Comments on the feasibility of LES for wings and on a hybrid, RANS/LES approach *Advances in DNS/LES: Proc. of 1st AFOSR*

Journal of Physics: Conference Series **579** (2015) 012004 doi:10.1088/1742-6596/579/1/012004

International Conference on DNS/LES In Lue C. and Lue Z. (eds) (Ruston, LA, /Greyden Press, Columbus, 1997 August, 4-8) pp137–47.

[12] Gavrilov A, Dekterev A, Sentyabov A, Minakov A and Platonov D 2012 Application of hybrid methods to calculations of vortex precession in swirling flows *Notes on Numerical Fluid Mechanics* **117** 449–59.

[13] Kuznetsov I, Zakharov A, Orekhov G, Minakov A, Dekterev A and Platonov D 2012 Investigation of free discharge through the hydro units of high head Francis turbine *IOP Conference Series: Earth and Environmental Science* 15 (PART 5)

[14] Sentyabov A V, Gavrilov A A, Dekterev A A and Minakov A V 2013 Analysis of RANS turbulence models by calculating the steady-state flow in the Turbine-99 draft tube *Computational Continuum Mechanics* **6**, № 1, 86–93

[15] Menter F R 1994 Two Equation Eddy Viscosity Turbulence Models for Engineering Applications *AIAA J.* **32**, № 8, 1598-1605

[16] Menter F R, Kuntz M and Langtry R 2003 Ten Years of Experience with the SST Turbulence Model *Turbulence, Heat and Mass Transfer* 4 (Begell House Inc.) In K. Hanjalic, Y. Nagano, and M. Tummers, editors 625–32

[17] Strelets M 2001 Detached Eddy Simulation of Massively Separated Flows. *AIIA Paper.* **2001-0879**

[18] http://www.ltu.se/research/subjects/Stromningslara/Konferenser/Francis-99?l=en

Journal of Physics: Conference Series **579** (2015) 012005 doi:10.1088/1742-6596/579/1/012005

Investigation of different simulation approaches on a high-head Francis turbine and comparison with model test data: Francis-99

Peter Mössinger, Roland Jester-Zürker, Alexander Jung
Voith Hydro Holding GmbH & Co KG, St Pöltener Straße 43, 89522 Heidenheim, Germany

E-mail: peter.moessinger@voith.com

Abstract. Numerical investigations of hydraulic turbo machines under steady-state conditions are state of the art in current product development processes. Nevertheless allow increasing computational resources refined discretization methods, more sophisticated turbulence models and therefore better predictions of results as well as the quantification of existing uncertainties. Single stage investigations are done using in-house tools for meshing and set-up procedure. Beside different model domains and a mesh study to reduce mesh dependencies, the variation of several eddy viscosity and Reynolds stress turbulence models are investigated. All obtained results are compared with available model test data. In addition to global values, measured magnitudes in the vaneless space, at runner blade and draft tube positions in term of pressure and velocity are considered. From there it is possible to estimate the influence and relevance of various model domains depending on different operating points and numerical variations. Good agreement can be found for pressure and velocity measurements with all model configurations and, except the BSL-RSM model, all turbulence models. At part load, deviations in hydraulic efficiency are at a large magnitude, whereas at best efficiency and high load operating point efficiencies are close to the measurement. A consideration of the runner side gap geometry as well as a refined mesh is able to improve the results either in relation to hydraulic efficiency or velocity distribution with the drawbacks of less stable numerics and increasing computational time.

1. Introduction

The increasing availability of computer resources allow the use of more accurate numerical schemes such as advanced turbulence models or higher discretization of the model domain. Therefore the Norwegian University of Science and Technology and Lulea University of Technology jointly are organizing a set of workshops to determine the state of the art of high head Francis turbine simulation. As the first of three upcoming workshops the present one is aimed at steady state simulations of a model turbine in three different operating points. In the pending workshops more sophisticated investigations like transient operating conditions and fluid structure interactions are investigated.

Voith Hydro, as one of the world's leading manufactures in the field of hydropower equipment and services, gathered a broad experience in hydraulic model test and numerical simulations. Increasing demands of computational fluid dynamics (CFD) to support the hydraulic development process lead to a continuous improvement of computer resources and simulation methods. To validate the current development process the provided geometry is used as an input for the in-house mesh and setup procedure. The numerical results are validated with experimental data and can also be compared with the results of other participants.

Journal of Physics: Conference Series **579** (2015) 012005 doi:10.1088/1742-6596/579/1/012005

2. Experimental Setup

2.1. model machine

The model machine used for this test case is the Tokke model, designed at the Norwegian University of Science and Technology (NTNU) and whose prototype turbine ($H = 377\ m$, $P = 110\ MW$) is operating at the Tokke power plant in Telemark, Norway. The turbine, operating at a test rig with open loop water circuit, is scaled down to $1 : 5,1$ model size [3]. Beside the runner with a total of 15 splitters and 15 full length blades the model includes 14 stay vanes inside the volute casing, 28 guide vanes and an elbow type draft tube. Figure 1 shows the meridional section of the model. The sketch also indicates the positions for pressure and velocity measurement.

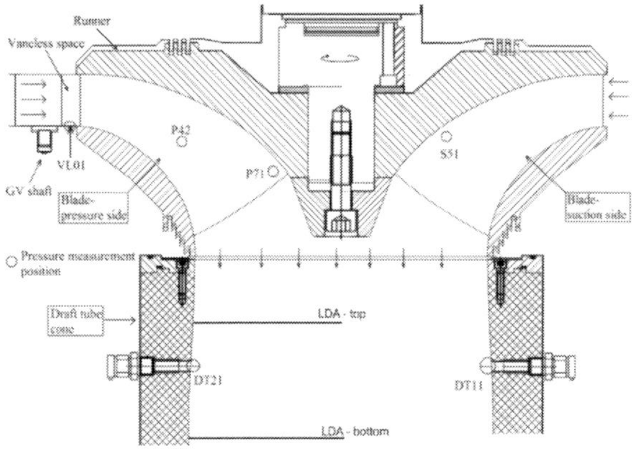

Figure 1: model machine, location of the pressure sensors and LDA measurement sections [3]

2.2. measurement

Three operating conditions were tested, which are visualized in the hill chart (cf. figure 2) and summarized in table 1. Beside the operation point for best efficiency (BEP), a point with very low discharge at part load (PL) and a high load point (HL) was chosen.

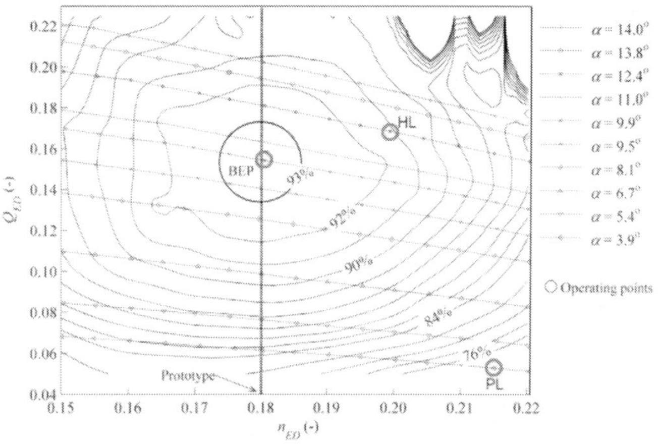

Figure 2: hill chart of the high head Francis turbine model [3]

Journal of Physics: Conference Series **579** (2015) 012005 doi:10.1088/1742-6596/579/1/012005

Table 1: operating points and experimental data for pressure and velocity (*) measurement

Parameter	symbol	unit	PL	BEP	BEP*	HL	HL*
Net head	H	m	12,29	11,91	12,77	11,84	12,61
Flow rate	Q	m^3/s	0,071	0,203	0,21	0,221	0,23
Runner angular speed	n	1/s	6,77	5,59	5,74	6,16	6,34
Guide vane angle	γ	°	3,91	9,84		12,44	
Efficiency	η_h	%	71,7/72,5*	92,6	92,4	90,7	91,0
Speed factor	n_{ed}	-	0,22	0,18		0,20	
Discharge factor	Q_{ed}	-	0,05	0,15		0,17	

Time dependent pressure measurement was realized in both the rotating and stationary frame at six positions according to figure 1. Additional velocity measurements of axial and tangential velocity components were carried out, using a laser Doppler anemometer (LDA) along two horizontal lines in the draft tube (cf. figure 1). Due to induced vibrations in the test rig at BEP and HL which caused interferences with the measurements it was necessary to change the absolute values of head, discharge and runner speed [3]. To maintain the required operating point the non-dimensional quantities n_{ed} and Q_{ed} were kept constant (cf. table 1).

3. Numerical Setting

3.1. solution setup

All simulations are performed with the software ANSYS CFX v15.0. CFX is an established and well validated fluid solver and allows very high accuracy in the prediction of complex turbulent flows [2]. It uses a finite volume based discretization scheme up to second order accuracy for convective fluxes and truly second order accuracy for diffusive fluxes. Due to the demanding geometry, particular the guide vane fillets and the runner splitter blades, which results in a reduced grid quality, a physical timescale control of $\Delta t = 10^{-3}$ s is necessary to gain numerical stability. Convergence is achieved with root mean square residuals of pressure, mass-momentum and turbulence parameters of $r \leq 10^{-3}$ and a total of iteration steps with elapsed pseudo time which exceeds the total domain advection time of factor 5 for simulations with spiral case and factor 10 for all other simulation runs.

During the analyses different kinds of two-equation eddy viscosity (k$-\epsilon$ and k$-\omega-$SST [5, 8, 13]) and Reynolds stress models (BSL-EARSM, BSL-RSM [1, 11]) are applied. As a main difference between these two approaches, the two-equation models using an eddy viscosity hypothesis for the Reynolds stresses, which relates them linearly to the mean velocity gradient:

$$\overline{u_i u_j} = \nu_T S_{ij} - \frac{2}{3}\delta_{ij} \tag{1}$$

while the explicit Reynolds stress model (EARSM) gives a nonlinear relation between the Reynolds stresses and the mean strain-rate, whereas the Reynolds stress model (RSM) solves all additional six terms.

3.2. mesh and boundary conditions

In steady-state simulations the interface (IF) between the stationary and the rotating frame can be modeled either using a circumferential averaging of the flow variables (*stage interface*) or by changing the reference frame while the relative orientation of the components across the interface is fixed (*frozen rotor*). The following computations are done using a stage interface frame

Journal of Physics: Conference Series **579** (2015) 012005 doi:10.1088/1742-6596/579/1/012005

change model between the stationary frame (guide vanes) and the runner. By circumferentially averaging the flow variables it is sufficient to model only one passage of stay vane, wicket gate and runner, applying rotational periodic interfaces. For the present case narrow gaps between stay and guide vanes impede a combined meshing of the two stationary blade rows in one domain. Meshing is done with support of an in-house tool, using scaled geometry and block-structured hexahedral cells, ensuring constant mesh conditions when varying the guide vane angle and furthermore allowing a rapid change of mesh parameters or interface positions. In this case an additional frozen rotor interface is applied between the two domains (cf. figure 3a). Figure 3b shows the respective mesh. While the runner is modeled in the same passage-periodic way, a

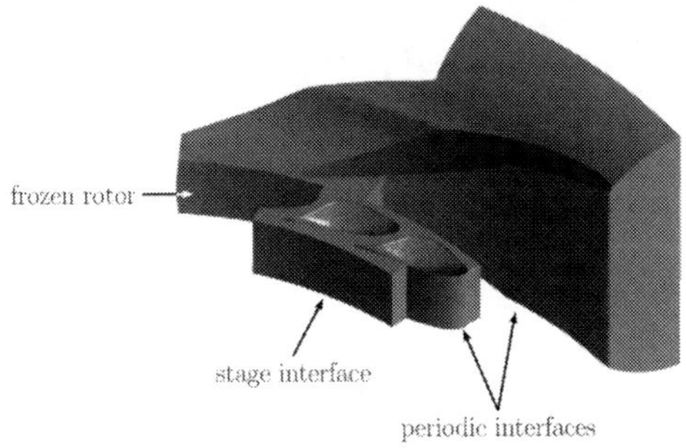

(a) boundary conditions

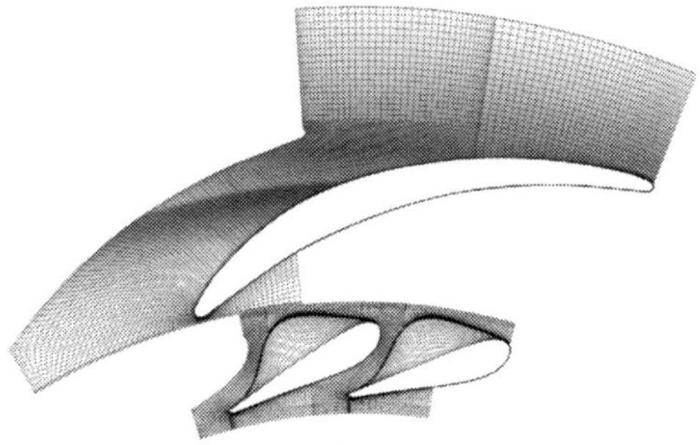

(b) mesh

Figure 3: numerical model of distributor (stay vane [SV] and wicket gate [WG])

full draft tube mesh is required (cf. figure 4b,c). The stay vanes in the present geometry extend far in the spiral case which prevents a separate modeling. Therefore the spiral case together with the stay vanes are meshed manually and, to avoid the frozen rotor interface, coupled with all 28 guide vanes by a general grid interface (*GGI*) for the two non-matching grids (cf. figure 4a). Hence four different domains lead to a variation of possible geometry combinations, summarized in table 2. As a basic geometry, G_1 contains all domains except for the spiral case. This configuration is expected to yield the best results for a numerical simulation. The influence of the spiral case is investigated with variant G_2. The draft tube plays a major role in terms of hydraulic efficiency and velocity distribution beneath the runner. Variant G_3 is used to investigate this influence. A transition from the runner passage to the draft tube is realized by circumferentially averaging the flow variables. Therefore in variant G_4 a shifted interface

(a) spiral case (SC)

(b) runner (RU)

(c) draft tube (DT)

Figure 4: numerical model of spiral case, runner and draft tube

Table 2: resulting domain variations

Model Type	G_1	G_2	G_3	G_4
spiral case	✗	✗	✓	✗
stay vane	✓	✓	✓	✓
wicket gate	✓	✓	✓	✓
runner	✓	✓	✓	✓
draft tube	✓	✗	✓	✓
shifted DT-IF	✗	✗	✗	✓
possible insight	reference	DT influence	SC influence	DT-IF influence

beneath both measurement planes allows the evaluation of a direct runner outflow. The meshed geometries result in grid densities which are summarized in table 3. Beside the basic coarse grid (M_2), with a total of 3,6 Mio. nodes the necessity of a combined spiral case, stay vane mesh results in a mesh M_1 with 20 Mio. nodes. To evaluate the grid sensitivity a refined mesh M_3 is used after the analysis of all basic calculations and having identified the best configuration. To allow the use of in-house pre- and post-processing procedures the runner rotational speed is kept constant at $n_{BEP} = 5,59$ s^{-1}. Flow rate and head are adjusted accordingly in order to achieve constant non-dimensional values of n_{ed} and Q_{ed}.

$$n_{ed} = \frac{nD}{\sqrt{gH}} \tag{2}$$

$$Q_{ed} = \frac{Q}{D^2\sqrt{gH}} \tag{3}$$

To adjust the particular operating point the measured head is multiplied with the efficiency, yielding to the so-called Euler head:

$$H_{eul,exp} = H \cdot \eta_{h,exp} \tag{4}$$

where the Euler head from the simulation is calculated with the flow velocities at the runner entrance and exit:

$$H_{eul,CFD} = \frac{1}{g}\left(u_2 c_{2u} - u_1 c_{1u}\right) \tag{5}$$

Table 3: used grid densities and resulting quality parameters

Grid type	M_1	M_2	M_3
Distributor+spiral case	17,969,250	-	-
Distributor	-	1,343,873	3,700,014
Runner	636,889	636,889	4,124,348
Draft tube	1,633,239	1,633,239	3,150,259
Total nodes ($\times 10^6$)	20,24	3,61	10,97
Min. element angle	8,3°	15,2°	13,3°
Max. aspect ratio	290	254	565
Max. expansion factor	789	101	166
Mean y^+	53	51	32

By correcting the guide vane angle in a manner to match the required Euler head within ±1% the similarity between the velocity fields before and after the runner for the experiment and simulation is ensured. As boundary condition a mass flow at the inflow according to table 1 and corrected with eq. 3 is applied. Additionally the inlet flow angle is set to 8.5° and turbulence parameters are likewise set to the fixed values of 5% intensity and an eddy length scale of 10^{-2} m. The domain outlet is defined with a zero static pressure condition allowing backflow by switching to total pressure based on the normal component of velocity.

4. Results

4.1. average pressure

Six pressure taps along the hydraulic waterway (cf. figure 1) measured the pressure at all three operating points. Due to the arbitrary pressure level of the numerical calculation the first sensor location at the runner blade pressure side P42 is used to identify the absolute pressure level. Figure 5 summarizes the percentage differences between the measured and the calculated pressure. The simulations are based on the basic domain (G_1) with the coarser grid (M_2) and the k-ω-SST turbulence model. In general a good agreement between the experimental

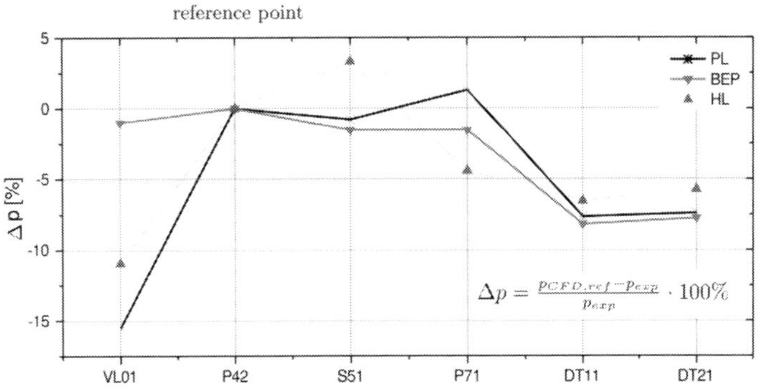

Figure 5: comparison between numerical and experimental pressure at five positions

and numerical results is found. For the blade pressure taps the differences amount to ±5%, whereas in the draft tube at all operating points the pressure is calculated about −7% lower than measured. In the vaneless space deviations up to −15% are found. A local mesh improvement has no considerable influence. The procedure of an adjusted wicket gate angle to reach the required Euler head leads to a slightly different pressure distribution in this area and thus may contribute to the deviations. It is expected that pressure levels are not varying in a significant manner for different model domains. The same applies by altering turbulence models, as already concluded by Trivedi et al. [10]. Furthermore, according to table 1, operating points for pressure measurements differ from those of the velocity measurements. Therefore the focus on proceeding numerical simulations lies in the operating conditions of the velocity measurements and comparing the resulting velocity distribution along the draft tube sections.

4.2. velocity distribution

Velocity measurements are carried out along two horizontal sections in the draft tube according to figure 1. The components available are in axial (c_{ax}) and tangential (c_u) direction. For a comparison with numerical data, velocities are normalized with the discharge at the respective operating point and the draft tube inlet area.

$$c_0 = \frac{Q_{in,DT}}{A_{in,DT}} \tag{6}$$

A first evaluation compares the different turbulence models. As in the previous chapter, simulations were done for the basic domain on the coarser grid. Further investigations evaluate the influence of different domains and a second, refined mesh. For part load operating conditions (cf. figure 6) it is found a satisfied agreement of the axial component between all turbulence models and the measurement data. Only near the wall slight deviations are visible, where

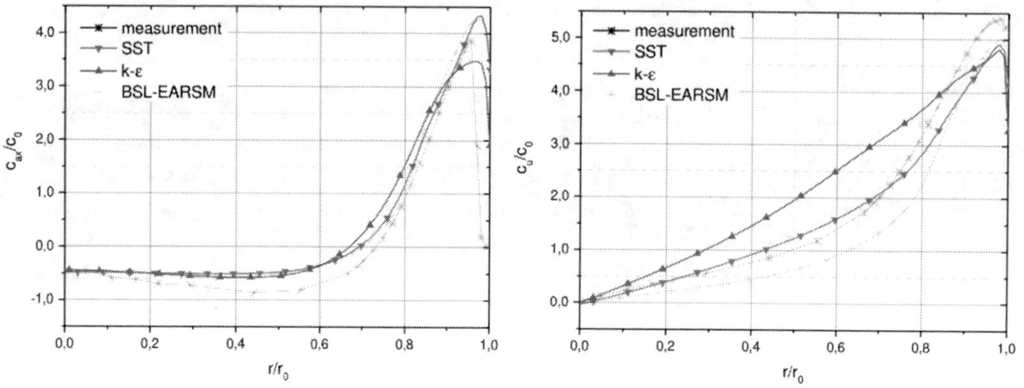

(a) top measurement plane

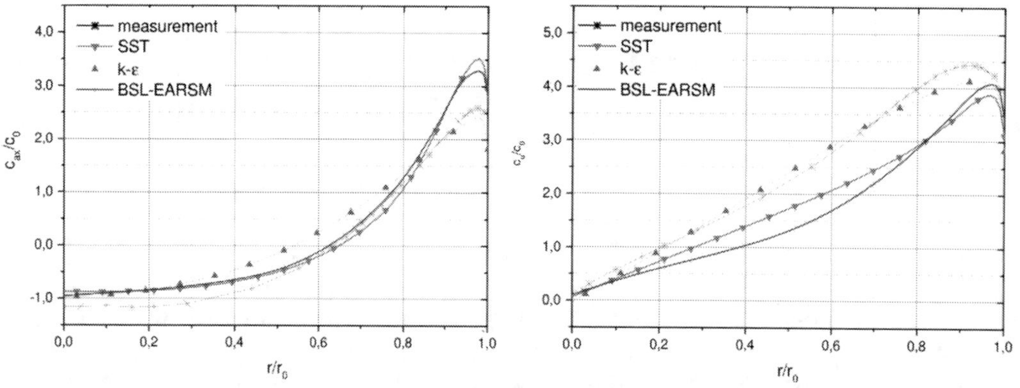

(b) bottom measurement plane

Figure 6: axial (left) and circumferential (right) velocity distribution at part load

the measurement shows a velocity drop at the top plane. Especially in this region a higher measurement resolution would be beneficial to evaluate this behavior in near-wall region. SST and BSL-EARSM model are over-predicting the velocity close to the walls, where the $k-\epsilon$ model exhibits a reduced axial component. The tangential velocity distributions are showing deviations of the numerical result at part load condition. Beside the simplification of turbulence anisotropy in the two-equation models which results in a reduced accuracy for swirl flows, also cavitation phenomena are not considered. However it is noticeable that the standard SST-model and the improved BSL-EARSM match especially the trend of the velocity gradient very well, yet under-predicting the swirl, where the $k-\epsilon$ model shows a higher tangential velocity without reproducing the increasing values near the wall due to a limited behavior in rotational flows and near-wall regions. For the BSL-RSM turbulence model it is not possible to gain a numerical stable solution at this part load operating point.

At best efficiency $k-\epsilon$, SST and BSL-EARSM show similar behavior, with increasing deviations for the axial flow near the center line and the draft tube wall (cf. figure 7). For the BSL-RSM approach larger discrepancies in axial velocities near the draft tube core occur. In this area the recirculation length is insufficiently predicted, resulting in increasing deviations near the draft tube center. Especially for the tangential velocity all turbulence models predict a drop near the center, indicating a rotational outflow. However, as the measurement confirms, the runner shows at the best efficiency point a nearly swirl-free outflow behavior.

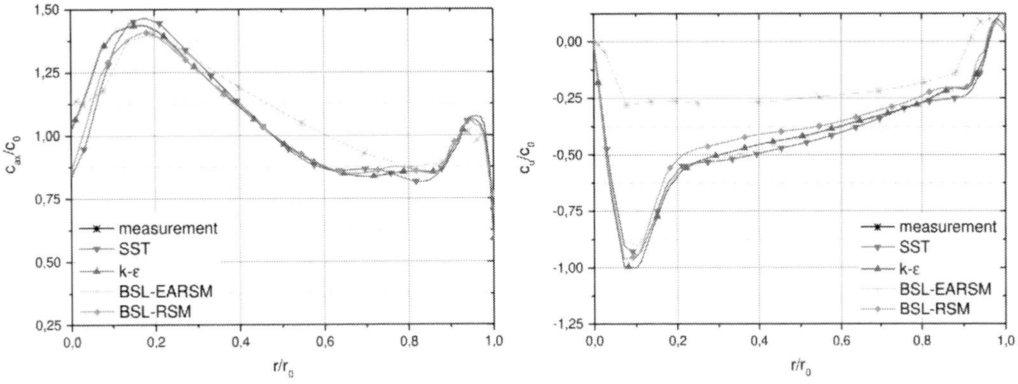

(a) top measurement plane

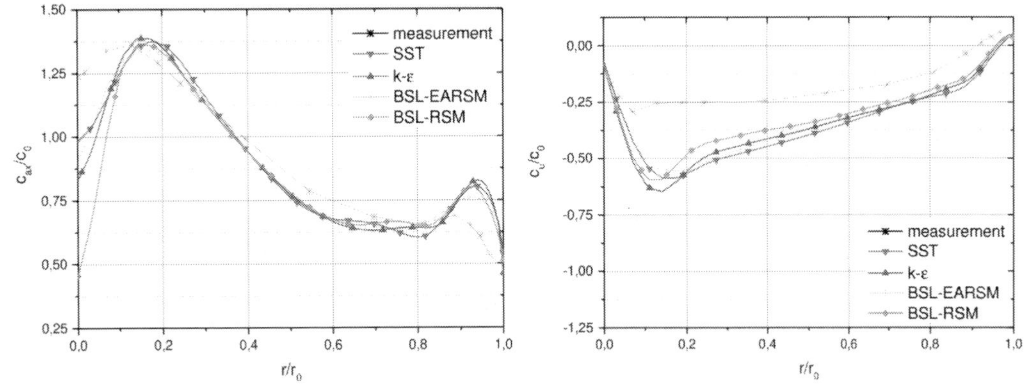

(b) bottom measurement plane

Figure 7: axial (left) and circumferential (right) velocity distribution at best efficiency

Similar results in comparison with experimental data for the axial velocity can be obtained at the high load operating point. The trend of the measurements are reproduced well by all turbulence models with remaining deviations in the core zone. At high load operating conditions a counter-rotating flow at the runner exit occurs which is captured by the simulation (cf. figure 8), although the decrease in streamwise direction is over-predicted. Again only minor differences between the turbulence models are visible. In general a best fit of the axial component is found at the part load whereas at other operating points some deviations can be observed especially in the core region of the draft tube. Beside some discrepancies in the core and near wall regions and although the use of improved turbulence models show a remarkably small influence, the accordance between experimental and numerical results are satisfying. However with the focus on stable, accurate and first of all efficient solutions the BSL-RSM approach seems not

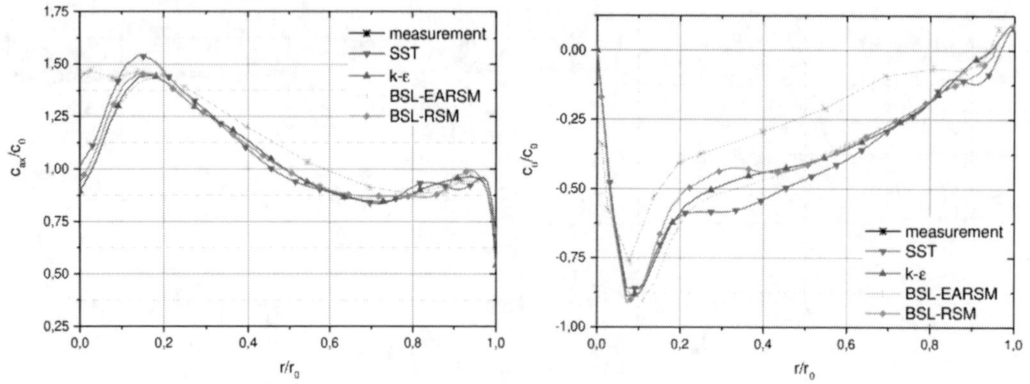

(a) top measurement plane

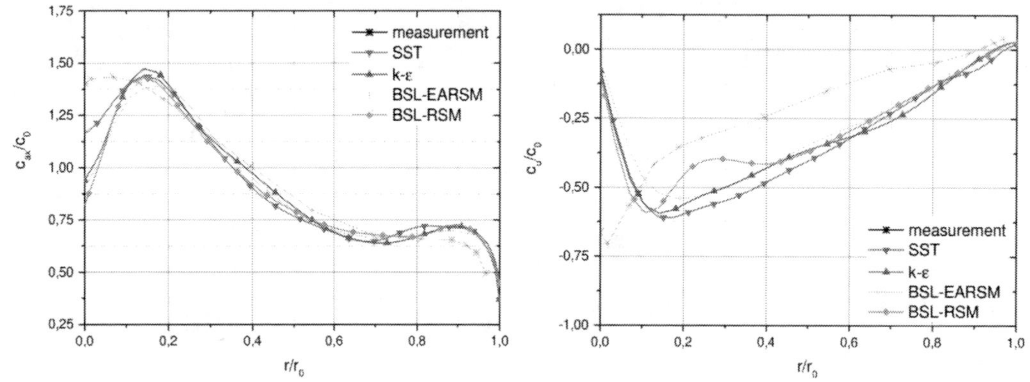

(b) bottom measurement plane

Figure 8: axial (left) and circumferential (right) velocity distribution at high load

to be appropriate for today development processes. In a further study adding and neglecting of different domains, such as spiral case and draft tube is investigated in order to indicate the significance for velocity distributions. To keep an overview only the evaluations with major influences are shown.

4.3. domain variation

According to table 2, there are four different domain modifications under investigation (G_1-G_4). Configuration G_1 is the already discussed reference model. In a first modification the draft tube of G_1 is shortened to the cone-shaped part of the draft tube (G_2). A downstream extension of $5 \times D_{DT}$ is necessary to avoid influences of the outlet boundary conditions on the measurement section. In variation G_3 the spiral case is added to the original domain. To circumvent an interface between stay vanes and guide vanes all 28 guide vanes are taken into account, leaving only one interface between guide vanes and runner. The last modification investigates the influence of circumferentially averaging the flow at the runner outlet on the draft tube velocity field. As shown in figure 9 the interface is shifted beneath the measurement region at the end of the conical section, allowing to evaluate the velocities in the rotating domain (G_4). However, as with geometry G_3, due to the curved periodical passage it is not possible to directly compare the flow at the straight measurement line. Therefore an average over the horizontal section is necessary. The results at different operating conditions are similar to previous investigations,

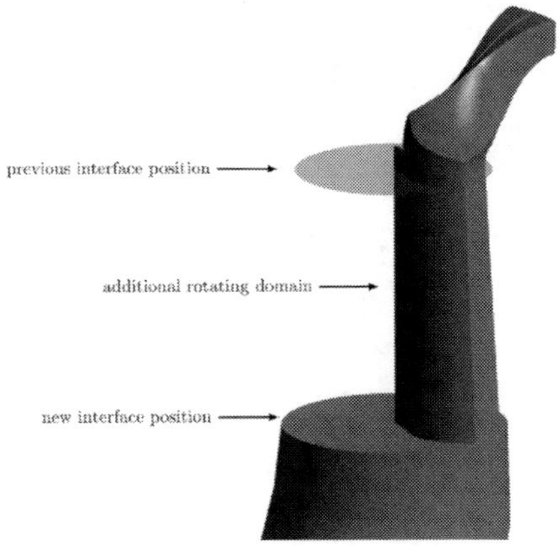

Figure 9: shifted interface in streamwise direction

though two characteristics are mentionable. According to figure 10 the neglected draft tube has an influence on the tangential velocity at part load only, whereas for all other configurations no deviations compared to the reference solution are found. However a drawback of this approach is a missing statement on the hydraulic efficiency, as the draft tube losses are unavailable. At best efficiency and high load the domain variations have minor influences, where the spiral case and reference model as well as the shifted interface and the neglected draft tube show almost identical characteristics (cf. figure 11). While the velocities of configuration G_1 and G_3 (+SC) are evaluated in the stationary frame of the draft tube, the velocities of G_2 (-DT) and G_4 (DT-IF) are calculated in the rotating frame. Therefore it can be stated that a circumferentially averaging of the velocities has a major effect compared to different domain modifications.

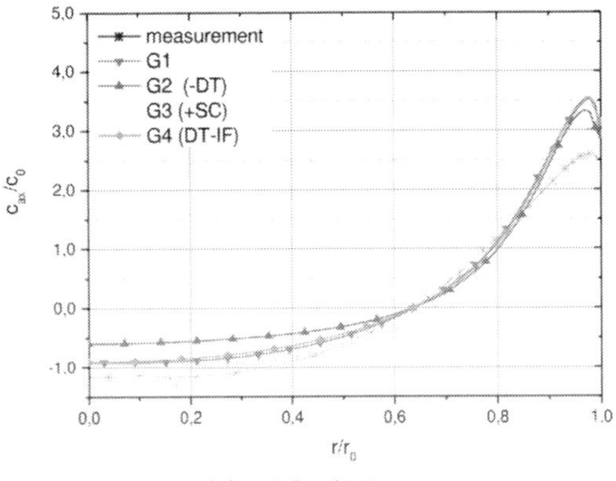

(a) axial velocity

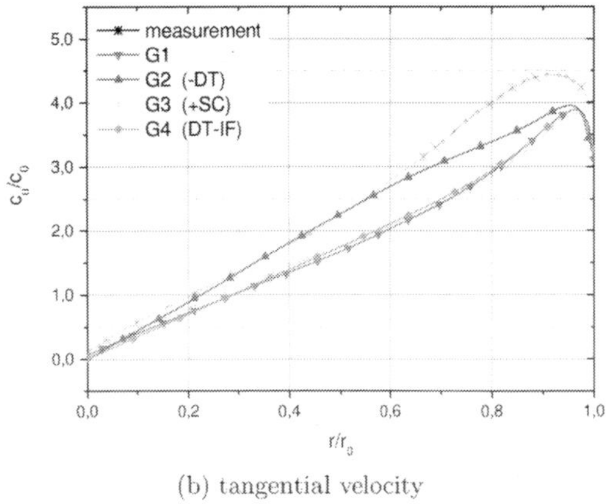

(b) tangential velocity

Figure 10: velocity distribution at part load for different domain models at bottom measurement plane

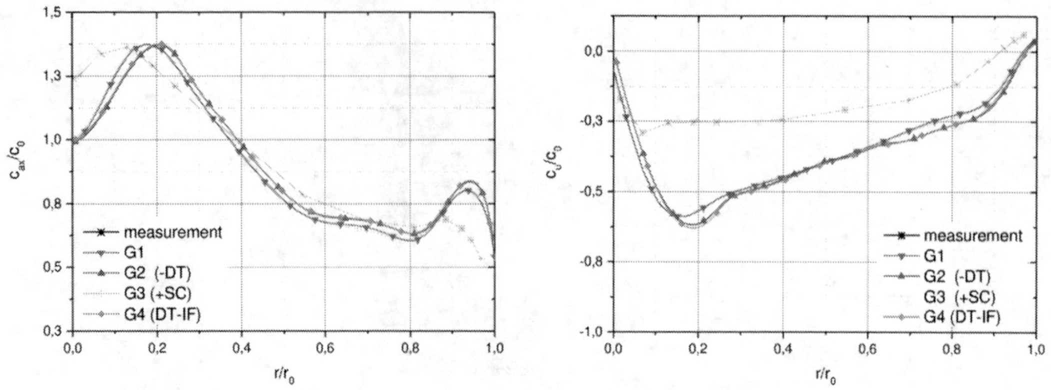

(a) axial velocity (b) tangential velocity

Figure 11: velocity distribution at best efficiency for different domain models at bottom measurement plane

4.4. hydraulic efficiency

According to IEC 60193 Annex N [6] the turbine efficiency is defined as:

$$\eta_h = \eta_E \eta_v \eta_r \tag{7}$$

where due to the small dimensions of the model machine buoyancy effects can be neglected and eq. 7 results in [6]:

$$\eta_h = \left(1 - \frac{1}{\rho g H}\sum_{k=c} \Delta p_{tot,k}\right)\left(1 - \frac{q' + q''}{Q}\right) - \frac{P_{Ld}}{\rho g H Q} \tag{8}$$

Here c stands for the components spiral case, guide vane, wicket gate, runner and draft tube. Seal leakage losses of crown (q') and band (q'') as well as disk friction losses (P_{Ld}) are estimated from a investigation of the runner side gap (cf. chapter 4.5). Spiral case losses are chosen accordingly to the simulations in chapter 4.3 (domain variation G_3) and considered for all test-cases. Figure 12 shows a comparison of the reference configuration with the measurement. All

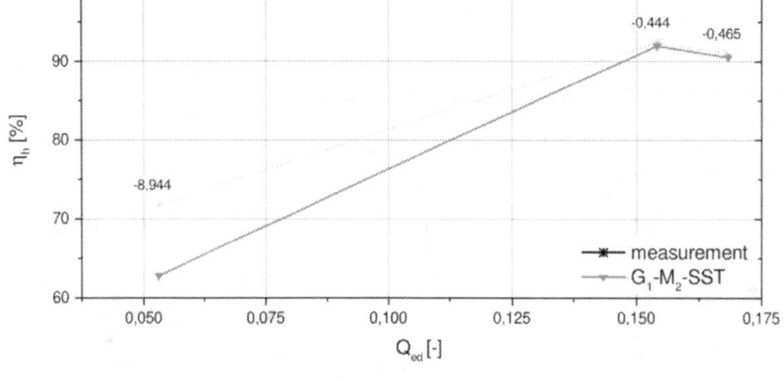

Figure 12: experimental and numerical hydraulic efficiency for three operating points

results are summarized in table 4. A negative percentage value refers to an under-predicted hydraulic efficiency of the simulation compared to the measurement, whereas a positive value means an over-prediction. Model G_2 is excluded in this evaluation, because the missing draft tube inhibits a statement of the total hydraulic losses. According to table 4 the same tendency for all calculations is observable. At part load conditions the efficiency is under predicted up to 9% for the SST turbulence model. Deviations up to 14% at low discharge operating conditions between numerical simulations and the measurement for this model case were already discussed by Trivedi et al. [10]. For the remaining two operating points, deviations of efficiency vary between -1,4% and 0,15%. Again absolute deviations of different variations regarding model domain or turbulence model are in a very small range. The BSL-RSM approach results in best accordance with the measurements. On the contrary at part load occurring stability issues and for BEP/HL an over-predicted recirculation area in the draft tube center is found.

Table 4: deviations of hydraulic efficiency for different turbulence models and model domains compared to measurements

OP absolute deviation	PL	BEP	HL
measurement η_h [%]	71,7	92,4	91,0
G_1-SST	-8,94	-0,44	-0,47
G_1-k-ϵ	-7,08	-0,70	-1,09
G_1-BSL-EARSM	-7,55	-0,69	-1,35
G_1-BSL-RSM	-	0,08	0,15
G_3-SST	-6,59	-0,28	0,01
G_4-SST	-6,94	-0,41	-0,15

4.5. runner side gap

Another area of investigation are the runner side gaps. With the input of basic geometry seal data, an in-house software is able to mesh the existing seals (cf. figure 13a). In a first approach these geometries of crown and band can be included in the numerical model for simulation (cf. figure 13b, *RSG-SST* in table 5). A second method considered the leakage losses with an analytical approach to calculate the seal leakage, disc friction losses as well as axial thrust based on formula from Gülich [4], Lorenz [7], Stampa [9] and Weber [12]. The resulting losses can be

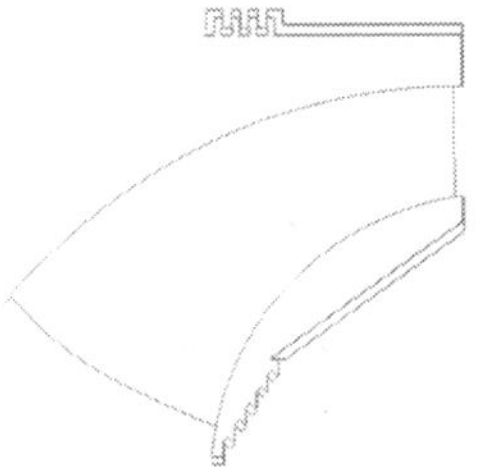

(a) model seals

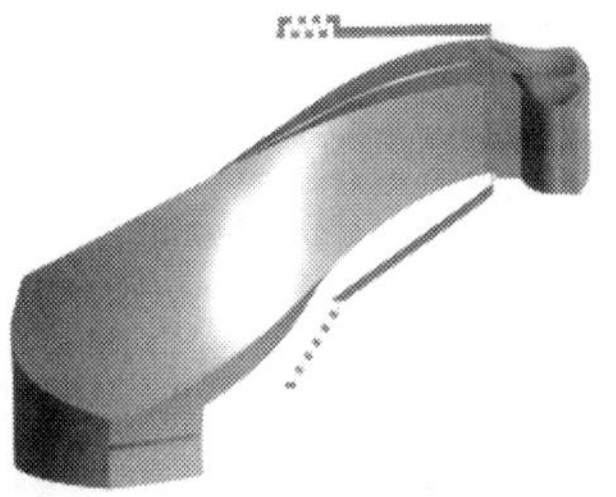

(b) seals in numerical model

Figure 13: seals and implementation in the numerical model

taken into account in the loss analysis for the hydraulic efficiency according to eq. 8 as done in all previous simulations (cf. table 4). For the hydraulic efficiency it is possible to improve the prediction as shown in table 5. As a result in this case the seal leakage and disc friction losses are over-predicted in the analytical approach. However a consideration of the three-dimensional seals in the numerical model leads to very small mesh sizes in streamwise direction, resulting in stability issues especially at operating points apart from best efficiency. Therefore it can be stated that if the seals are not under special investigation it is sufficient, especially in terms of numerical stability and efficiency, to consider the seal leakage in an analytical way and take it into account for the resulting hydraulic efficiency.

Table 5: deviations of hydraulic efficiency for a runner side gap consideration compared to measurements

OP absolute deviation	PL	BEP	HL
measurement η_h [%]	71,7	92,4	91,0
G_1-SST	-8,94	-0,44	-0,47
RSG-SST	-6,45	0,00	-0,20

4.6. optimized model

Since there are no relevant improvements by using higher sophisticated turbulence models, at least not in a order to justify the increasing complexity, numerical instability and computational time, the $k-\omega-$SST turbulence model is used for the following configuration. Furthermore the spiral case has no significant influence on the velocity distribution, whereas modeling the draft tube is vital, especially at part load conditions. Therefore the domain variation under investigation is G_1 with the refined mesh M_3. Although the total mesh size is tripled, the improvements vary within a very small range. At part load no change of the velocity distribution is observable, whereas at best efficiency (cf. figure 14) and high load a slightly shift towards the measurement occurs. The axial velocity field remains similar, while the appearing swirl near the runner hub is decreasing. With a higher mesh density better predictions of the hydraulic

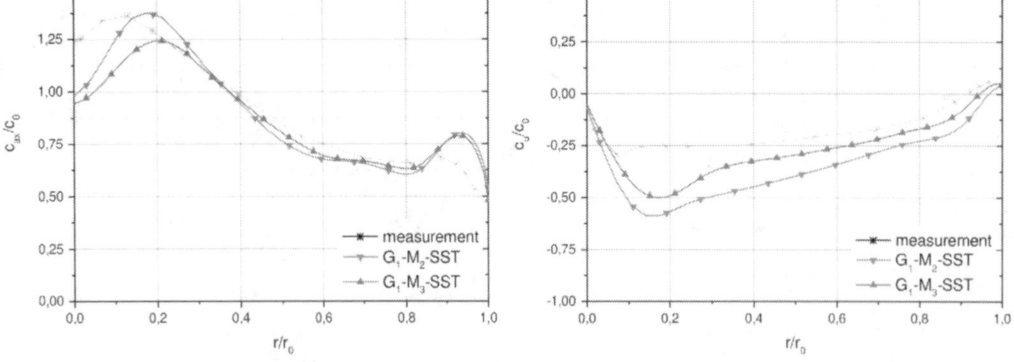

(a) axial velocity (b) tangential velocity

Figure 14: velocity distribution at best efficiency with a refined mesh at bottom measurement plane

Journal of Physics: Conference Series **579** (2015) 012005 doi:10.1088/1742-6596/579/1/012005

losses are possible. This results in a reduced hydraulic efficiency which in this case, as the efficiency is already underrated for BEP and HL, leads to a greater deviation in comparison with measurement data (cf. table 6).

Table 6: deviations of hydraulic efficiency for a refined mesh model compared to measurements

OP absolute deviation	PL	BEP	HL
measurement η_h [%]	71,7	92,4	91,0
G_1-M_2-SST	-7,98	-0,44	-0,47
G_1-M_3-SST	-5,89	-0,82	-0,65

5. Discussion

Increasing computational resources allow the use of more advanced turbulence models as well as refined grid densities and the simulation within reasonable time spans. A high head Francis model turbine is modeled for one passage of distributor and runner, applying periodical boundary conditions, with a full draft tube. Three operating points at part load, best efficiency and high load are under investigation. With the in-house procedure a fast, reliable meshing as well as setup-procedure and post-processing is possible. Pressures at six locations in the rotating and stationary frame are analyzed and differences are found to $\pm5\%$ in the runner, -7% in the draft tube and up to -15% in the vaneless space when compared to the experimental data. The application of different two-equation and Reynolds stress turbulence models showed minor influences on the velocity distribution in the draft tube beneath the runner. However with the BSL-RSM turbulence model the recirculation area beneath the runner is over-predicted which led to deviations for the velocity field in the draft tube in comparison with other turbulence models. Overall the $k-\omega-$SST turbulence model confirmed the best solution in terms of robustness, stability and accuracy. Domain variations indicated an influence of the draft tube at part load conditions on the occurring swirl, where at other operating points the interface position at which a circumferential averaging of the flow variables is carried out showed the major effect. Hydraulic efficiency is well predicted at the operating points BEP and HL with a tendency to smaller values, where at PL larger deviations occurred, which were also discussed by other authors. The included runner side gaps improved the prediction of the hydraulic efficiency with the drawbacks of increasing computational time and reduced numerical stability. A refined mesh reduced discrepancies of the tangential velocity at best efficiency while the increasing losses resulted in larger deviation of the hydraulic efficiency. Although the results showed good agreement in terms of velocity and hydraulic efficiency, further investigations for efficiencies at part load conditions and velocity distribution at best efficiency operating point are necessary. Further mesh refinements or the detailed trailing edge meshing will influence the over-predicted swirl at best efficiency. Also the use of scale resolving turbulence models as large eddy simulations (LES) could improve the simulation results.

Nomenclature

Term	*Symbol*	*Unit*	**Indices**	
Velocity, absolute	c	m/s	0	normalized
Diameter	D	m	1	runner inlet
Gravity acceleration	g	m/s^2	2	runner outlet
Hydraulic head	H	m	ax	axial
Turbulent kinetic energy	k	m^2/s^2	eul	Euler

Term	*Symbol*	*Unit*
Runner angular speed	n	1/s
Speed factor	n_{ed}	-
Pressure	p	Pa
Disk friction losses	P_{Ld}	W
Leakage losses	q', q''	m^3/s
Discharge	Q	m^3/s
Discharge factor	Q_{ed}	-
Residuum	r	-
Mean strain rate tensor	S_{ij}	1/s
Time	t	s
Velocity, components in CCS	u_i, u_j	m/s
Velocity, circumferential	u	m/s
Dimensionless wall distance	y^+	-
Guide vane angle	γ	°
Kronecker delta	δ_{ij}	-
Difference	Δ	-
Efficiency	η	%
Turbulence eddy viscosity	ν_T	m^2/s
Density	ρ	kg/m^3

Indices

exp	experimental
E	specific energy
in	inlet
h	hydraulic
ref	reference
tot	total
u	circumferential
v	volumetric

Abbreviations

BSL	Baseline turbulence model
CCS	Cartesian coordinate system
CFD	Computational fluid dynamics
BEP	Best efficiency
DT	Draft tube
EARSM	Explicit algebraic RSM
G_1-G_4	Domain variation
HL	High load
IF	Interface
LDA	Laser Doppler Anemometer
M_1-M_3	Mesh variation
PL	Part load
RSG	Runner side gap
RSM	Reynolds stress model
RU	Runner
SC	Spiral case
SST	Shear stress turbulence model
SV	Stay vanes
WG	Guide vanes

References

[1] ANSYS Inc: *ANSYS CFX-Solver Theory Guide, release 15.0*, 2011
[2] Aschenbrenner, T.; Otto, A.; Moser, W.:, *Classification of vortex and cavitation phenomena and assessment of CFD prediction capabilities*, 23rd IAHR Symposium on Hydraulic Machinery and Systems, 2006
[3] Cervantes, M.: *Francis-99 Workshop*, Lulea University of Technology, http://www.ltu.se/research/subjects/Stromningslara/Konferenser/Francis-99, accessed August 15, 2014
[4] Gülich, J.F.: *Disk friction losses of closed turbomachine impellers*, Forschung im Ingenieurwesen, vol. 68, p. 87-95, 2003
[5] Hoffmann, K.; Chiang S.: *Computational Fluid Dynamics*, vol. 3, Engineering Education System, 2000
[6] International Electrotechnical Comission; *IEC 60193: Hydraulic turbines, storage pumps and pump turbines - Model acceptance test*, International Electrotechnical Comission, Geneva, Switzerland, 1999
[7] Lorenz, J.: *Untersuchung von Spaltdichtungen für hydraulische Strömungsmaschinen*, Master-Thesis, University of Konstanz 1986
[8] Menter, F.: *Improved two equation k-ω turbulence model for aerodynamic flows*, NASA technical memorandum 103975, 1992
[9] Stampa, B.: *Experimentelle Untersuchung an axial durchströmten Ringspalten*, PhD-Thesis, Technische Universität Braunschweig, 1971
[10] Trivedi, C.; Cervantes, J.M.; Ghandi, B.K.; Dahlhaug, O.G.: *Experimental and numerical studies for a high head Francis turbine at several operating points*, Journal of Fluid Engineering, vol. 135, 2013
[11] Wallin, S; Johansson A.: *An explicit algebraic Reynolds stress model for incompressible and compressible turbulent flows*, Journal of Fluid Mechanics, vol. 403, pp. 89-132, 2000
[12] Weber, D.: *Experimentelle Untersuchung an axial durchströmten kreisringförmigen Spaltdichtungen für Kreiselpumpen*, PhD-Thesis, Technische Universität Braunschweig, 1971
[13] Wilcox, D.: *Turbulence Modelling for CFD*, DCW Industries, Inc., 1994

Numerical simulation of flow in a high head Francis turbine with prediction of efficiency, rotor stator interaction and vortex structures in the draft tube

D Jošt[1], A Škerlavaj[1], M Morgut[1], P Mežnar[1] and E Nobile[2]

[1] Turboinštitut, Ljubljana, Slovenia
[2] University of Trieste, Department of Engineering and Architecture, Italy

E-mail: dragica.jost@turboinstitut.si

Abstract. The paper presents numerical simulations of flow in a model of a high head Francis turbine and comparison of results to the measurements. Numerical simulations were done by two CFD (Computational Fluid Dynamics) codes, Ansys CFX and OpenFOAM. Steady-state simulations were performed by k-ε and SST model, while for transient simulations the SAS SST ZLES model was used. With proper grid refinement in distributor and runner and with taking into account losses in labyrinth seals very accurate prediction of torque on the shaft, head and efficiency was obtained. Calculated axial and circumferential velocity components on two planes in the draft tube matched well with experimental results.

1. Introduction

Nowadays, numerical simulations of Francis turbines are quite accurate. In our experience (using a commercial CFD code, namely ANSYS-CFX), the difference between experimentally and numerically obtained values of efficiency is usually less than one percent [1]. Furthermore, differences in numerical results due to geometry modifications follow the same trend as obtained with measurements. The results can be further improved by performing simulations of labyrinth seals.

Three workshops for Francis turbines are announced by the Norwegian University of Science and Technology (NTNU), Norway, and Luleå University of Technology (LTU), Sweden, in order to further validate the capabilities of the CFD technologies. As a matter of fact, the goal of the first workshop is to determine the state-of-the-art of numerical predictions for steady operating conditions. The second and the third workshop will be focused on transient operation of turbines and fluid-structure interaction, respectively. In all workshops, a high-head Francis turbine model (Tokke model) will be employed as a reference test case. The model was designed and experimentally measured at Water Power Laboratory at NTNU [2].

In order to further improve its expertise in CFD, University of Trieste and Turboinštitut joined in the ACCUSIM (Accurate Simulations in Hydro-Machinery and Marine Propellers) project with primary aim to develop reliable, high fidelity methods for the accurate prediction, and optimization, of the performances of hydro-machinery and marine propellers. This paper is a result of the synergic collaboration of both partners at the first Francis 99 workshop.

The paper is focused on efficiency prediction but also a rotor stator-interaction and prediction of vortex structures in the draft tube will be investigated. Besides predicted efficiency also the values of torque, head (or flow rate if head is input data) and pressure fluctuations are compared to the measured values. Details of flow field where the effects of the rotor-stator interaction and vortex structures are significant are presented.

The capabilities of two different well-known CFD codes were further evaluated. The simulations were carried out using ANSYS-CFX 15.0, a commercial CFD solver, and OpenFOAM-2.1.1, an open source CFD toolbox.

2. Numerical strategy

2.1. Numerical models

Steady-state simulations wire performed with the most widely used turbulence models, standard *k*-ε and shear-stress-transport (SST) [3, 4]. The SST became a standard in turbo machinery. Both numerical codes, OpenFOAM and CFX, follow the same versions of the turbulence models, with the same values of constants.

Due to our positive past experience [5, 6], unsteady simulations were performed with zonal large-eddy-simulation (ZLES) [7, 8]. The main idea of the ZLES model is to resolve the flow inside a predefined zone with the LES model, and the rest of the domain with the Reynolds-averaged Navier-Stokes (RANS) model. In the presented simulations, the zone of the ZLES model was defined within the scale-adaptive-simulation SST (SAS SST) [9] turbulence model. The zone started just after the interface between the runner and the draft tube, and it is extended to the outlet of the computational domain.

In our research of Kaplan turbines [5, 6] it was concluded that CC and KL improve the accuracy of predictions. KL reduces flow energy losses at stagnation points, therefore its effect was significant in stay and guide vane cascades. CC acted mostly in the draft tube, where it reduced flow energy losses, especially for large discharge. Due to positive effects of CC and KL for Kaplan turbines also in this research most of the simulations with CFX were performed in combination with curvature correction (CC) [10] and Kato Launder limiter of production term (KL) [11] (see table 1). In simulations with OpenFOAM code CC and KL were not included.

For the *k*-ε model scalable wall functions were used. ω-based turbulence models (SST, SAS SST) use automatic near-wall treatment that allows a gradual switch between wall functions and low-Reynolds number method. The LES zone uses a WMLES formulation [12] at walls.

As far as the discretization of the advection term in steady-state simulations is concerned, the 'high resolution' (in CFX) scheme and linearUpwindscheme (in OpenFOAM) were used. In ZLES simulations the bounded central difference scheme was used.

Table 1. List of simulations performed

Case Nr.	Software	Grid (basic/new)	Turb. Model	CC+KL	Bound. Cond. (Q/H)	Labyrinth seals source/sink
01	CFX	BG	SST	Y	Q	
02	CFX	BG	SST	N	Q	
03	CFX	BG	SST	Y	H	
04	CFX	BG	*k*-ε	Y	Q	
05	CFX	BG	*k*-ε	N	Q	
06	CFX	BG	*k*-ε	Y	H	
07	CFX	BG	ZLES	Y	Q	
08	CFX	NG	SST	Y	Q	
09	CFX	NG	SST	N	Q	
10	CFX	NG	SST	Y	H	
11	CFX	NG	SST	Y	Q	Y (Part Load)
12	CFX	NG	*k*-ε	Y	Q	
13	CFX	NG	*k*-ε	N	Q	
14	CFX	NG	*k*-ε	Y	H	
15	CFX	NG	ZLES	Y	Q	
16	CFX	NG	ZLES	Y	H	
17	OF	BG2	SST	N	Q	
18	OF	BG2	*k*-ε	N	Q	

2.2. Computational grids

The Tokke turbine consists of a spiral casing, 14 stay vanes, 28 guide vanes, a runner with 15 full length blades and 15 splitters, and a draft tube. Simulations were performed on model size of the turbine on block-structured grids. Computational grids that were kindly provided by the organizers of Francis 99 workshop are denoted as basic grids (BG). The set of grids that were prepared by the authors, for CFX simulations, are denoted as new grids (NG). A combination of basic grids provided by the organizer and new grids done by the authors was not possible due to different approach in meshing. The BG consisted of three domains: distributer (spiral casing, stay vanes and guide vanes) with 3.6 million nodes, runner with 5.9 million nodes and draft tube with 3.6 million nodes. The geometry of the NG is the same as that of BG. The NG (figure 1) consisted of six domains: spiral casing, stay vanes, guide vanes, runner, draft tube and an extension of the draft tube. The extension was added in order to move the outlet boundary away from the region of interest. Losses in the draft tube extension were not included in calculation of head and efficiency. In case of OpenFOAM simulations BG2 grid was used. BG2 is the same as the BG, only the grid in the draft tube region was modified. The modification was necessary because of solver stability problems. It seems that the problem was related to the discretization of the rotor – draft tube interface. Number of nodes in all grids can be seen in table 2. A difference between number of nodes in BG and in NG is insignificant, but NG is more refined near the walls, especially at stay and guide vanes and runner blades. Therefore NG is more suitable for ω-based turbulence models SST and ZLES. An advantage of BG is that it has, contrary to the NG, no general grid interfaces (GGI) between spiral casing and stay vane cascade and between stay and guide vane cascades.

In BG an averaged values of y^+ (in ANSYS CFX, at BEP) in distributor and runner are equal to 185 and 49.3, respectively. In NG averaged values of y^+ in spiral casing, stay vane cascade and guide vane cascade are equal to 70.9, 22.8 and 17.2, respectively, while in the runner averaged value of y^+ is 6.1.

With OpenFOAM, values of y^+ on same grids in the distributor and runner were much smaller than in CFX. This is related to the fact that OpenFOAM is based on a Cell-Centered Finite Volume Method while CFX employs the node-centered finite volume method (more precisely the Control Volume-Based Finite Element Method-VFEM)

Figure 1. New grid (NG) with draft tube extension

Table 2. Number of nodes in computational grids

	BG	BG2	NG
Spiral casing			1,461,359
Stay vanes			3,318,588
Distributer	3,607,016	3,607,016	
Guide vanes			887,852
Runner	5,850,470	5,850,470	3,818,400
Draft tube	3,639,241	4,186,580	4,533,681
Extension			574,685
Total	13,096,727	13,644,066	14,594,565

Journal of Physics: Conference Series **579** (2015) 012006 doi:10.1088/1742-6596/579/1/012006

A coupled simulation of labyrinth seals and flow in a turbine is not reasonable because of high computational costs. Therefore, de-coupled simulations of flow in labyrinth seals were performed. Block-structured numerical grids of an eight-degree periodic section in circumferential direction were created. In case of labyrinth seal at shroud, a basic grid with 9 million elements and a refined grid with 63.6 million elements were used. Since the results were nearly the same on both grids, only a basic grid with 10 million nodes was created for labyrinth seal at hub.

2.3. Boundary conditions

Most of the simulations were done with flow rate prescribed at the inlet and average pressure at the outlet. For steady state simulations frozen rotor condition was prescribed between the distributor and the runner and between the runner and the draft tube while transient rotor stator condition was used for transient simulations. When measured value of flow rate was input data (besides geometry and rotating speed), values of head, torque on the shaft and efficiency were calculated from numerical results. Prescribed flow rate is suitable for comparison of calculated velocity distribution with measured results.

In a design process of a new turbine, a flow rate corresponding to a certain guide vane opening is not known in advance. Therefore it is important to know how accurate numerical prediction is when a value of head is input data and a value of flow rate is a result of numerical simulation. For that reason numerical simulations were done also with total pressure prescribed at the inlet. In this case head is input data, while a value of flow rate is a result of numerical simulation.

Simulations of flow in labyrinth seals were performed in order to obtain volumetric and torque losses. At the inlet and outlet of the bottom labyrinth seal, pressure for opening type boundary condition was prescribed. Pressure values were determined from results of the turbine simulations with the SST turbulence model on BG. Discharge through labyrinth seal and torque on the rotating walls were the results of the simulations. For each operating point two simulations for different values of pressure at the inlet and outlet were performed. The volumetric and torque losses for the other boundary conditions (pressure values) were obtained by interpolations.

In case of labyrinth seal at hub, torque on rotating walls is the result of the simulation. It depends only on rotational speed, therefore only one numerical simulation per operating point was performed.

3. Discussion of results

Numerical simulations were performed and compared to the measurements at three operating points: Part Load (PL), Best Efficiency Point (BEP) and High Load (HL).

3.1. Comparison of calculated and measured velocity field in the draft tube

Organizers of Francis-99 workshop performed measurements of mean and fluctuating axial (U and u') and circumferential (V and v') velocity with laser Doppler anemometry (LDA). Measurements were performed along Line 1 (at Z = -0.2434 m) and Line 2 (at Z = -0.5614 m) in the draft tube. Positive U is defined in streamwise direction, whereas V is positive in the runner rotational direction.

Measured mean axial and circumferential velocity components along Line 1 and Line 2 were compared with numerical results obtained with solvers CFX and OF, using different turbulence models and different grids. At first all simulations with CFX solver were performed with Curvature Correction (CC) and Kato Launder limiter (KL). In simulations with OpenFOAM CC and KL were not included. In order to correctly compare the results of CFX and OF, steady state simulations with CFX were repeated without CC and KL. In all simulations with the ZLES model CC and KL were included.

The comparison of axial and circumferential velocity components with experimental results is presented in figures 2, 3a, 3b and 3c. For steady state simulations the results were analysed at every 50 iterations and it was found out that velocity distribution on Line 1 and Line 2 stabilized after sufficient number of iterations, except at PL when SST CC KL was used. The variation of velocity components during the simulation with SST CC KL at PL is presented in figure 2. It has to be mentioned that in case of steady-state simulations results depend upon the relative position of Line 1 and Line 2 to the runner blades. Namely, due to frozen rotor condition the position of wakes behind the blades is fixed during the simulation and no mixing due to runner rotation is taken into account. For transient simulations with the ZLES model in figures 3a, 3b and 3c, velocities were obtained from statistical averaging over last 7 runner revolutions.

At PL distribution of velocity components along Line 1 and Line 2 depends on turbulent model being used, while the differences due to different grids (BG, BG2, NG) and different solvers (CFX and OF) are negligible. The difference due to using CC and KL is significant. Therefore, all steady state results can be classified into four groups: *k*-ε, *k*-ε CC KL, SST and SST CC KL. At PL in an additional steady state simulation volumetric losses due to labyrinth seals were taken into account in such a way that before the runner a mass sink and behind the runner a mass source were prescribed (see subsection 3.4). This simulation was performed on slightly modified NG with SST CC KL turbulence model. The effect of volumetric losses on axial and circumferential velocity components is visible, but small.

At BEP and HL velocity distribution along Line 1 and Line 2 no longer depends only on turbulence model, different results were obtained also due to different grid and different solver. All simulations predicted too small axial velocity in the middle of the draft tube (at $r = 0$). The region of disagreement in the middle of the draft tube is wider when OF solver was used. Circumferential velocity component at the BEP and at the HL is most accurately predicted by the ZLES model. All other models predicted too large circumferential velocity in opposite direction of runner rotation at the region in the middle of the draft tube.

It is difficult to conclude which turbulence model provided the best results. For example, at PL, the axial velocity component obtained with *k*-ε model agrees well with measurement near the wall but the agreement is much worse in the middle of the draft tube. On the contrary, SST and ZLES predicted much too large axial velocity component near the wall but they are more successful in the middle of the draft tube. For circumferential velocity component on Line 1 the best agreement was obtained with *k*-ε CC KL and with SST, while on Line 2 the best agreement was obtained with *k*-ε.

Velocity fluctuations are presented in figure 4a and 4b. Fluctuating velocities predicted with ZLES agree with the experimental results well. Larger disagreement was obtained only at PL for v'.

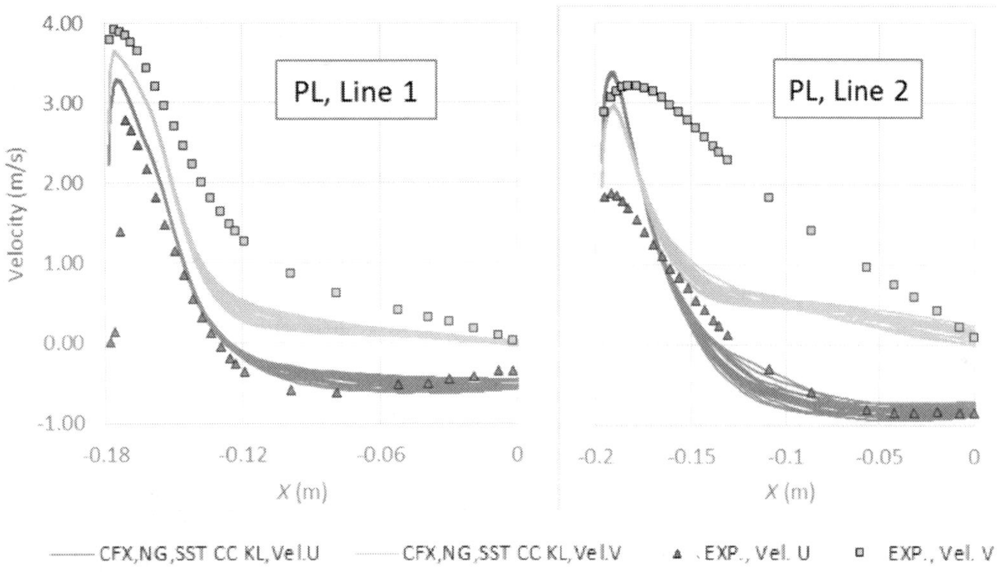

Figure 2. Velocities U and V during steady state simulations with SST CC KL at PL.

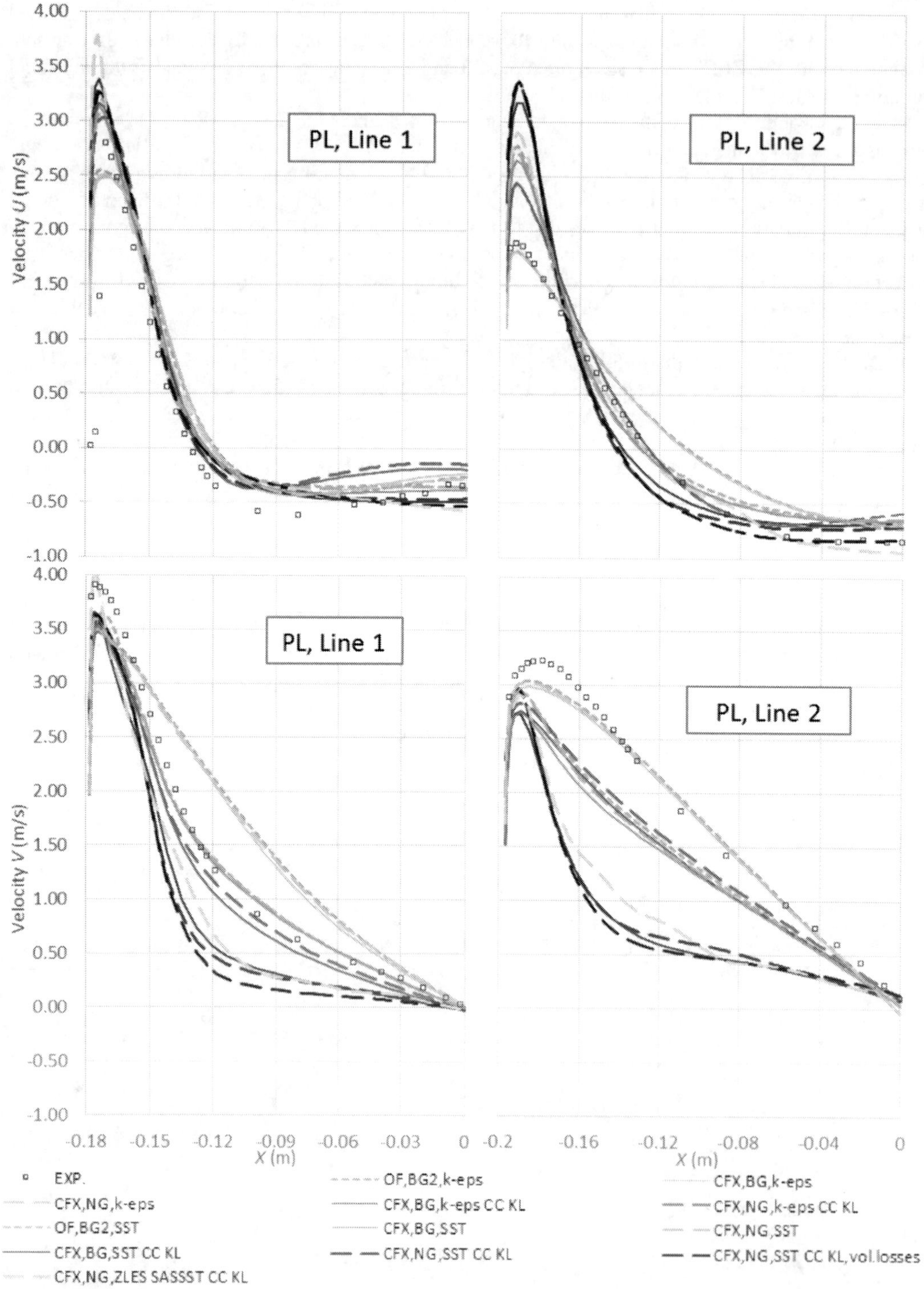

Figure 3a. Comparison of velocities U and V at PL. Numerical results obtained using OF and CFX in combination with different computational grids and turbulence models.

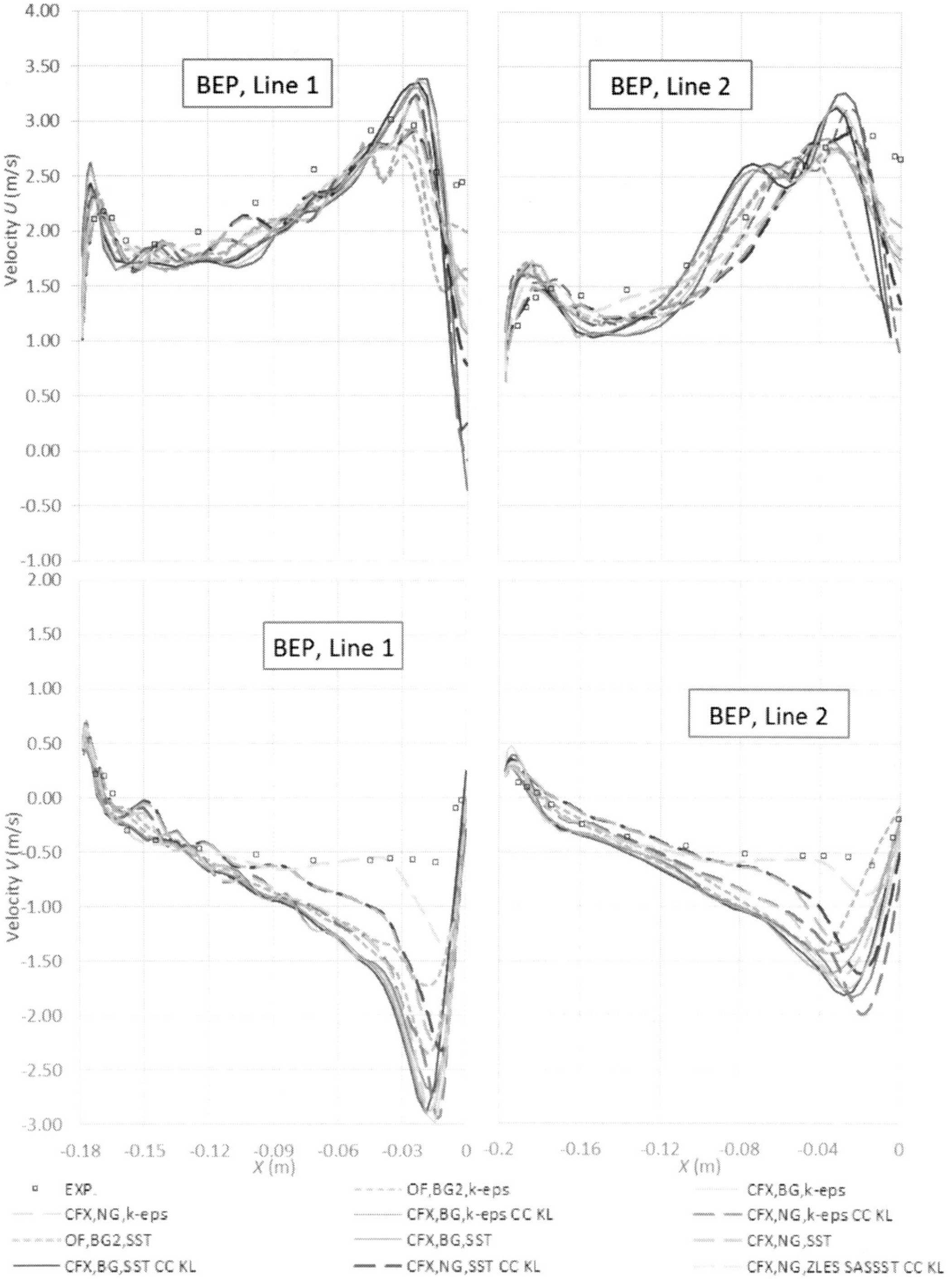

Figure 3b. Comparison of velocities U and V at BEP. Numerical results obtained using OF and CFX in combination with different computational grids and turbulence models.

Journal of Physics: Conference Series **579** (2015) 012006 doi:10.1088/1742-6596/579/1/012006

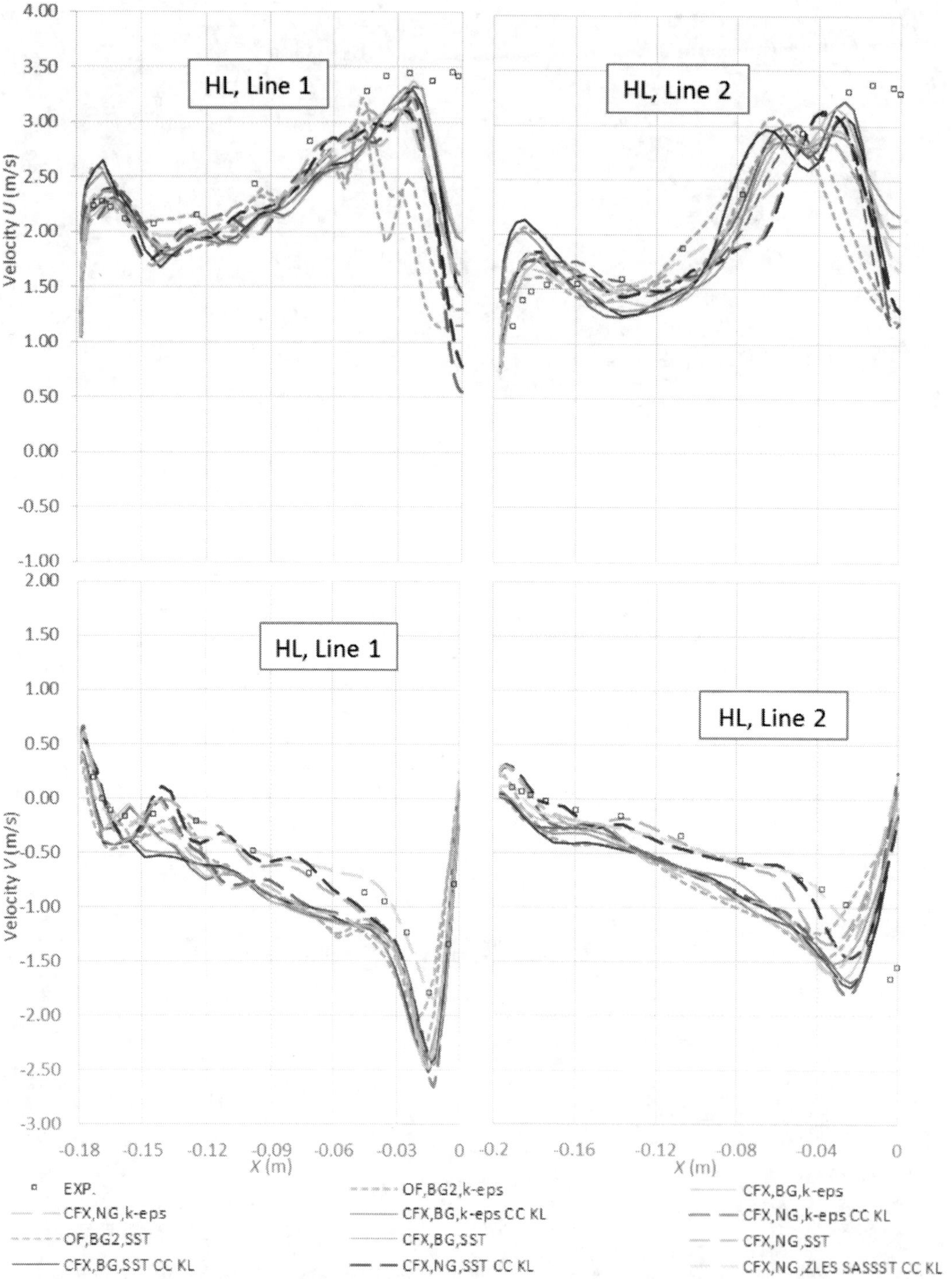

Figure 3c. Comparison of velocities U and V at HL. Numerical results obtained using OF and CFX in combination with different computational grids and turbulence models.

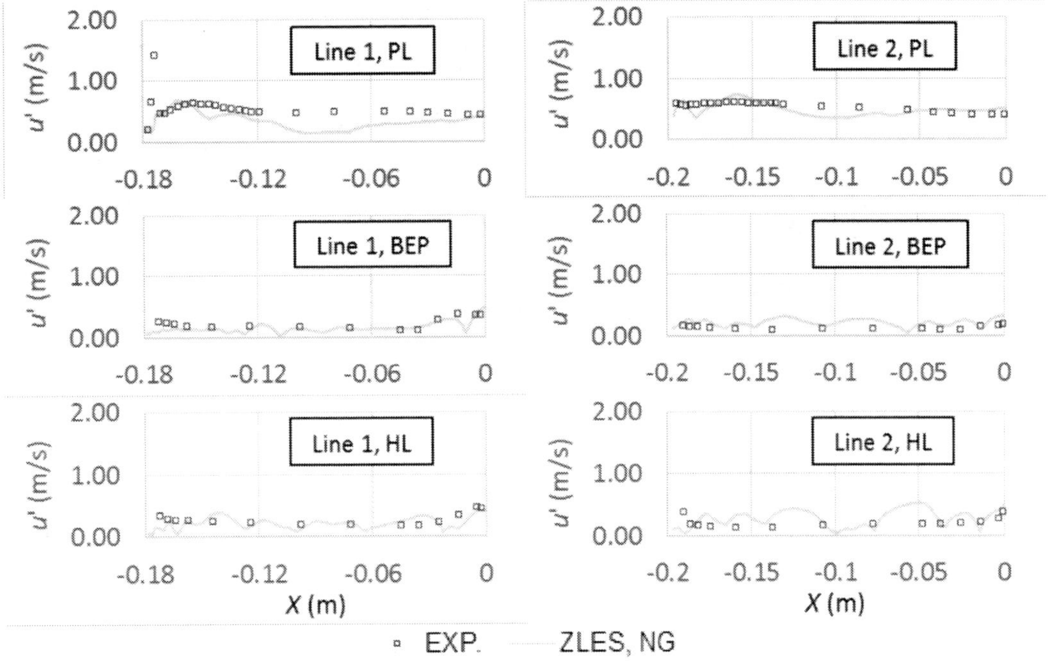

Figure 4a. Fluctuating axial velocity at Line 1 and Line 2. Comparison of experimental and CFX results on NG obtained with ZLES model.

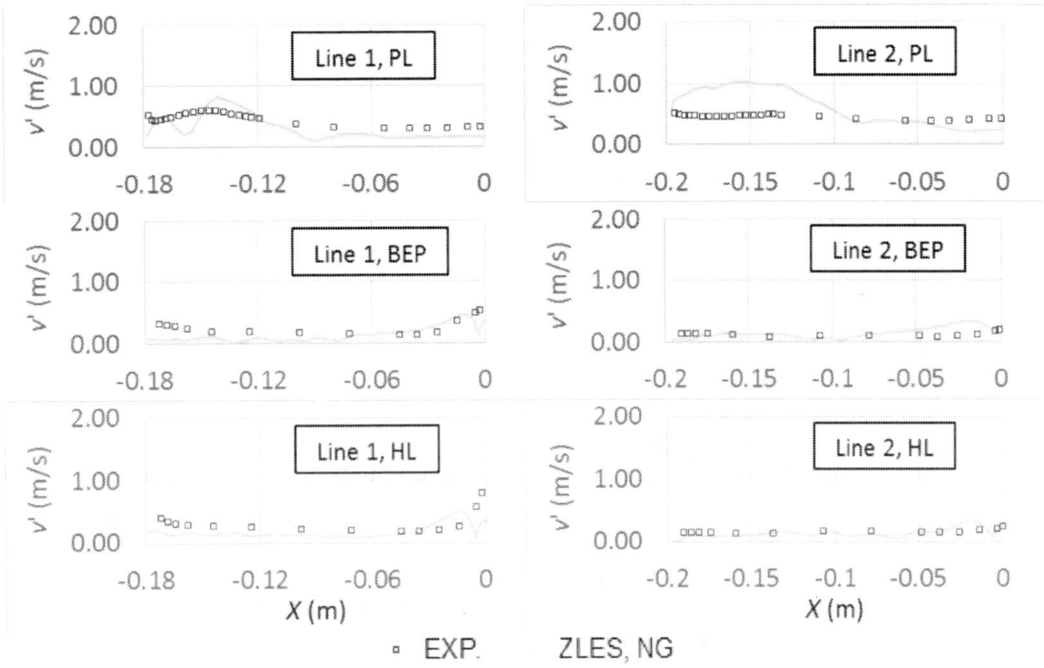

Figure 4b. Fluctuating circumferential velocity at Line 1 and Line 2. Comparison of experimental and CFX results on NG obtained with ZLES model.

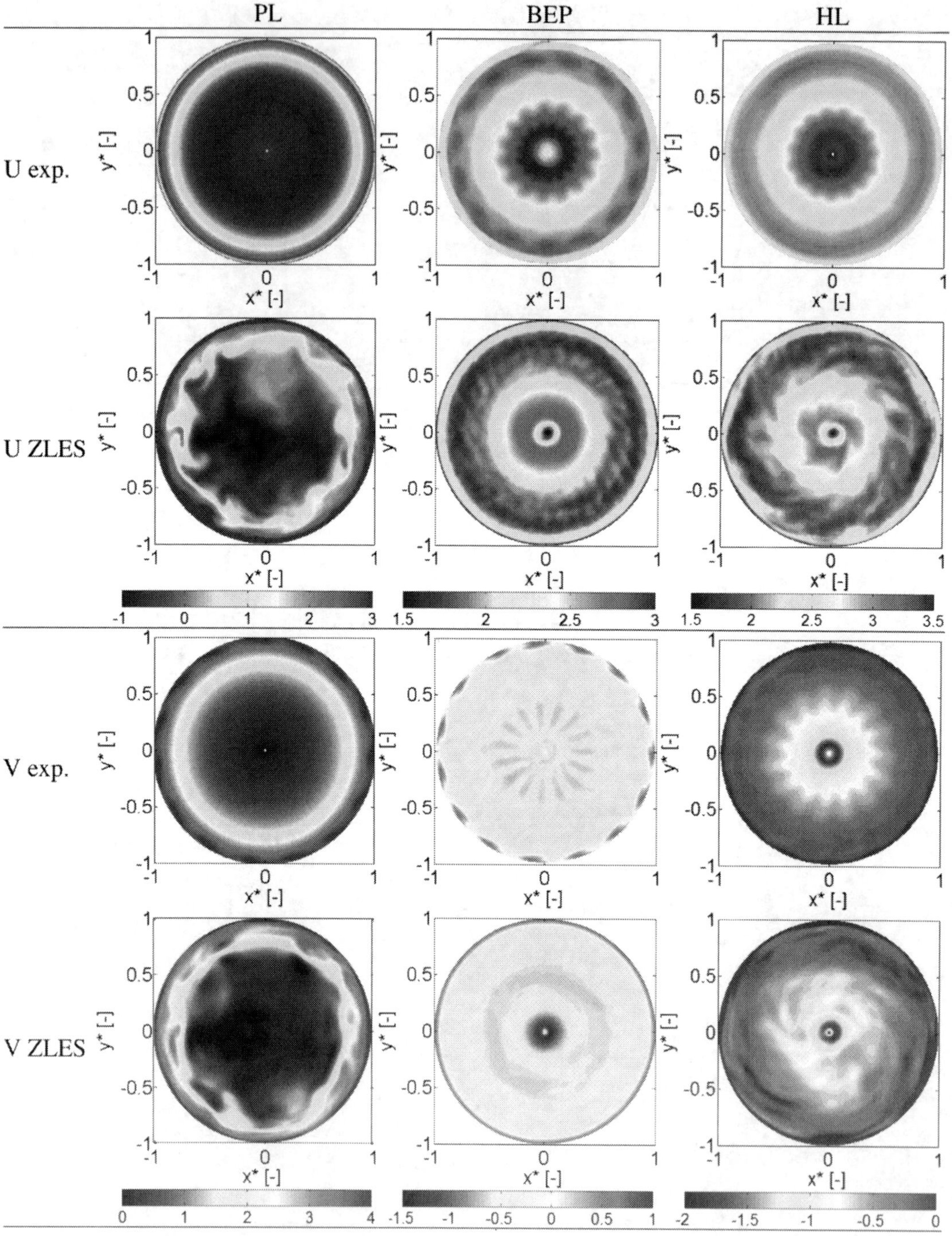

Figure 5. Axial (U) and circumferential (V) velocities at Plane 1, comparison of experimental and CFX results on NG with ZLES model.

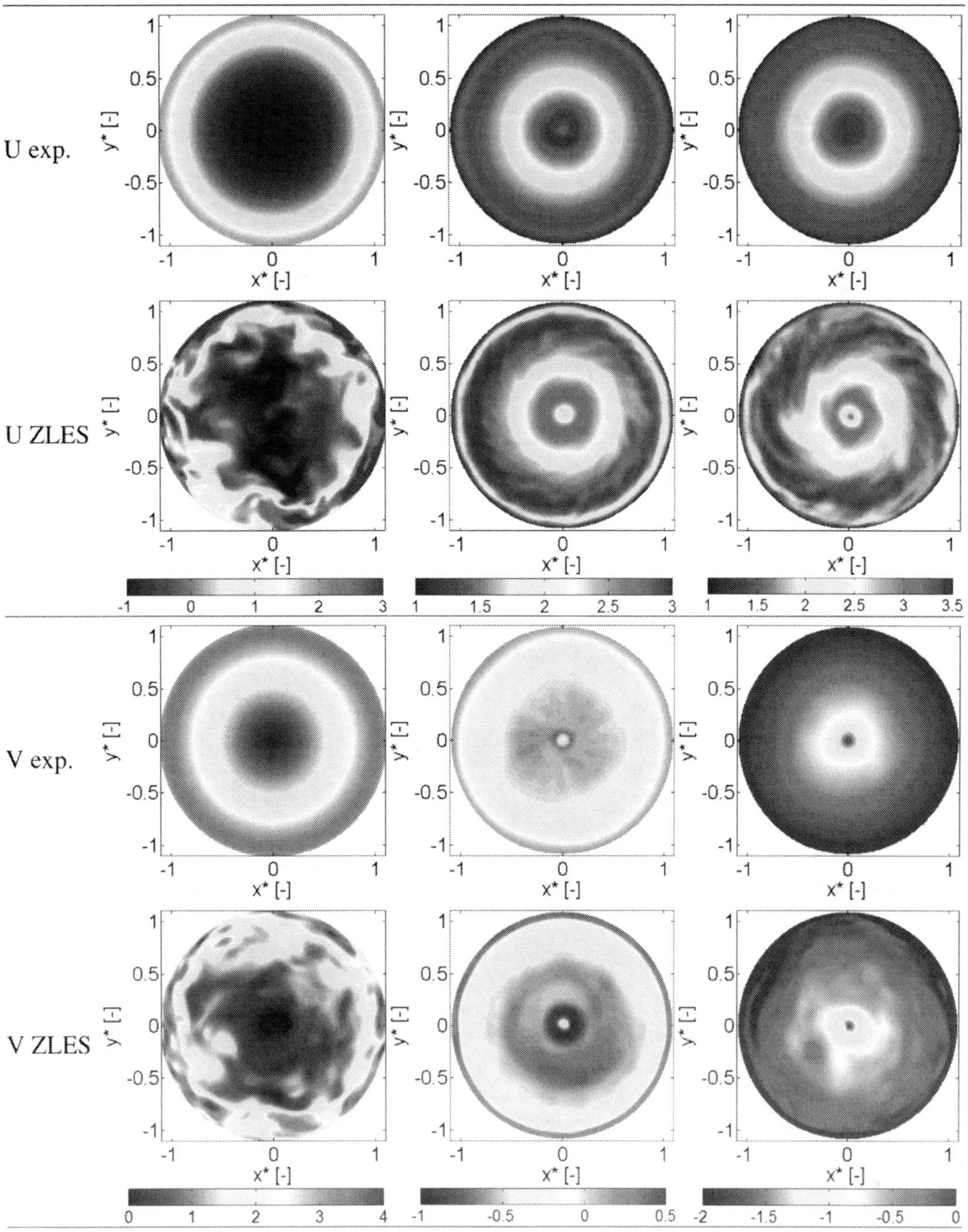

Figure 6. Axial (U) and circumferential (V) velocities at Plane 2, comparison of experimental and CFX results on NG with ZLES model.

Instantaneous velocity values of ZLES NG on Plane 1 and 2 are compared to phase-resolved values of experimental data on corresponding Line 1 and Line 2 (figures 5 and 6). Phase-resolved experimental data was averaged over many runner rotations. Such post-processing can capture wakes behind runner blades, whereas all other phenomena in the draft tube are circumferentially averaged. From such standpoint it can be concluded that ZLES NG simulation predicts range of velocity field in draft tube well (figures 5 and 6). Wakes behind runner blades on Plane 1 are in comparison with the experiment less visible.

3.2. Pressure pulsation

The organizers of Turbine-99 workshop performed time-dependent pressure measurements. Pressure sensors, located in stationary frames and presented in this paper are VL01 (in vaneless space between

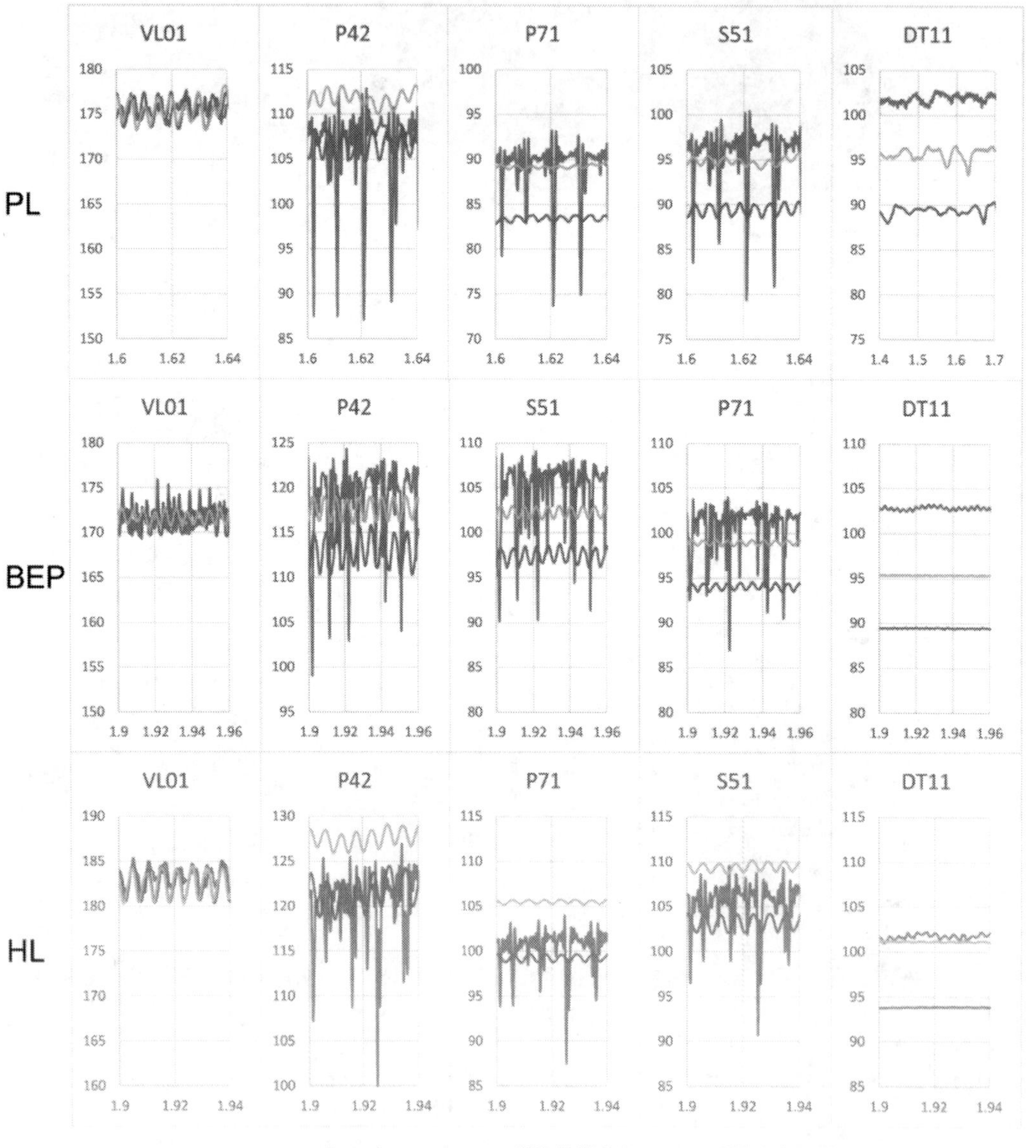

Figure 7. Pressure pulsation at five locations at Part Load (PL), BEP and High Load (HL).

guide vanes and the runner) and DT11 (in draft tube, at *Z*=-0.3058 m). Pressure sensors, located in rotating domain, are P42 (centre of blade on pressure side of blade), P71 (near trailing edge and hub on pressure side of blade) and S51 (on suction side of blade, two thirds along the blade, near hub).

Results of ZLES BG and ZLES NG simulations were compared to experimental data. At first, all numerically obtained pressure levels were shifted in order to set on sensor VL01 the same pressure level as it was obtained in measurements (figure 7).

From figure 7 it is possible to observe that in case of ZLES BG pressure drop between VL01 and P42 is approximately 7 kPa larger than in case of ZLES NG simulation at PL, BEP and HL. This is in agreement with larger runner head in case of BG. Large pressure spikes in the runner observed in experimental data were numerically not predicted. Pressure signal at VL01 and DT11 seems similar in numerical results and experimental data.

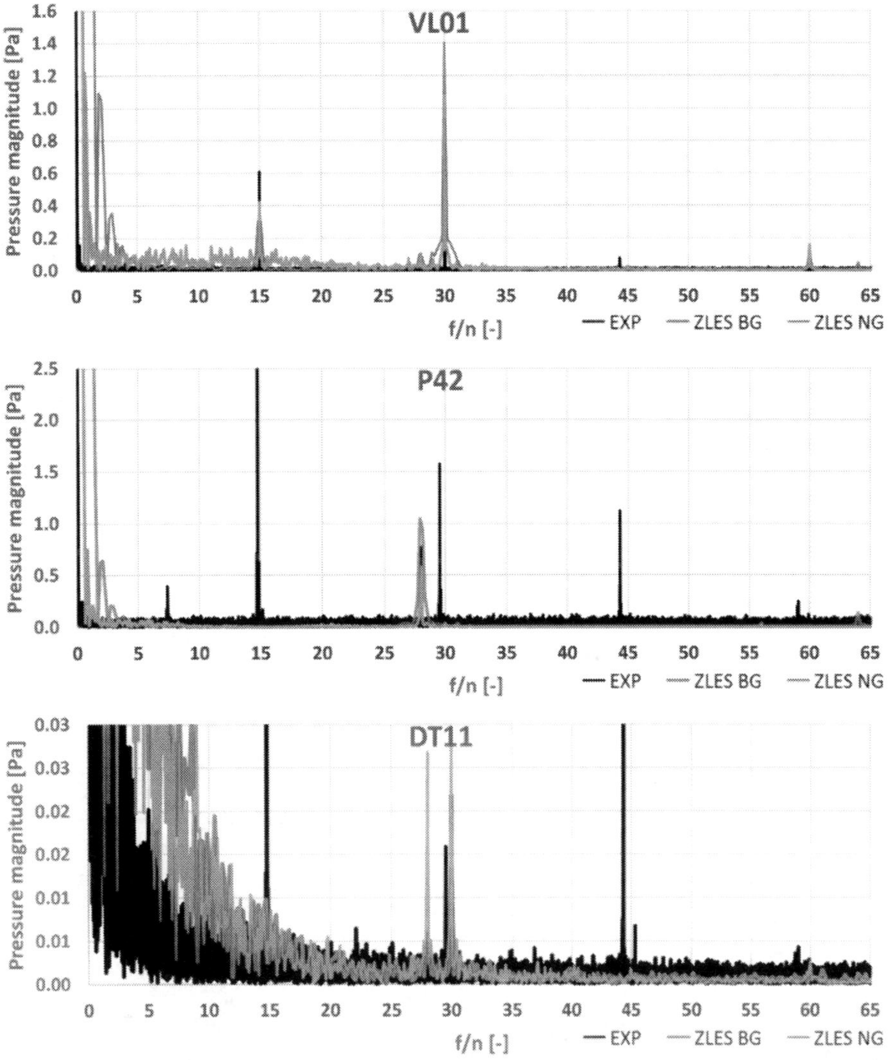

Figure 8. Pressure pulsation frequencies, FFT, Part Load (PL).

Fourier analysis of pressure signals at PL was carried out in figure 8. Two sources of peaks can be observed in charts. The first source of peaks is guide vane cascade with 28 blades, at 189.56 Hz (189.56 s^{-1} / 6.77 s^{-1} = 28). The peak can be observed in runner (e.g., P42) and in draft tube (DT11). The second source is runner with 30 blades, with peak at 203.1 Hz (203.1 s^{-1} / 6.77 s^{-1} = 30). It can be observed at VL01 and at DT11. There is also a peak at half of the latter frequency (at 101.55 Hz), probably because of the differences between 15 splitter and 15 full-length blades. There are also some peaks at multiplicators of 101.55 Hz (e.g., in VL01). All of these peaks were predicted by ZLES BG and ZLES NG simulations.

Some of the peaks in figure 8 were not predicted by numerical simulations. For instance, at P42 there are peaks with large magnitude, resembling pressure peaks due to runner blades in VL01. However, the analysis reveals that their frequency is somewhat lower. For instance, a pressure peak in P42 and in DT11 is present at 200.06 Hz instead of at 203.1 Hz.

3.3. Flow energy losses in distributor, runner and draft tube

From numerical results flow energy losses in all turbine parts were calculated as a difference of total pressure divided by density and gravity. Flow energy losses in the spiral casing and especially in the stay and guide vane cascade depend greatly on computational grid. Proper grid refinement near the vanes is crucial for accurate prediction of friction losses. In figure 9, where velocity field on mid cross section is illustrated, it can be seen that BG is not properly refined near the stay and guide vanes and therefore boundary layer is not well resolved. Therefore, flow energy losses calculated on BG were larger than those calculated on NG. The influence of grid refinement on losses was more significant at PL where guide vane opening was small and velocity large. Flow energy losses in distributor are at PL in range 4.6-6.9% of head, at BEP around 4% and at HL around 3% of head.

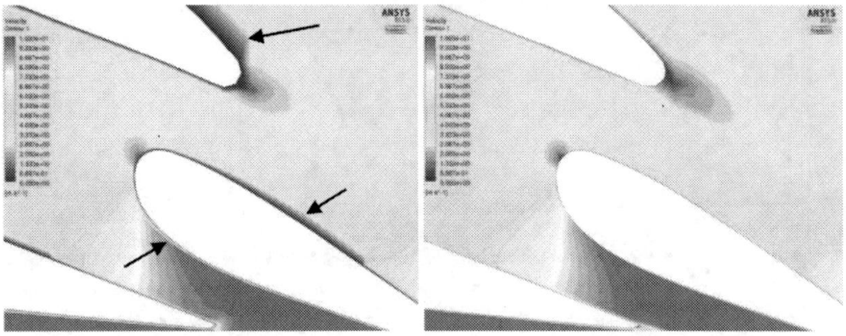

Figure 9. Velocity distribution on mid cross section at BEP, obtained with CFX using SST CC KL model on BG (left) and on NG (right)

Grid refinement near runner blades has great influence on pressure field in the runner. Consequently with too large values of y^+ much too large values of runner head were obtained. On the other hand, with large values of y^+ on runner blades too large values of torque on the shaft were obtained. Due to both effects smaller losses in the runner were obtained on BG than on NG. The difference is quite large at PL while at BEP and HL it is insignificant.

Flow in the draft tube depends on the operating point. At PL a circumferential velocity component at the runner outlet is large and consequently there are large swirls in the draft tube (see figure 10a). As expected at such extreme part load (flow rate was less than 35% of flow rate at BEP) no vortex rope was obtained, but some incoherent vortex structures can be observed (see figure 10b). Calculated flow energy losses depend on turbulence model being used. The smallest losses were obtained by the k-ε model, and the largest by the ZLES model. With the same turbulence model larger losses were obtained on NG than on BG. Flow energy losses in the draft tube are between 6.65% (k-ε model on BG) and 8.9% (ZLES on NG) of head. At BEP circumferential velocity at the draft tube inlet is small, flow is equally distributed between right and left part of the draft tube. Below the hub, vortex structures are rolled around strong vertical central vortex. Flow energy losses are between 0.46 - 0.6% of head, depending on turbulence model and grid. At HL, flow in the draft tube looks very similar as at BEP. Also vortex structures are similar. Flow energy losses are between 0.56 - 0.65% of head.

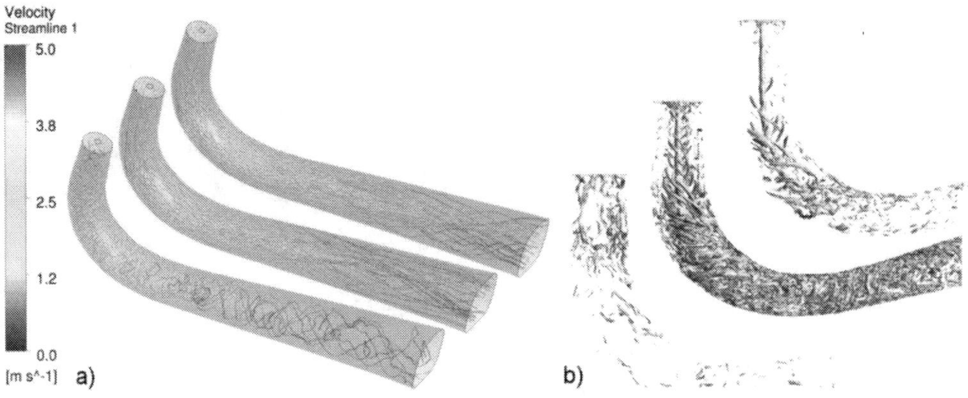

Figure 10. Flow in the draft tube at Part Load (bottom), BEP (middle) and High Load (top): a) streamlines; b) vortex structures. Results of CFX on NG with the ZLES turbulence model.

3.4. Labyrinth seals

The results of flow simulation in the bottom labyrinth seal are volumetric losses and torque losses. They depend on rotational speed of the runner and also on pressure difference between labyrinth seal inlet (before the runner) and outlet (behind the runner). These values of pressure were obtained from the results of flow simulations in the turbine and differed due to used turbulence model and grid. Volumetric losses are approximately the same for all three operating points, but they are, due to small flow rate, more important at PL than at the other two operating points. Torque losses are the largest at PL mainly due to the highest rotational speed. Torque losses in the top labyrinth seal depend only on runner rotating speed and therefore they are the largest at PL and the smallest at the BEP. The results for the bottom and the top labyrinth seals at all three operating points are presented in table 3.

Table 3. Volumetric and torque losses in labyrinth seals.

Operating point	Rotating Speed (s^{-1})	Volumetric losses in bottom labyrinth seal (l/s)	Torque losses in bottom labyrinth seal (Nm)	Torque losses in top labyrinth seal (Nm)
PL	6.77	0.435	8.86	7.335
BEP	5.59	0.426	6.10	5.355
HL	6.16	0.47	7.40	6.3

The sum of torque losses in bottom and top labyrinth seals were subtracted from the torque on the shaft calculated from the results of flow simulations in the turbine. In such a way corrected torque better agreed with the measurements.

Relative values of volumetric losses (volumetric losses divided by the discharge value) are the largest at PL, therefore at this operating point a flow simulation in the turbine was performed with prescribed mass sink before the runner and mass source behind the runner. This simulation was performed only with CFX on slightly modified NG mesh with SST CC KL turbulence model. Due to smaller discharge in the runner, calculated head and torque on the shaft slightly decreased. Due to volumetric losses also the efficiency value decreased. The effect of volumetric losses on results is presented in figure 11. It can be expected that for the other operating points the effect of volumetric losses would be even smaller.

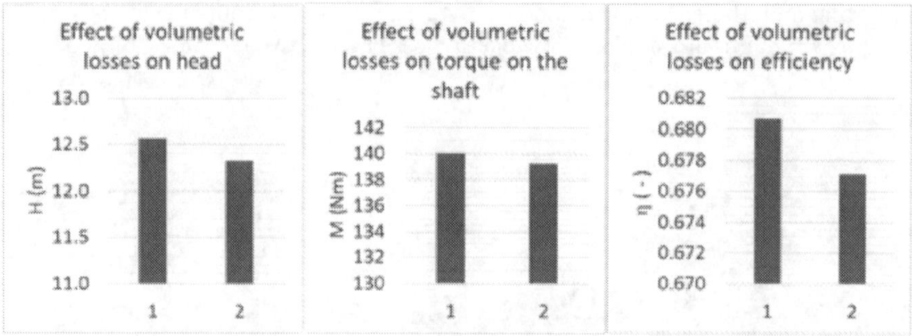

Figure 11. Effect of volumetric losses on torque on the shaft and efficiency at Part Load, simulation with ANSYS CFX, on NG mesh with SST turbulence model (1 - without volumetric losses; 2 - with volumetric losses).

3.5. Prediction of head (flow rate), torque and efficiency

In figures 12a, 12b and 12c results of integral quantities (head, flow rate, torque and efficiency) are presented. At first, simulations were done with flow rate as input data, the results were head, torque on the shaft and efficiency. At second, input data was head, the results were flow rate, torque on the shaft and efficiency. With head as input data, simulations were performed only with CFX solver using k-ε and SST models (on BG and NG) and ZLES on NG mesh. Steady-state simulations were performed with and without CC KL. The effect of CC and KL on efficiency is small, mostly less than 0.2%. An exception is simulation with SST on NG at PL, where efficiency is 1% lower due to included CC and KL (not presented in figure 12).

Efficiency is calculated from values of head, flow rate and torque on the shaft. If head and torque (or head and flow rate) are wrong for approximately the same percent, predicted efficiency can agree well with measurements even when numerical simulation is not sufficiently accurate. A quality of numerical prediction can be seen only if besides efficiency also predicted values of flow rate, head and torque are presented as in figures 12a, 12b and 12c. For each operating point all values were divided by the measured values at that operating point. Values of torque and efficiency obtained without and with torque losses in labyrinth seals are presented. With black vertical line the results obtained with prescribed flow rate Q are separated from the results obtained with prescribed head H.

At BEP and at PL predicted values of head (when flow rate was input data) were larger than the measured values. Discrepancy was very large, even 10-15% when the BG was used. At HL predicted head was too large when CFX on BG was used, while with OpenFOAM on BG2 and CFX on NG predicted values of head were smaller than the measured one. At PL and at BEP for all steady-state simulations with head as input data, the predicted flow rate was too small. The discrepancy was smaller on NG than on BG. The smallest discrepancies (2.4% at PL and -2.94% at BEP) were obtained when SST CC KL model was used on NG. More accurate values of flow rate were obtained with ZLES model where at PL the discrepancy was less than 0.1%, while at BEP it was less than 2.6%. At HL on BG too small values of flow rate were obtained, while on NG predicted values were higher than the measured one.

With flow rate as input data predicted values of torque on the shaft obtained on BG were too large even when torque losses in labyrinth seals were taken into account. Discrepancies were very large, at PL, when SST and ZLES were used, even more than 20%. With more refined grid near blades (NG) the results improved and for all simulations discrepancies to the measurements were less than 5%. With head as input data, due to too small values of calculated flow rate at PL and at BEP, also predicted torque was too small, the discrepancy was the smallest when ZLES model on NG was

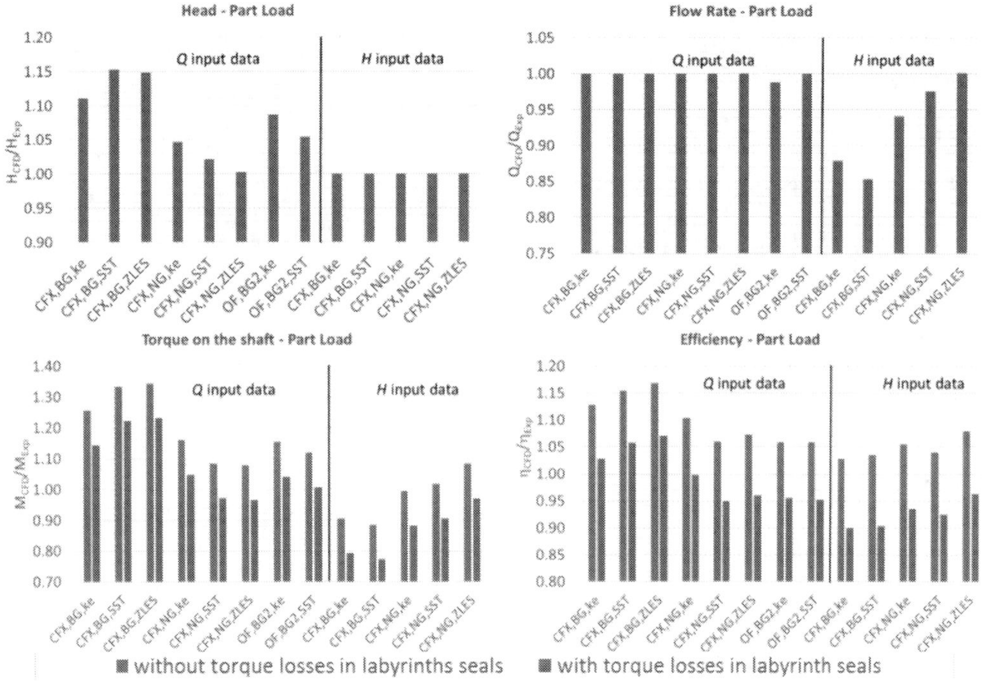

Figure 12a. Comparison of numerical results to the experimental values, Part Load. CFX simulations were performed with CC and KL.

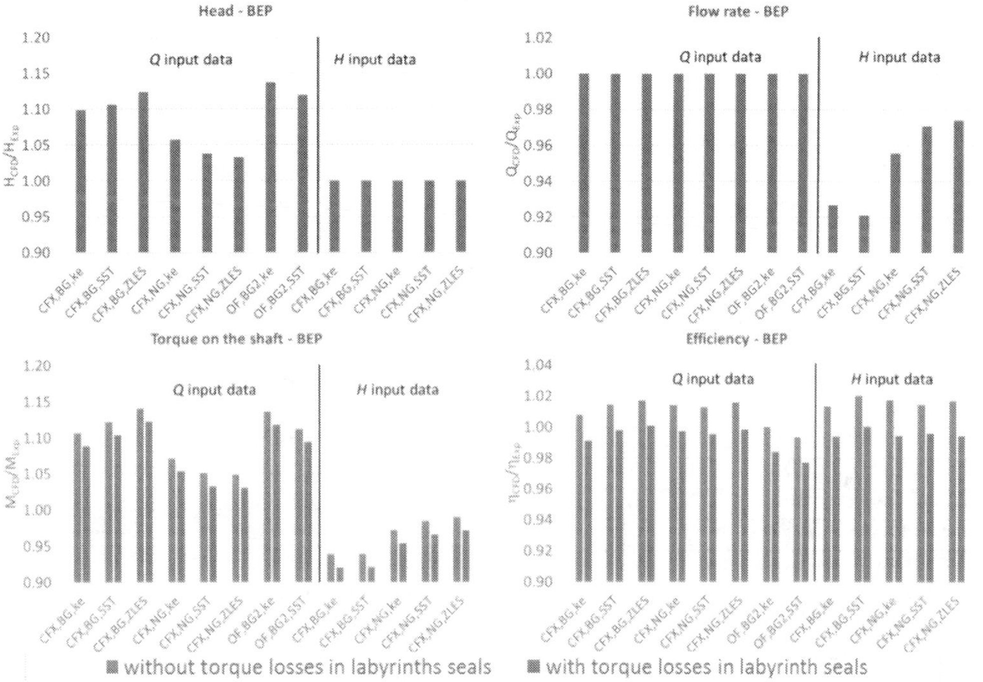

Figure 12b. Comparison of numerical results to the experimental values, BEP. CFX simulations were performed with CC and KL.

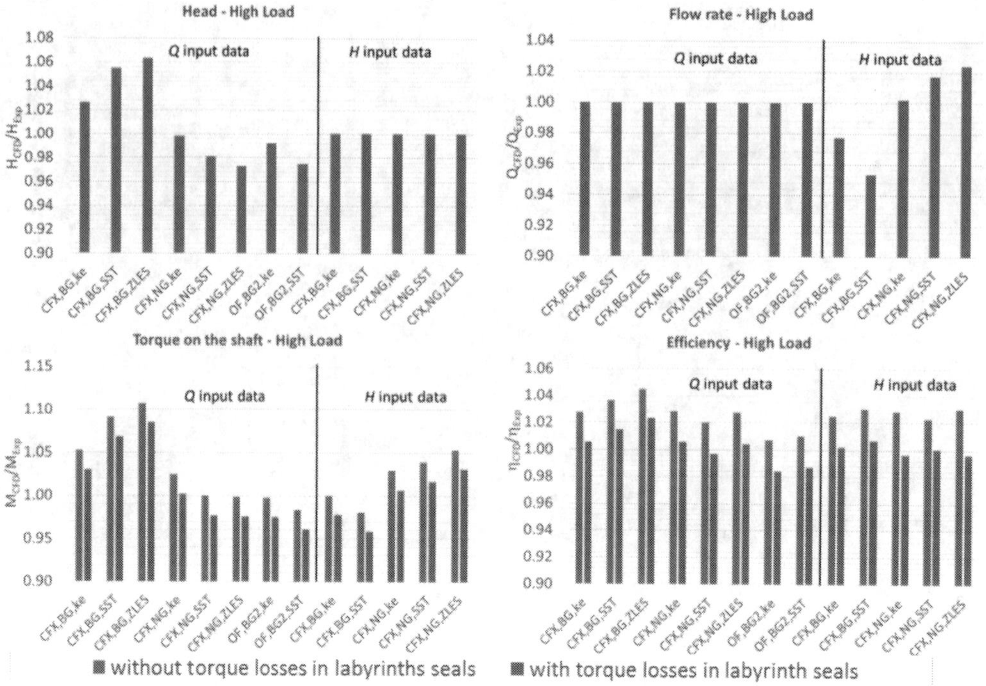

Figure 12c. Comparison of numerical results to the experimental values, High Load. CFX simulations were performed with CC and KL.

used. At HL the predicted values of torque were too small when BG was used, while on NG too large values were obtained.

Efficiency at BEP was very accurately predicted by all simulations with CFX, a discrepancy to the measured values was less than 0.7%. With OpenFOAM only steady state simulations on BG were performed, discrepancy to the measured value was in case of the *k*-ε model about 1.6% and in case of the SST model about 2.3%. At HL efficiency prediction with CFX on NG was also quite accurate (discrepancy smaller than 0.7%), while on BG discrepancy to the measured value was larger, especially when SST (discrepancy 1.5%) and ZLES (discrepancy 2.3%) were used. With OpenFOAM too small values were obtained. At HL efficiency prediction was less accurate, especially for simulations on BG. The best results were obtained by *k*-ε on NG, the discrepancy was only 0.27%. But the agreement with measurements was so good only because head and torque were both wrong for nearly the same percent, about 4.6%. When head was input data and flow rate a result of simulation, efficiency prediction at PL was very poor, discrepancy was between 3.7% (ZLES, NG) and -10% (*k*-ε, BG), depending on grid and turbulence model being used.

Even more than absolute value of efficiency it is important to get the right shape of efficiency curve and correct position of the best efficiency point. Three operating points are not enough to obtain the efficiency curve, but at least it can be checked whether the efficiency value is the highest at the BEP and the differences between efficiency values at BEP, HL and PL are approximately the same as the measured ones. Efficiency curves, with flow rate as input data, obtained by simulations on BG and NG, with different turbulent models (*k*-ε, SST, ZLES) and with both solvers (CFX, OpenFOAM), taking into account torque losses in labyrinth seals, are presented in figure 13, left. It can be seen that the shape of efficiency curves agrees well with the measured curve except in case of simulations with CFX on BG using SST and ZLES models, where efficiency values at HL and PL are too high. When head was input data and flow rate a result of simulations (figure 13, right), positions of all three operating points are slightly shifted due to a bit inaccurate numerical values of flow rate, and predicted efficiency values at PL are underestimated, otherwise a shape of the efficiency curves is well captured.

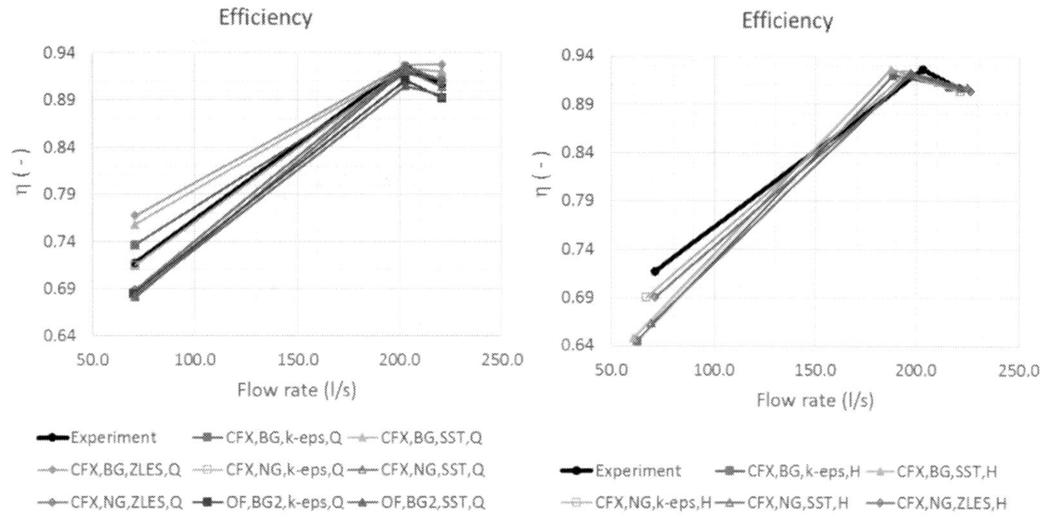

Figure 13. Turbine efficiency, numerical and experimental results. Torque losses in labyrinth seals are taken into account. Left: flow rate as input data. Right: head as input data. CFX simulations were performed with CC and KL.

4. Conclusions

Flow in a high-head Francis turbine was analysed by two CFD codes, CFX and OpenFOAM, using three turbulence models (k-ε, SST and ZLES) and three computational grids. Numerical results were compared with available experimental data. The following conclusions can be drawn:

- Predicted axial and circumferential velocity components on Line 1 and Line 2 in the conical part of the draft tube agree satisfactorily with the experimental data, but no numerical setup was the best at all three operating points. ZLES model agreed much better with experimental data than steady-state simulations at HL and BEP.
- Axial and circumferential velocity components on two planes in the conical part of the draft tube, predicted with ZLES CC KL generally agree well with experimental data, but wakes behind runner blades on Plane 1 are in comparison with experiment less distinct.
- Numerically predicted pressure pulsations at five locations were compared to the measured results. CFD predicted frequency of pressure pulsations correctly, but did not detect some pressure peaks in the runner.
- Torque losses in labyrinth seals have to be taken into account to get proper values of torque on the shaft.
- Proper grid refinement near stay and guide vanes and especially near runner blades is crucial for accurate prediction of torque on the shaft, head and efficiency. With too coarse grid near runner blades significantly overestimated values of head and torque were obtained.
- In general, the CC and KL options did not have any significant effect on results.
- Steady-state simulations with k-ε and SST turbulence models on basic grid (BG2) were performed with open source code OpenFOAM. The values of efficiency as well as local field details (velocity components) obtained with OpenFOAM were in line with those predicted with CFX

Acknowledgments
The research was partly funded by the European Commission - FP7-PEOPLE-2013-IAPP - Grant No. 612279 and by the Slovenian Research Agency ARRS - Contract No. 1000-09-160263.

References

[1] Jošt D, Lipej A, Mežnar P 2008 Numerical prediction of efficiency, cavitation and unsteady phenomena in water Turbines, *9th Biennial ASME Conf. on Eng. Sys. Design and Analysis, ESDA08 (Haifa, Israel)*

[2] Chirag T, Cervantes M J, Gandhi B K and Dahlhaug O G 2013 Experimental and numerical studies for a high head Francis turbine at several operating points *J. Fluids Eng.* **135** 111102

[3] Menter F R, Kuntz M and Langtry R 2003 Ten years of industrial experience with the SST turbulence model Turbulence, Heat and Mass Transfer 4 eds. K Hanjalić, Y Nagano and M Tummers (New York: Begell House) pp 625–632

[4] Menter F R and Esch T 2001 Elements of industrial heat transfer prediction *16th Brazilian Congr. of Mech. Eng. (Uberlandia, Brazil)*

[5] Jošt D, Škerlavaj A, Lipej A 2012 Numerical flow simulation and efficiency prediction for axial turbines by advanced turbulence models *IOP Conf. Ser.: Earth Environ. Sci.* **15**(6) p 062016

[6] Jošt D, Škerlavaj A, Lipej A 2014 Improvement of efficiency prediction for a Kaplan turbine with advanced turbulence models *J. Mech. Eng.* **60**(2) pp 124–34

[7] Menter F R, Gabaruk A, Smirnov P, Cokljat D and Mathey F 2010 Scale-adaptive simulation with artificial forcing *Progress in Hybrid RANS-LES Modelling* ed. S-H Peng, P Doerffer and W Haase (Berlin: Springer) pp 235–46

[8] Adamian D and Travin A 2011 An efficient generator of synthetic turbulence at RANS–LES interface in embedded LES of wall-bounded and free shear flows *Computational Fluid Dynamics 2010* ed. A Kuzmin (Berlin: Springer) 739–44

[9] Egorov Y and Menter F 2008 Development and application of SST-SAS turbulence model in the DESIDER project *Advances in Hybrid RANS-LES Modelling* ed. S-H Peng and W Haase (Heidelberg: Springer) pp 261–70

[10] Smirnov P E and Menter F 2009 Sensitization of the SST turbulence model to rotation and curvature by applying the Spalart-Shur correction term *J. Turbomach.* **131**(4) 041010

[11] Kato M and Launder B E 1993 The modelling of turbulent flow around stationary and vibrating square cylinders *Proc. 9th Symp. on Turbulent Shear Flows* pp 10.4.1–10.4.6.

[12] Shur M L, Spalart P R, Strelets M K and Travin A K 2008 A hybrid RANS-LES approach with delayed-DES and wall-modeled LES capabilities *Int. J. Heat Fluid Flow* **29**(6) pp 1638–49

Journal of Physics: Conference Series **579** (2015) 012007 doi:10.1088/1742-6596/579/1/012007

The influence of disc friction losses and labyrinth losses on efficiency of high head Francis turbine

D Čelič[1] and H Ondráčka[2]

[1] Independent Researcher, Damjan Čelič, s.p., Novo mesto, Slovenia
[2] Hydraulic Designer, Scandic Energy, Rakkestad, Norway

E-mail: damjan.celic@gmail.com

Abstract. CFD is a very powerful tool, which significantly reduces the need for model testing of Francis turbines. Numerical models are constructed so as to cover all the essential geometric and physical properties of the physical models. However, for the sake of simplicity, some geometric details are often neglected. In case of Francis turbines, labyrinths are usually not included in the numerical model and consequently the disc friction losses and labyrinth losses are ignored. This may lead to inaccurate prediction of turbine efficiency. In our study we have investigated the importance of considering all the geometrical details of labyrinth for the accurate prediction of efficiency of high head Francis turbine. The research was performed by using commercial software Numeca FINE^{TM}/Turbo.

1. Introduction

Labyrinth seals are the most common type of non-contact seal between the runner and the turbine covers, which is widely used in turbo-machinery. Water leaking through a labyrinth will cause reduced efficiency since it will not be utilized by the runner. The leakage flow through the seals depends upon the size of the gap [1]. For a new turbine the labyrinth gap will be small and the leakage will be low. As the seals wear, the gap increases and so does the leakage. The labyrinth consists of two parts, a static seal connected to the covers and a rotating part connected to the runner.

In the past, much effort has been put into determination of volumetric and mechanical efficiencies and leakage behavior of Francis turbine [2]. Mechanical and volumetric efficiencies have been expressed as a function of specific speed. The empirical formulae have been obtained by fitting the model test data available at that time.

According to international standard for rehabilitation and performance improvement of Hydraulic turbines [3], the internal mechanical losses may significantly lower down the efficiency of Francis turbine operating at low specific speed. Figure 1 represents a plot of loss distribution at peak efficiency depending on the specific speed N_q for a wide range of model Francis turbines in 2005. This plot gives a good idea of what one may expect in the way of performance for a totally new unit at that point of time. It can be seen that the labyrinth losses and disc friction losses contribute significantly to the turbine hydraulic losses in case when the specific speed N_q of a turbine is low.

Turbine output is decreased due to labyrinth losses (friction losses and volumetric losses - water leakage) by the nearly constant value (kW) for the whole operating range of the turbine. Their relative importance in % decreases with the flow. If labyrinth losses represent e.g. 3% at full load, they can be higher than 6% at part load.

The Tokke Model, which is under this investigation, is a high head Francis turbine with the specific speed of $N_q = 23.6$. Roughly estimated from Figure 1, the labyrinth losses and disc friction losses at this N_q are almost as high as spiral case losses, stay vane losses and guide vane losses all together. Consequently, if the internal mechanical losses are not taken into account in the CFD analysis, a higher difference between the experimental and numerical hydraulic efficiency is expected.

In previous experimental and numerical studies of the Tokke Model [4], the difference between experimental and numerical hydraulic efficiencies at best efficiency point (BEP) was rather small (0.85%), while the difference between experimental and numerical efficiencies at lower discharge turbine operation was quite significant (14%).

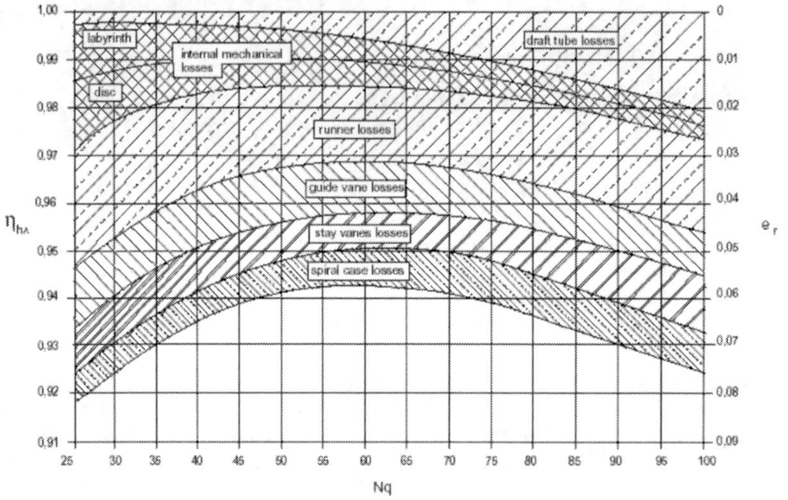

Figure 1. Efficiency and distribution of losses versus specific speed for Francis turbine [3].

This research is mainly focused on the CFD analysis of a detailed Tokke Model in order to get more accurate prediction of its hydraulic efficiency.

2. Numerical model and mesh

Numerical analyses are useful tools for making predictions about the properties of real-world problems. The accuracy of those predictions strongly depends on quality of numerical models and on simplifications adopted during the analysis in order to ensure the feasibility of a solution.

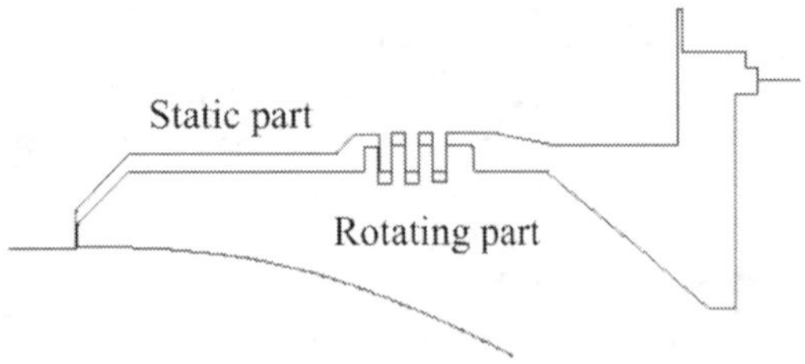

Figure 2. Runner seal on crown side.

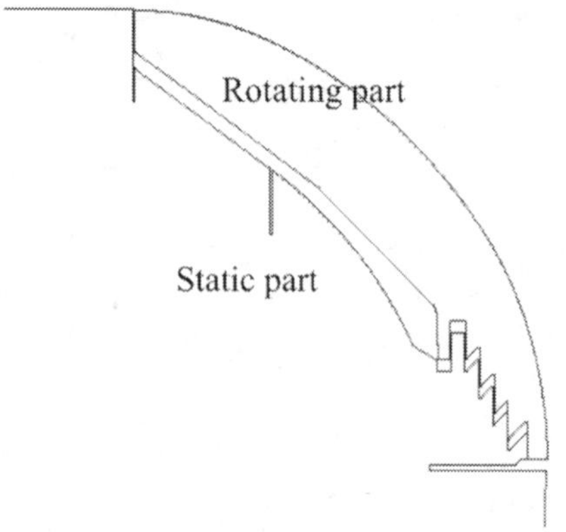

Figure 3. Runner seal on band side.

To get more accurate solution, our numerical model is built in a way to represent the conditions at test site as good as possible. The geometry of the model starts at the inlet of spiral casing and ends at the outlet of draft tube. Both runner seals as depicted in figure 2 and figure 3 are also taken into consideration. The runner seal geometry corresponds to the geometry at the test site. The only difference is that the edges inside the labyrinth are not rounded but sharp. This simplification helps to maintain the mesh size within manageable limits.

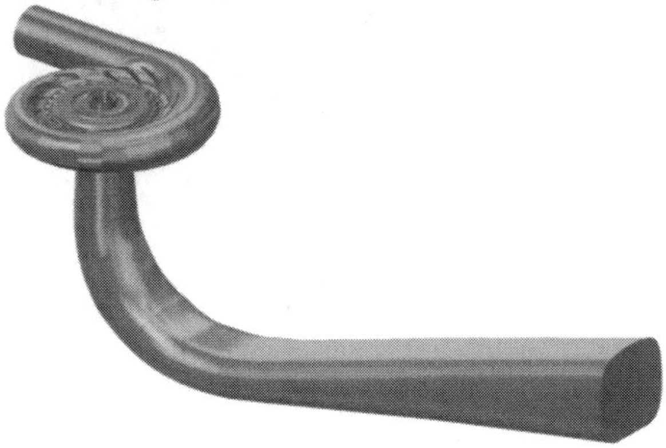

Figure 4. Computational domain.

The computational domain consists of the Spiral casing with Stay vanes, the Guide vane cascade, the Runner with labyrinths, and the Draft tube (see figure 4). The spatial discretization of the flow domain of the Spiral casing and the Draft tube was performed by IGG™, which is an interactive geometry modeler and a grid generation system for multi-block structured grids. The spatial discretization of the flow domain of the Guide vane and the Runner with labyrinths was performed by AutoGrid5™ which is an automatic meshing system for turbo-machinery configurations.

Table 1. Grid size.

Part of the turbine	App. number of nodes
Spiral casing with a Stay vane	2.600.000
Guide vane	6.000.000
Runner	8.700.000
Labyrinths	30.100.000
Draft tube	1.000.000
Total	48.400.000

The full size 360° mesh was generated to cover potential unevenness of the flow distribution inside the turbine. Figure 5 shows the computational grid of the whole turbine, while figure 6 depictures some details about the mesh of the runner, the guide vane and the labyrinth. Information about the grid size is gathered in table 1.

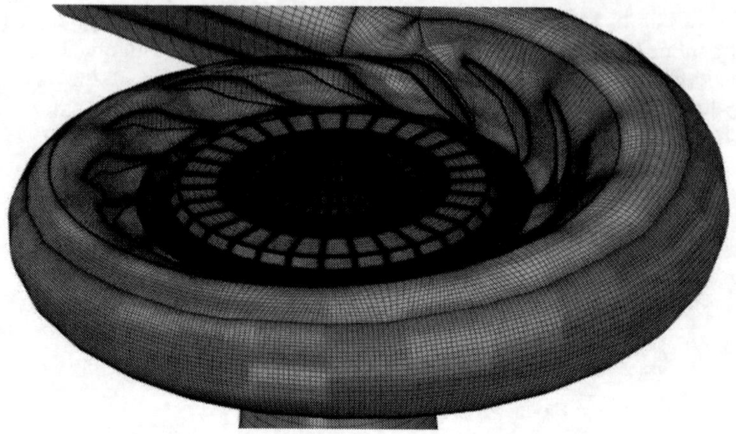

Figure 5. Computational grid.

A great care and consideration was needed while generating the mesh inside the labyrinths. The size of the mesh should be as small as possible to lower down the computational cost. However, all the significant geometrical details need to be captured with the mesh. Additionally, also the close-to-wall boundary layer has to be set correctly in order to capture the velocity gradient at the solid walls.

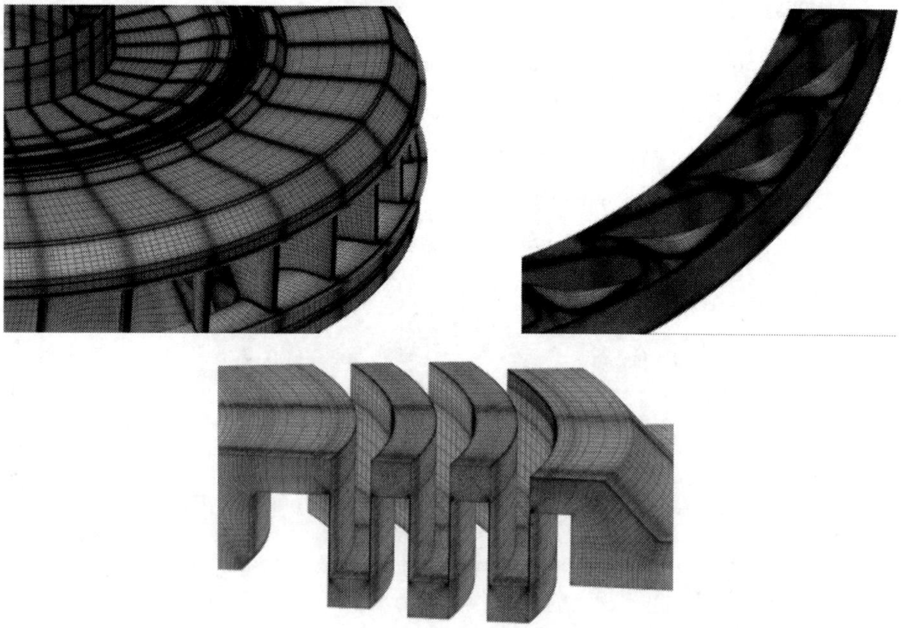

Figure 6. Computational grid in details.

3. Flow analysis

For the flow analysis the computer code Numeca FINE™/Turbo 9.02 has been used. The working fluid is water and the flow inside the turbine is assumed to be steady. The Reynolds-averaged Navier-Stokes equations for incompressible turbulent flow are adopted. The k-ω SST (Shear Stress Transport) turbulence model with extended wall function was used to provide a link between the turbulent transport of momentum and energy and mean flow properties.

A central spatial discretization scheme and a Merkle preconditioning were used. For the purpose of efficient computation and fast convergence a multigrid strategy "V-cycle" was employed [5]. Three multigrid levels with a linear progression of smoothing sweeps on successive grid levels and 100 cycles per grid level have been chosen. The amount of sweeps is the amount of times the Runge-Kutta operator is applied.

The simulation was performed for three given operating points:

- part load (PL);
- the best efficiency point (BEP);
- high load (HL).

Table 2 summarizes the operating conditions of a turbine at a test site.

Table 2. Test conditions corresponding to the operating points.

Operating point	Head (*m*)	Flow rate (m^3/s)	Runner speed (*Hz*)	Hydraulic efficiency (%)
PL	12.29	0.07	6.77	72.5
BEP	12.77	0.21	5.74	92.4
HL	12.61	0.23	6.34	91.0

At the inlet of the computational domain the absolute total pressure was prescribed in accordance with the turbine head at the test site for a given operating point. At the outlet of the computational domain the average static pressure was prescribed.

The steady analyses were performed on a PC with an eight-core Intel i7 processor and 64GiB of RAM, running on a 64-bit Windows 7 OS. For the same computational domain including runner seals, two different Rotor/Stator approaches were taken into consideration: the frozen rotor approach (FR) and the mixing plane approach (MP).

Figure 7 and figure 8 show the y^+ at the runner and at the runner seals, respectively.

Figure 7. y^+ at the runner.

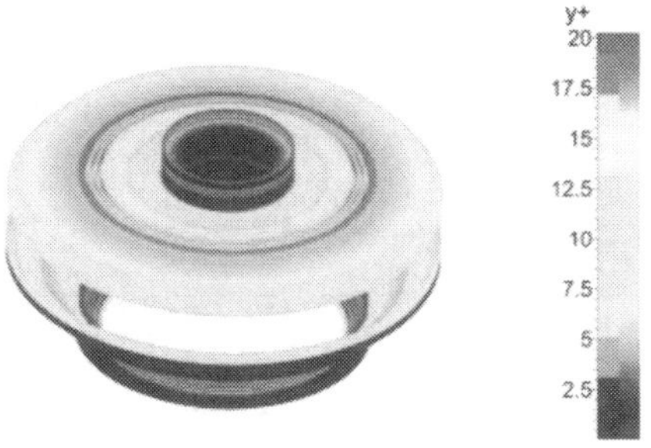

Figure 8. y^+ at the runner seals.

Our numerical model does not have balancing holes, so the flow leakage is possible only through the runner seal on a band side (see figure 3). Table 3 represents the quantities of band-seal leakage for given three operating points. The flow through the labyrinth is almost the same for all three operation points. The leakage flow is driven only by the pressure difference, which is for all three operating points almost the same, since the head of a turbine is more or less the same.

Table 3. Mass flow through the runner seal on a band side.

	Mass flow at the Inlet of a turbine (*kg/s*)	Mass flow through the labyrinth (*kg/s*)	Proportion of the flow through the labyrinth (*%*)
PL	72.3	0.318	0.44
BEP	207.7	0.329	0.16
HL	237.3	0.367	0.15

As expected, the proportion of the flow lost through the labyrinth is rather small, only 0.44% at a PL operation and even smaller at BEP and HL operation. Velocity conditions inside the runner and inside the labyrinth are shown in figure 9 and in figure 10, respectively.

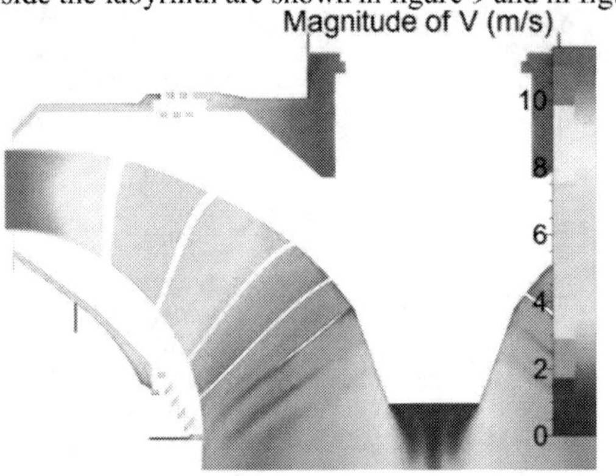

Figure 9. Velocity conditions inside the runner at a BEP operation.

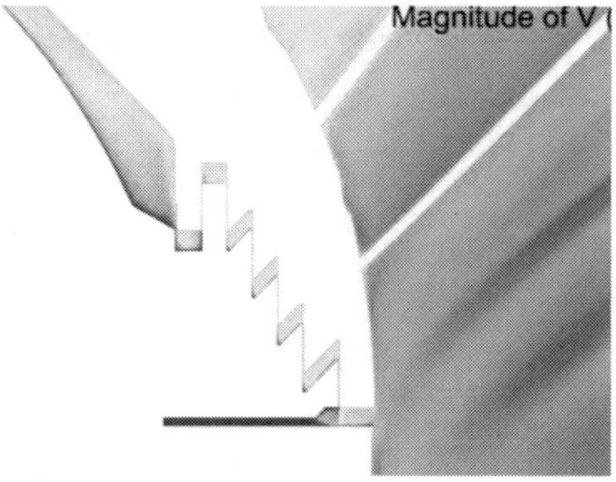

Figure 10. Velocity conditions inside the labyrinth at a BEP operation.

3.1. *Pressure data validation*

Time dependent pressure measurements in the stationary and rotating frames are available from the Tokke model test rig. The position of the pressure measurements are given in table 4. Our numerical analyses are stationary and, therefore, the information about the pressure fluctuations is not available. To compare our calculated pressure values to the time dependent pressure measurements, some statistical manipulation of experimental data is needed. A median pressure was calculated for the purpose of comparison with the CFD. In order to maintain the information about the pressure oscillation, the 25th percentile and the 75th percentile were calculated and shown together with the median pressure as some kind of a confidence interval.

Table 4. Position of the pressure sensors.

	VL01	DT11	DT21	P42	P71	S51
x (m)	0.2623	-0.0904	0.0904	7.16E-5	-0.0666	-0.0800
y (m)	0.1935	0.1566	-0.1566	0.1794	0.0423	0.0838
z (m)	-0.0296	-0.3058	-0.3058	-0.0529	-0.0860	-0.0509

The comparison of pressure data for each particular operation point is given in figure 11, figure 12 and figure 13, respectively. The CFD pressure data are available for both the FR approach and the MP approach.

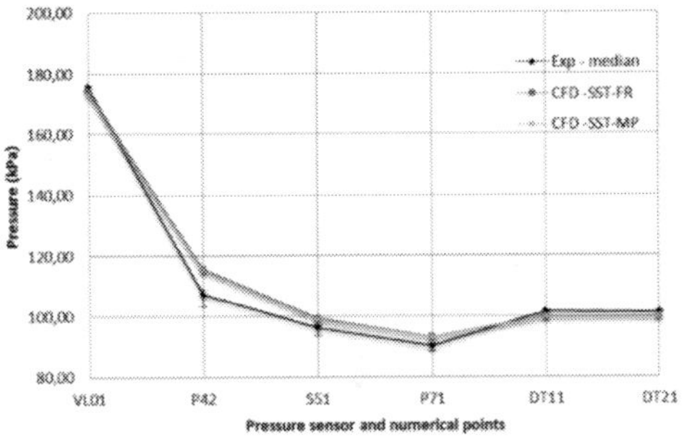

Figure 11. Comparison of experimental and numerical pressure at different locations in the turbine for PL operation.

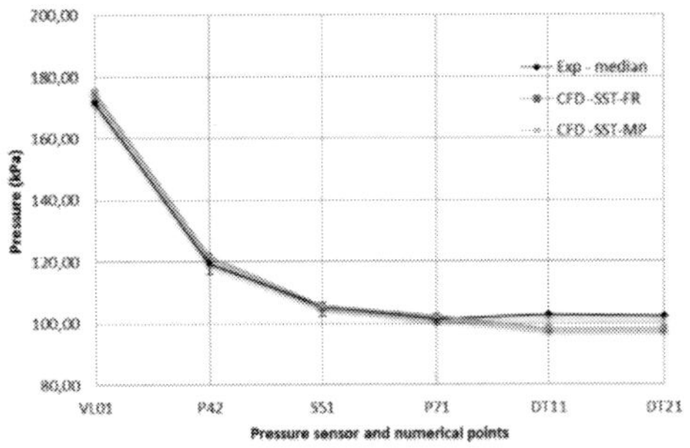

Figure 12. Comparison of experimental and numerical pressure at different locations in the turbine for BEP operation.

In the case of PL operation (see figure 11), the differences between the calculated pressure values and measured median values are the highest at the pressure side of a blade (P42, P71) and at the suction side of a blade (S51). However, at the same three positions also the pressure fluctuations were the highest. The maximal difference observed between the measured median values and numerical results was 7.8% at position P42 in the case of PL operation.

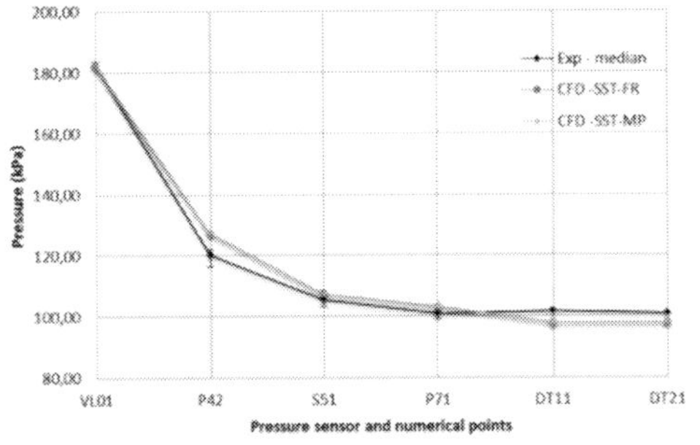

Figure 13. Comparison of experimental and numerical pressure at different locations in the turbine for HL operation.

For all three operation regimes the same trend can be observed: the pressure at points located on runner blade (P42, P71, S51) is always over-predicted, while the pressure in the draft tube (DT11, DT21) is always under-predicted. As can be seen from figure 11, figure 12 and figure 13, the pressure predictions of FR approach are in a very good agreement with the pressure predictions of MP approach.

3.2. Velocity data validation

Velocity measurements (axial and tangential) were performed with a laser Doppler anemometer (LDA) along two horizontal lines in the draft tube. The coordinates are available in table 5.

Table 5. Location of velocity measurement.

	Line 1		Line 2	
	Point 1	Point 2	Point 1	Point 2
x (m)	-0.1789	0	-0.1965	0
y (m)	0	0	0	0
z (m)	-0.2434	-0.2434	-0.5614	-0.5614

Journal of Physics: Conference Series **579** (2015) 012007 doi:10.1088/1742-6596/579/1/012007

Positive axial velocity is defined in the stream-wise direction and the tangential velocity is positive in the runner rotational direction. A comparison of experimental and numerical velocities is given in figure 14 - figure 25.

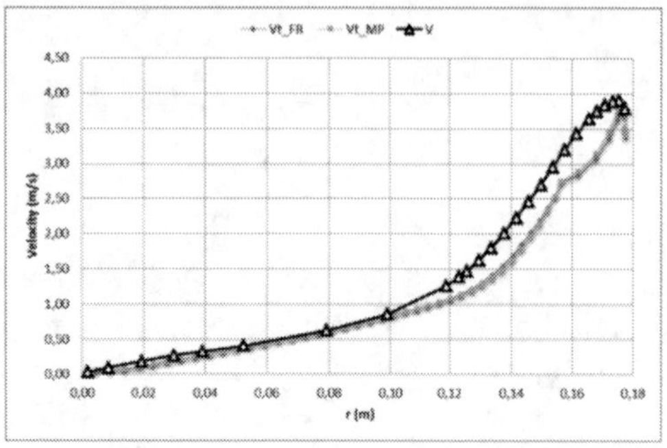

Figure 14. PL: Tangential vel. - upper line.

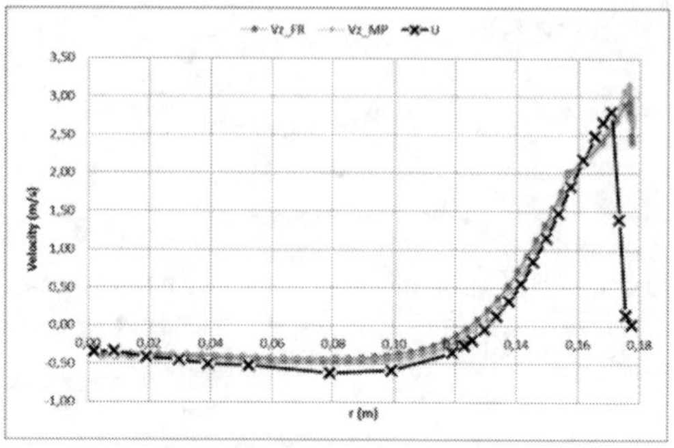

Figure 15. PL: Axial velocity – upper line.

In the case of PL operation, a very good agreement between CFD predictions and measurements is achieved, except for the tangential velocity at lower line (see figure 16). The differences between FR solution and MP solution are very small.

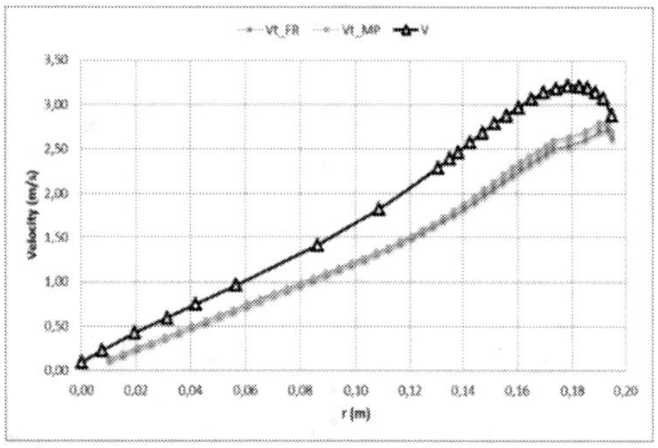

Figure 16. PL: Tangential vel. - lower line.

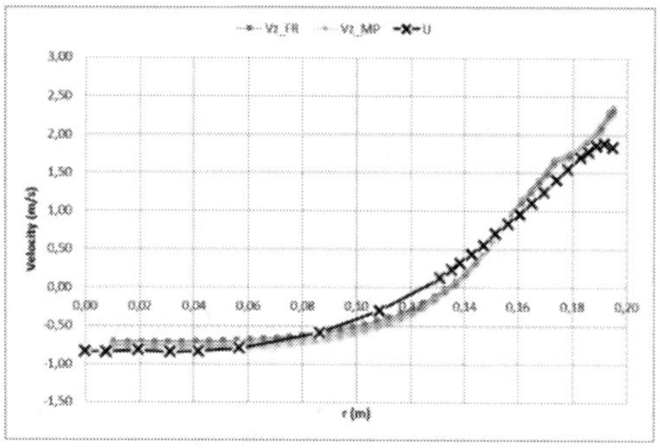

Figure 17. PL: Axial velocity – lower line.

In case of BEP operation, there is a noticeable difference between FR and MP solution. The latter generally better predicts the shape of both tangential velocity profile and axial velocity profile. The predictions of tangential velocity are better in the close-wall region (r > 0.1m) and worse in the center region (r < 0.1m).

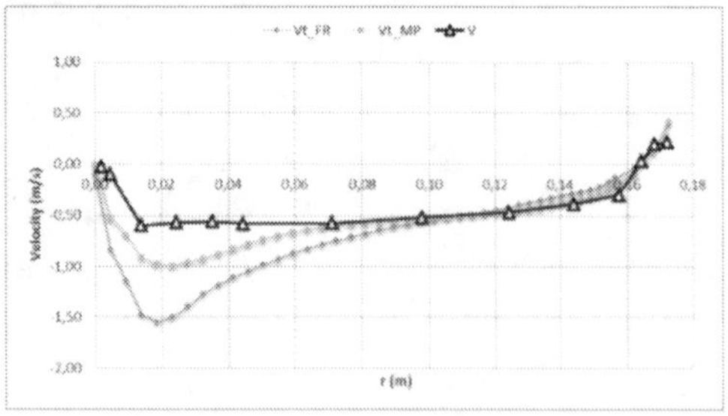

Figure 18. BEP: Tangential vel. - upper line.

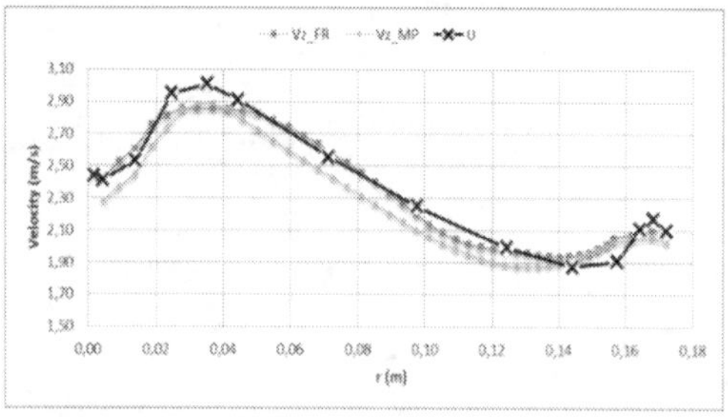

Figure 19. BEP: Axial velocity – upper line.

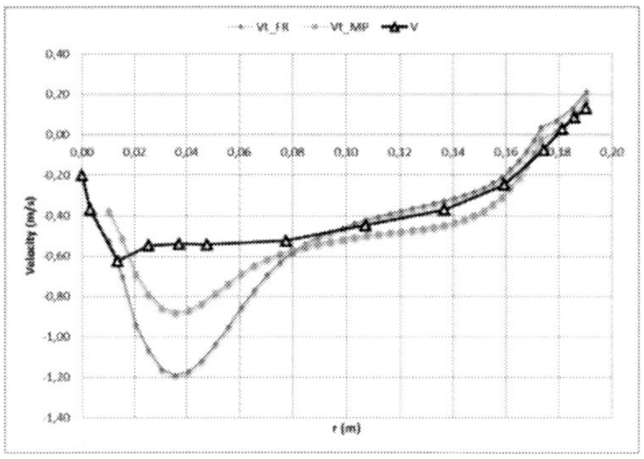

Figure 20. BEP: Tangential vel. - lower line.

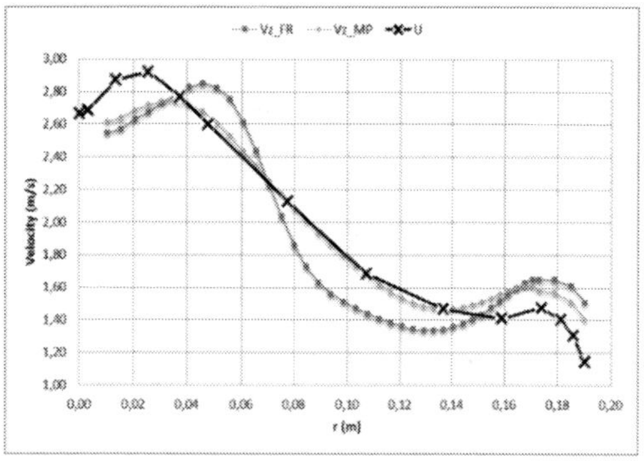

Figure 21. BEP: Axial velocity – lower line.

In a HL operation case, the predictions of tangential velocity are not as good as they were in case of BEP operation. In general, the predictions at upper line are for all three operation cases better than the predictions at lower line. Higher discrepancies between measured and calculated velocities could be the consequence of a fact that the SST turbulence model in Numeca FINE^{TM}/Turbo 9.02 does not include curvature correction.

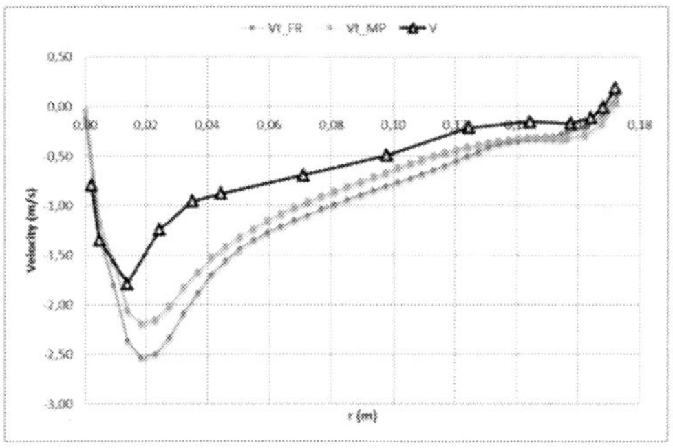

Figure 22. HL: Tangential vel. - upper line.

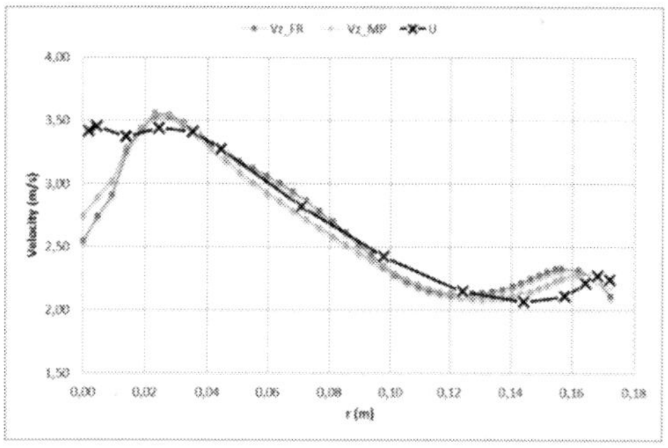

Figure 23. HL: Axial velocity – upper line.

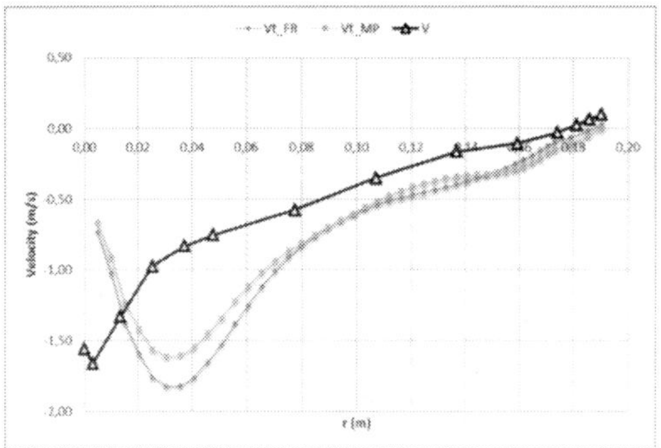

Figure 24. HL: Tangential vel. - lower line.

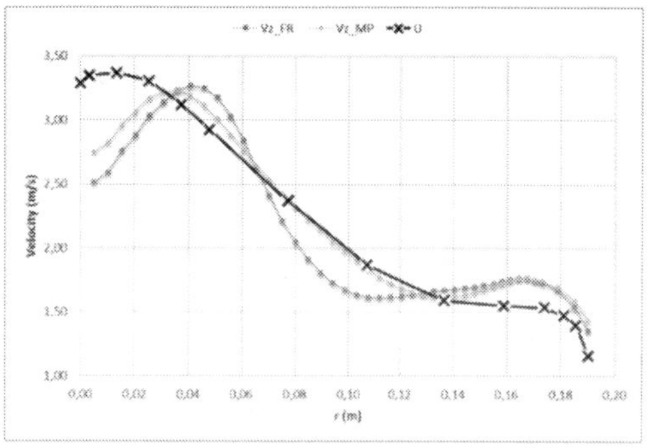

Figure 25. HL: Axial velocity – lower line.

3.3. Efficiency and torque

As a final result of CFD simulations, for each of the three given operating points the hydraulic efficiency of the turbine was calculated. The comparison of measured and predicted efficiencies is

Journal of Physics: Conference Series **579** (2015) 012007 doi:10.1088/1742-6596/579/1/012007

given in figure 26. The notation "NoLab" represent the solution, which is identical to FR solution, except that no runner seals are taken into account.

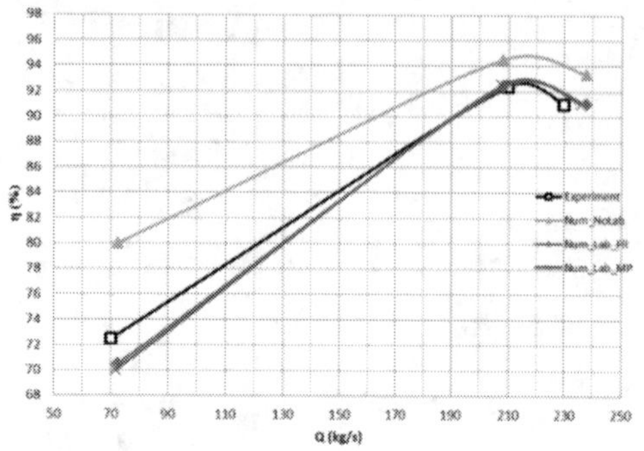

Figure 26. Hydraulic efficiency of a turbine at given operating points.

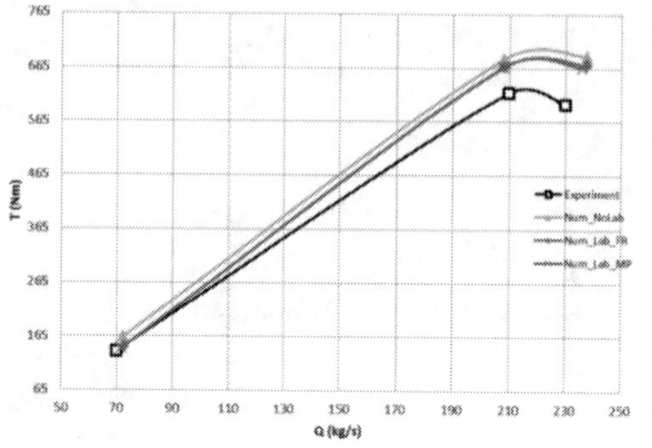

Figure 27. Torque of a turbine at given operating points.

From figure 26 it is evident that with the consideration of labyrinths much better prediction of hydraulic efficiency is achieved. By including the labyrinths to the numerical solution, the maximum difference between the experimental and numerical efficiencies at PL turbine operation lowers down from 7% to 2%. The efficiency predictions at BEP and HL operation points are in that case within 0.1% consistent with measurements. It is rather interesting that the efficiency predictions at PL operation are even under-predicted.

The comparison of measured and calculated torque is shown in figure 27. The best match between measured and predicted torque is achieved at PL operation. At BEP and HL operation the torque is over-estimated. There is a very good agreement between FR and MP solution. The torque curve of a "NoLab" solution is parallel to FR and MP solution curves. That indicates the influence of runner seals on the turbine torque.

4. Conclusion

In present study, a CFD investigation of a high head model Francis turbine was carried out. All geometrical details of a Tokke turbine, including runner seals, were taken into consideration. Steady state numerical simulations were performed at three representative operating points and validated with available experimental data. Except for a small change in the shape of velocity profile curves, there is no noticeable difference between FR and MP solution. A very good agreement between measured and predicted pressure values was achieved. The maximum difference occurred at the same point where the pressure fluctuations in a measured time dependent signal were the highest. Also the validated tangential and axial velocities in the draft tube were not far from measured values. However, the predictions of the velocities at upper line position were better than those located at lower line.

Including the runner seals, a very good prediction of the hydraulic efficiency compared to the experimental data was achieved. Instead of 7% over-prediction at PL operation, a 2% under-prediction was obtained. The influence of runner seals on the turbine torque was rather small. Also the proportion of the flow which leaked through the labyrinth was small, with the maximum amount of 0.44% at PL operation. Therefore, the main reason for the efficiency drop was in the labyrinth losses and disc friction losses. Especially at part load operation those losses significantly lowered down the turbine hydraulic efficiency. The influence of labyrinth loses on the turbine hydraulic efficiency decreased with the increase of the flow through the turbine.

Acknowledgements

The authors gratefully acknowledge the sponsorship of Scandic Energy.

References

[1] Zhao W, Nielsen T K and Billdal J T, 2010 Effects of cavity on leakage loss in straight-through labyrinth seals *IOP Conf. Series: Earth and Environmental Science* **12**

[2] Kurokawa J and Kitahora T, 1994 Accurate determination of volumetric and mechanical efficiencies and leakage behaviour of Francis turbine and Francis pump-turbine *Proc. XVII IAHR (Beijing)*

[3] IEC 62256:2008 83-84

[4] Chirag T, Cervantes M J, Gandhi B K and Dahlhaug O G, 2013 Experimental and numerical studies for a high head Francis turbine at several operating points *Journal of Fluid Engineering* **135**

[5] FINE™/Turbo v9.0, 2013 Discretization and Solution Theory *Theoretical manual* 5-11

Numerical investigation of a high head Francis turbine under steady operating conditions using foam-extend

M. Lenarcic, M. Eichhorn, S. J. Schoder and Ch. Bauer

Institute for Energy Systems and Thermodynamics, Vienna University of Technology, Getreidemarkt 9, 1060 Vienna, Austria

E-mail: markus.lenarcic@tuwien.ac.at

Abstract.
In this work the incompressible turbulent flow in a high head Francis turbine under steady operating conditions is investigated using the open source CFD software package FOAM-extend-3.1. By varying computational domains (cyclic model, full model), coupling methods between stationary and rotating frames (mixing-plane, frozen-rotor) and turbulence models ($k\omega$-SST, $k\epsilon$), numerical flow simulations are performed at the best efficiency point as well as at operating points in part load and high load. The discretization is adjusted according the y^+-criterion with $y^+_{mean} > 30$. A grid independence study quantifies the discretization error and the corresponding computational costs for the appropriate simulations, reaching a GCI $< 1\%$ for the chosen grid. Specific quantities such as efficiency, head, runner shaft torque as well as static pressure and velocity components are computed and compared with experimental data and commercial code. Focusing on the computed results of integral quantities and static pressures, the highest level of accuracy is obtained using FOAM in combination with the full model discretization, the mixing-plane coupling method and the $k\omega$-SST turbulence model. The corresponding relative deviations regarding the efficiency reach values of $\Delta\eta_{rel} \sim 7\%$ at part load, $\Delta\eta_{rel} \sim 0.5\%$ at best efficiency point and $\Delta\eta_{rel} \sim 5.6\%$ at high load. The computed static pressures deviate from the measurements by a maximum of $\Delta p_{rel} = 9.3\%$ at part load, $\Delta p_{rel} = 4.3\%$ at best efficiency point and $\Delta p_{rel} = 6.7\%$ at high load. Commercial code in turn yields slightly better predictions for the velocity components in the draft tube cone, reaching a good accordance with the measurements at part load. Although FOAM also shows an adequate correspondence to the experimental data at part load, local effects near the runner hub are captured less accurate at best efficiency point and high load. Nevertheless, FOAM is a reasonable alternative to commercial code that makes it possible to predict integral quantities and local parameters under steady operating conditions adequately.

1. Introduction

Numerical flow simulations are still the most applicable way for determining the operating conditions of hydraulic machines. Keck et al. [1] chronologically recapitulate the main steps of ordinary CFD towards its specific application in turbomachinery by implementing rotor-stator coupling methods like mixing-plane and frozen-rotor interfaces. In this context, the open source CFD software package OpenFOAM® is noted, which was commercialized and distributed by NABLA Ltd. from 2000 till 2004 [2]. Since then, a lot of effort has been applied in adapting and generating necessary features for turbomachinery purposes. The corresponding main interests

Journal of Physics: Conference Series **579** (2015) 012008 doi:10.1088/1742-6596/579/1/012008

and extensions in this field are still coordinated by the OpenFOAM® Turbomachinery Working Group [3], which continuously organizes workshops to forward appropriate code innovations [4]-[7]. The implementation of the General Grid Interface (GGI), the mixing-plane approach as well as the frozen-rotor coupling method is described in [8]-[9] and part of the new release called FOAM-extend-3.1 [10]. The capabilities of this mixing-plane coupling method are investigated in [11], wherein the computed results correlate well with the expected main flow characteristics. Further CFD simulations in hydraulic turbines have been carried out with OpenFOAM® in [12]-[15].

In order to evaluate the accuracy of state-of-the-art CFD simulations under steady operating conditions, the first Francis-99 workshop has been introduced by the NTNU in Trondheim. Therefore, appropriate measurements at different operating points were provided, which prescribe the numerical boundary conditions, validate the computed results and reveal the current possibilities of numerics and powerful computers. The investigated hydraulic machine is a scaled down model of Tokke power plant in Norway - the associated test rig is located at the NTNU. The model consists of 14 stay vanes (including two different types) embedded inside a spiral case, 28 guide vanes, a runner with 15 splitter and 15 full length blades and an elbow draft tube. Measurements have been carried out at part load (PL), best efficiency point (BEP) and high load (HL) with different guide vane angles, discharges and rotational speeds. Steady-state as well as unsteady flow simulations have already been done on this model by Trivedi et al. [16] and Moritz [17], leading to good agreements with experimental data at BEP and HL but to significant differences at PL. Based on these publications, further numerical investigations have to be performed. The corresponding conditions are summarized in table 1, which promote the validation of numerical simulations over a wide range of possible turbine operating conditions.

Table 1. Operating conditions proposed for the Francis-99 workshop; Operating conditions for the LDA measurements (*)

Objective	Unit	PL	BEP	HL	$PL^{\star}_{LDA}$	$BEP^{\star}_{LDA}$	$HL^{\star}_{LDA}$
H	m	12.290	11.910	11.840	12.290	12.77	12.61
Q	m^3/s	0.071	0.203	0.221	0.071	0.21	0.23
n	1/s	6.770	5.590	6.160	6.770	5.74	6.34
α	deg	3.910	9.840	12.440	3.910	9.84	12.44
η_M	%	71.690	92.610	90.660	72.500	92.40	91.00
T	Nm	137.520	619.560	597.990	-	-	-
ρ	kg/m^3	999.230	999.190	999.200	999.230	999.19	999.20

In the course of this work, numerical flow simulations of the mentioned scaled down model are done using FOAM-extend-3.1 and commercial code. The geometry is spatially discretized with Ansys ICEM® 15.0 according to the y^+-criterion with $y^+_{mean} > 30$ and composed to a cyclic model and a full 360° model with integrated spiral case (see section 3.2). Effects of different turbulence models ($k\omega$-SST, $k\epsilon$) and coupling methods at the interfaces between stationary and rotating frames (mixing-plane, frozen-rotor) are outlined as well.

Various integral quantities such as the head, efficiency and runner torque as well as static pressure and velocity components are computed and compared with experimental data. Therefore, pressure measurements have been performed at six different locations in the vaneless space, the runner and the draft tube (see figure 1). Furthermore, LDA measurements for the axial- and circumferential velocity components have been done at several points along two horizontal lines in the draft tube cone (see table 2). The relative positions of the pressure sensors P41, P71 and S51 are fixed within the runner domain, rotating with the appropriate rotational speed n.

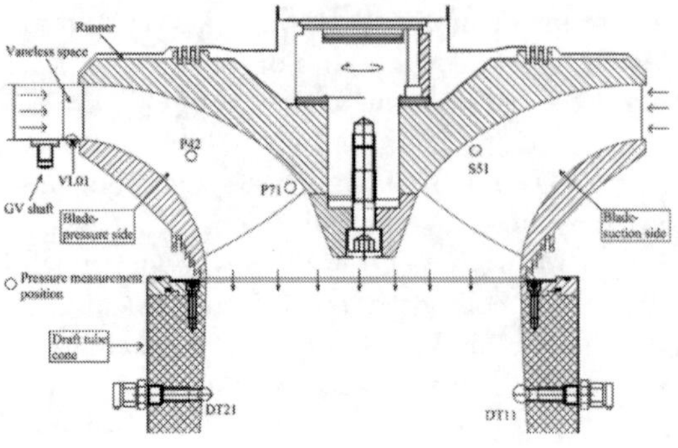

Figure 1. Monitoring points of the experimental pressure measurements [18]

Table 2. Cartesian coordinates of LDA monitoring points

Coord.	Unit	Top Line		Bottom Line	
-	-	Start	End	Start	End
x	mm	0	-178.9	0	-196.5
y	mm	0	0	0	0
z	mm	-243.4	-243.4	-561.4	-561.4

2. Numerical Setup

The computational domain includes all relevant components between spiral case inlet and draft tube outlet, whereby the distance between inlet and actual model is chosen sufficiently large in order to keep the influence of the boundary condition moderate. Stationary parts like spiral case (SC), stay vanes (SV) and guide vanes (GV) are connected to each other by General Grid Interfaces (GGI). The numerical couplings between guide vanes, runner (RN) and draft tube (DT) are achieved by mixing-plane and frozen-rotor approaches respectively.

A constant discharge is set as inlet boundary condition for full models and a corresponding velocity distribution with an incidence angle of $\beta = 9°$ is prescribed for cyclic models. The turbulent quantities are computed by specifying a turbulent intensity of $I = 0.05$ and a uniform turbulent mixing length of $l_t = 0.01m$. In order to consider rotational effects of the runner, a constant rotational speed is defined depending on the appropriate operating conditions. A so-called opening boundary condition with a prescribed value for the mean static pressure enables a re-entering flow and reduces numerical oscillations at the outlet. Values of discharge and rotational speed are set according to table 1. Considering an incompressible single phase flow based on the Reynolds-averaged Navier-Stokes (RANS) method, steady operating conditions are prescribed for all numerical simulations. Convective terms are approximated by 2nd-order interpolation schemes. The maximum number of iteration steps is limited by an abortion criterion for the absolute values of residuals and the evaluated quantities are averaged over a certain iteration step period. In order to improve the numerical stability during the calculation process, moderate under-relaxation factors are chosen. The SIMPLE-algorithm executes the steady-state pressure-velocity coupling in FOAM, whereas commercial code in turn uses a fully coupled solver. With regard to the mixing-plane approach integrated in FOAM, pressure and velocity components are area-averaged at the interface, whereas the turbulence quantities are flux-averaged. Commercial code operates with a constraint of constant total pressure on both sides of the interface. But unless otherwise noted, the computed results are obtained with FOAM using the mixing-plane approach and the kω-SST turbulence model.

3. Discretization

3.1. Grid independence study and quality

In order to quantify the discretization error, a grid independence study is carried out using the full model at three different grid refinements ($h_i = 1, 1.3, 1.7$). Therefore, the efficiency η is computed with commercial code at BEP and compared with an estimated value $f_{h_i=0}$ from the Richardson Extrapolation [19]:

$$f_{h_i=0} \approx f_{h_i=1} + \frac{f_{h_i=1} - f_{h_i=1.3}}{r^p - 1} \tag{1}$$

Therein, $f_{h_i=1,1.3}$ are the computed values, r represents the grid refinement with a value of $r \geq 1.3$ according to [20] and p is the formal order of accuracy. Referring to this equation, the value of the quantity $f_{h_i=0}$ can be approximated if the normalized grid spacing h_i approaches zero. The corresponding data set is given in table 3 and figure 2, where a monotonic convergence regime is clearly visible.

Table 3. Normalized efficiency η as quantity of f_{h_i}

Grid	h_i	r	$f_{h_i}/f_{h_i=1.7}$
-	-	-	-
Fine	1.0	1.3	0.995
Mean	1.3	1.3	0.997
Coarse	1.7	-	1.000

Table 4. Grid Convergence Index (GCI) at different mesh refinements

ϵ_{21}	ϵ_{32}	p	GCI$_{21}$	GCI$_{32}$
-	-	-	%	%
0.2	0.3	1.57	0.52	0.79

The Grid Convergence Index (GCI) represents an empirical error band around the asymptotic numerical value. The GCI of the fine grid is defined as

$$GCI_{21} = \frac{F_s|\epsilon_{21}|}{r^p - 1} \tag{2}$$

and the GCI of the coarse grid as $GCI_{32} = F_s|\epsilon_{32}|r^p/(r^p - 1)$. The safety factor is assumed with $F_s = 1.25$ according to [21]. The computed results of the grid independence study show for either GCI comparatively small deviations from the asymptotic value (see table 4).

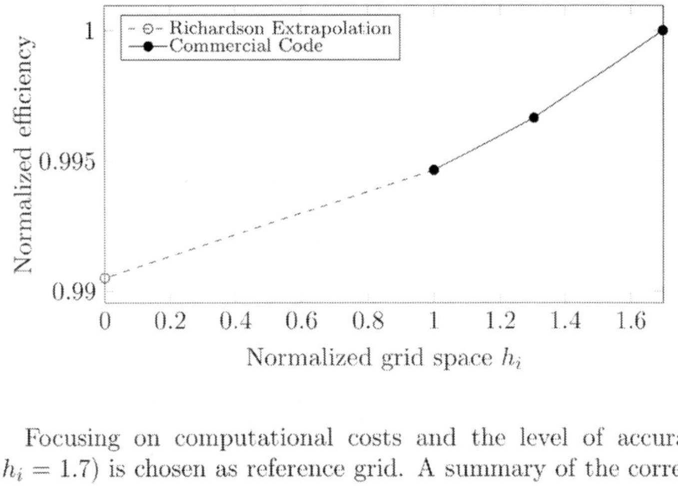

Figure 2. Computed normalized efficiency at three different grid refinements ($h_i = 1, 1.3, 1.7$) using commercial code with the full model; Comparison with the asymptotic estimation from the Richardson Extrapolation

Focusing on computational costs and the level of accuracy, the hexahedral coarse grid ($h_i = 1.7$) is chosen as reference grid. A summary of the corresponding grid quality is given in table 5. The standard wall function, as a suitable compromise between computational time and numerical accuracy, is considered for all further computations, as long as the dimensionless wall distance exceeds values of $y^+ > 30$.

Journal of Physics: Conference Series **579** (2015) 012008 doi:10.1088/1742-6596/579/1/012008

Table 5. Grid quality; Quantities are averaged from SV type 1 and SV type 2 integrated in the spiral case (*)

Objective	Unit	SC	SV*	GV	RN	DT
Min. Determinant	-	0.41	0.35	0.36	0.18	0.66
Min. Angle	deg	14	23	21	11	42
y^+_{mean}	-	> 30	> 30	> 100	> 100	> 100

3.2. Physical models

The influence of the computational domains on the computed results are carried out by considering two different types of physical models: A cyclic model with a simplified extension of the inlet domain, which includes the whole 360° geometry except the spiral case (see figure 3) and a full model, which represents the spatial discretization of the entire hydraulic machine with an extension of the inlet domain at the spiral case (see figure 4).

Table 6. Composition of the physical models and their total number of nodes

Model	SC	SV Type 1	SV Type 2	GV	RN	DT	Total Number of Nodes
-	-	-	-	-	-	-	10^6
Cyclic	-	180°	180°	360°	360°	360°	4.3
Full	360°	180°	180°	360°	360°	360°	4.2

Detailed information about the model compositions and their total number of nodes can be seen in table 6.

Figure 3. Cyclic model

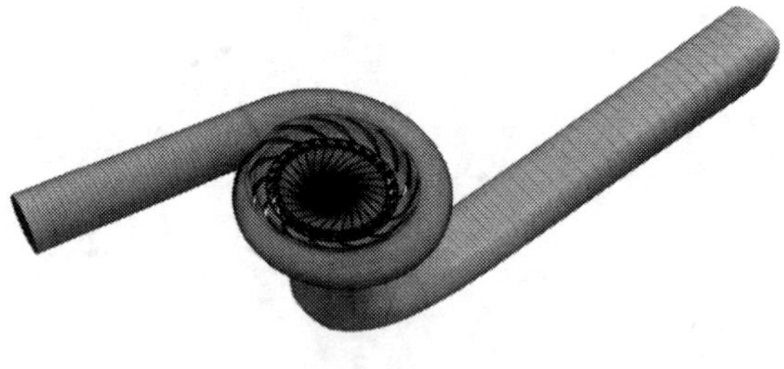

Figure 4. Full model

4. Results

By varying computational domains, coupling methods between stationary and rotating frames, turbulence models and solvers, numerical flow simulations are performed for the operating points PL, BEP and HL. The prescribed operating conditions and monitoring points are given in table 1 and table 2. Regarding the results, table 7 and table 8 illustrate the relative deviations of the computed integral quantities and static pressure results with the corresponding measurements. The evaluation of the velocity components are shown from figure 5 to figure 10. Therein, a radius ratio of $r/r_0 = 0$ represents the starting point on the rotational axis and $r/r_0 = 1$ the end point at the draft tube wall.

Journal of Physics: Conference Series **579** (2015) 012008 doi:10.1088/1742-6596/579/1/012008

4.1. Cyclic models

Table 7 shows the relative deviations between computed results and experimental data concerning the integral quantities and the static pressures. By considering the cyclic model it can be seen that FOAM yields a high level of accuracy at all investigated operating points. Since the mixing-plane interface causes significant differences at PL concerning the runner shaft torque ($\Delta T_{rel} \sim 37\%$), the frozen-rotor coupling method has been additionally applied to evaluate the influence of either approach. Despite of reaching comparable values for the static pressure and the head H, the runner shaft torque is differently captured and more precise using the frozen-rotor interface. This can be lead back to the underlying mathematical algorithm and the different treatment of simplified mass transfer at this specific operating condition. However, the computed results obtained with commercial code are slightly affected by the different types of coupling methods, except for the runner shaft torque at PL. So by comparing the computed results with the measurements, FOAM yields the best agreement using frozen-rotor and - despite of the significant discrepancy at PL - acceptable results using the mixing-plane coupling method. Although commercial code predicts the integral quantities more or less insufficiently, the computed values for the static pressure are comparable with those obtained with FOAM.

Table 7. Relative deviations of integral quantities and static pressure between computed results and measurements using the cyclic model, the $k\omega$-SST turbulence model, FOAM/commercial code (CC) and the mixing-plane (MP)/frozen-rotor (FR) coupling method

OP	Config	Unit	H	T	η	VL01	DT11	DT21	P42	P71	S51
PL	FOAM, MP	%	0.9	-36.8	-28.8	0.7	1.3	1.6	11.2	8.0	6.4
	FOAM, FR	%	0.6	13.3	5.3	2.8	0.1	0.4	11.8	5.5	6.0
	CC, MP	%	13.4	19.3	3.7	2.4	-1.8	-1.7	8.5	11.7	4.5
	CC, FR	%	14.1	28.0	8.7	3.4	-1.8	-1.7	8.8	11.8	3.9
BEP	FOAM, MP	%	3.7	1.2	-3.0	1.2	-2.6	-2.2	4.3	3.8	3.0
	FOAM, FR	%	4.4	2.9	-2.2	2.1	-2.6	-2.2	4.3	3.7	3.1
	CC, MP	%	19.8	13.4	-4.4	5.3	-2.6	-2.2	3.9	1.4	8.0
	CC, FR	%	19.5	13.9	-3.9	4.6	-2.5	-2.1	3.8	1.5	8.4
HL	FOAM, MP	%	1.4	-6.9	-7.8	-0.4	-1.7	-0.9	6.3	5.1	3.7
	FOAM, FR	%	1.3	-0.6	-2.1	-0.6	-1.8	-1.0	7.2	4.9	3.7
	CC, MP	%	13.7	8.5	-3.4	-0.7	-7.3	-6.5	2.4	-3.6	4.4
	CC, FR	%	13.2	8.1	-3.4	-1.2	-7.3	-6.5	2.6	-3.5	4.3

The computed velocity components at PL show an adequate correspondence to the measurements for both the axial velocities c_u and the circumferential velocities c_θ (see figure 5), whereas enhanced differences occur at the draft tube wall. Those discrepancies are attributable to the applied standard wall function, which limits the accuracy when the flow condition differs too much from the ideal one and large gradients of physics occur in the boundary layer. However, the computed results at BEP and HL are more or less comparable, showing the maximum deviations near the runner hub, where existing flow structures are not captured well (see figure 6 and figure 7). But although discharge and swirl are incorrectly predicted in this region, the experimental data are mainly reproduced yet. A comparison between the computed velocities obtained with both solvers and coupling methods shows no clear preference, since the quantities are qualitatively similar. But, the accuracy tends to decrease when the mixing-plane approach is applied with FOAM, whereas using the frozen-rotor coupling method yields an adequate correspondence to the measurements.

Journal of Physics: Conference Series **579** (2015) 012008 doi:10.1088/1742-6596/579/1/012008

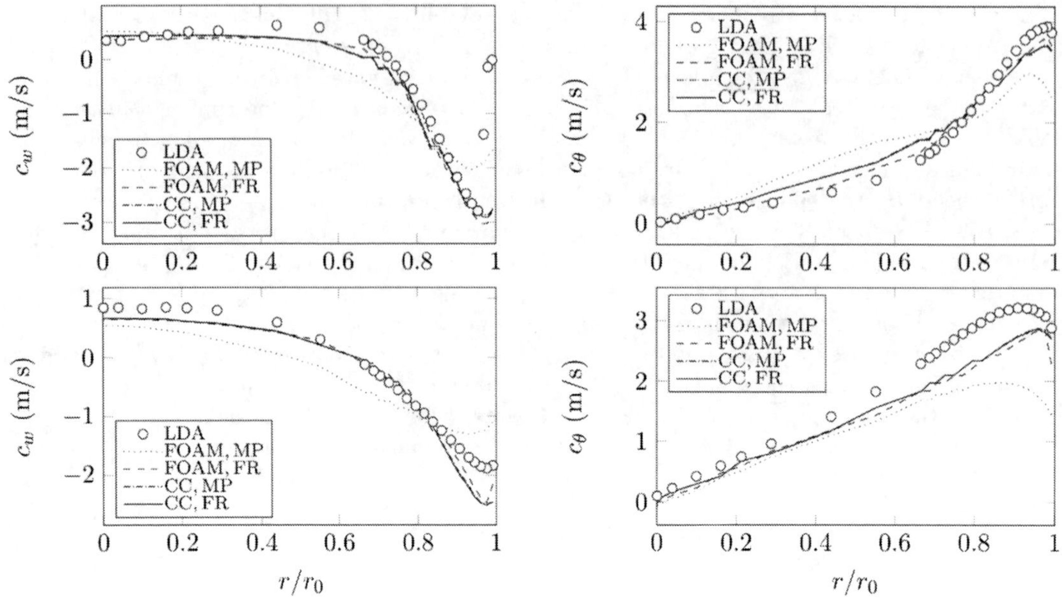

Figure 5. Axial velocity (left) and circumferential velocity (right) on the top line (top) and the bottom line (bottom) at operating point PL using the cyclic model, the kω-SST model, FOAM/commercial code (CC) and the mixing-plane (MP)/frozen-rotor (FR) coupling method

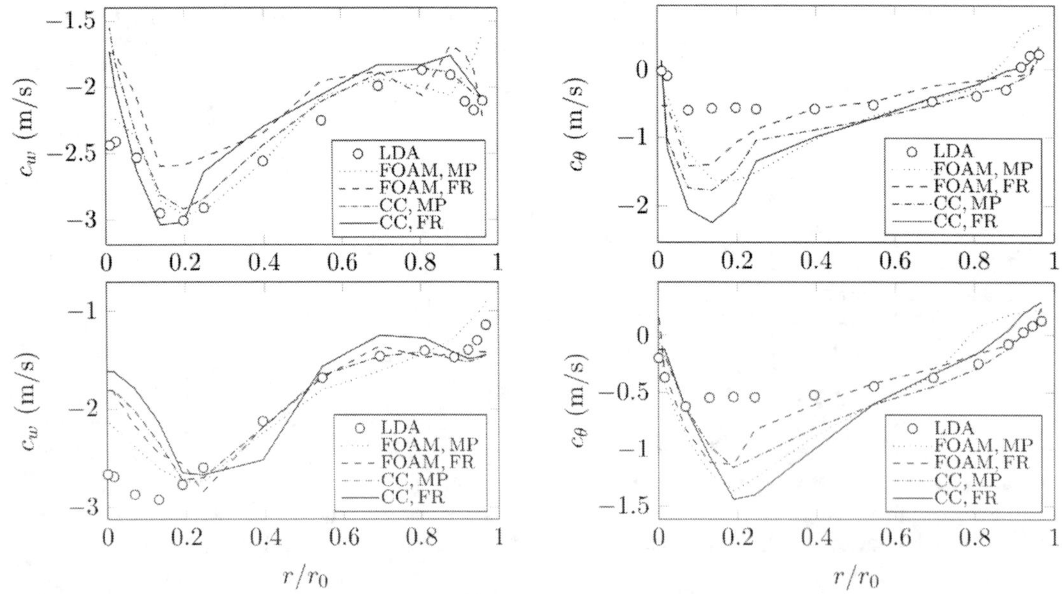

Figure 6. Axial velocity (left) and circumferential velocity (right) on the top line (top) and the bottom line (bottom) at operating point BEP using the cyclic model, the kω-SST model, FOAM/commercial code (CC) and the mixing-plane (MP)/frozen-rotor (FR) coupling method

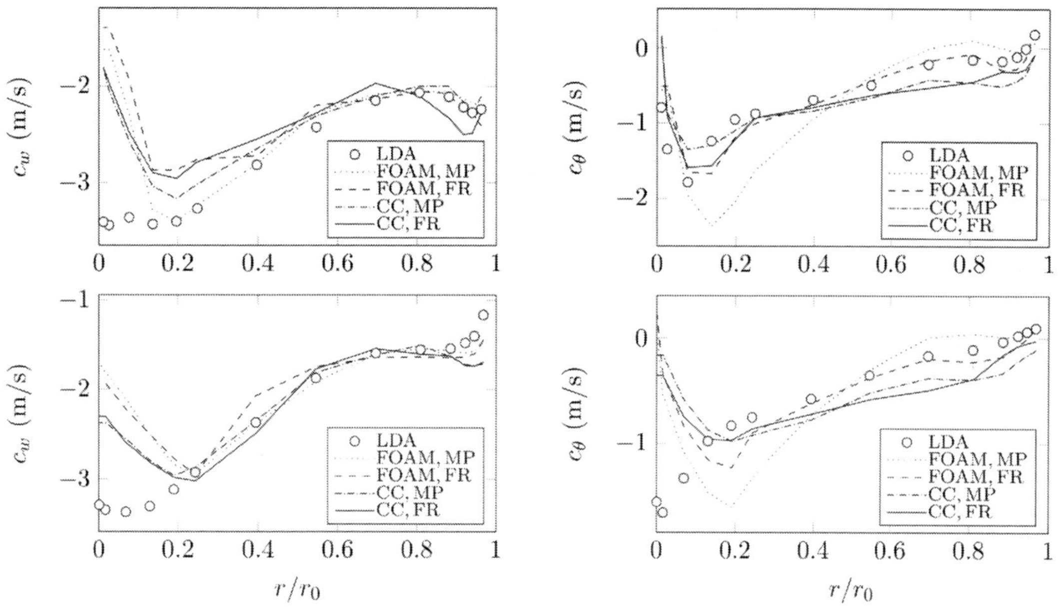

Figure 7. Axial velocity (left) and circumferential velocity (right) on the top line (top) and the bottom line (bottom) at operating point HL using the cyclic model, the kω-SST model, FOAM/commercial code (CC) and the mixing-plane (MP)/frozen-rotor (FR) coupling method

4.2. Full models

Table 8 illustrates the relative deviations between computed results and measurements regarding integral quantities and static pressures. Therefore, the $k\omega$-SST and the $k\epsilon$ turbulence models are applied using the full model with both coupling methods. Considering the frozen-rotor interface does not significantly improve the accuracy, so the corresponding results are not further discussed in this work. Nevertheless, the computed results obtained with FOAM and the $k\omega$-SST turbulence model show a high accuracy concerning the integral quantities at BEP and even adequate values at HL and PL. Compared to the computed results using the cyclic model, considering the full model yields a higher precision regarding the head H at BEP and regarding the runner shaft torque T at PL - the corresponding values are similar at HL. Using the full model positively affects the computation of the integral quantities, which can be lead back to the possibility of considering an affected incident flow as consequence of the different operating conditions. Using the $k\epsilon$ turbulence model with FOAM leads to significant deviations between numerics and experimental data at each operating point, whereas an improving trend is evident when using commercial code. This can be attributed to its underlying numerics and wall treatment, preventing a high level of accuracy in appearance of complex flow structures and near-wall effects. Generally, the head H is computed more precise than the runner shaft torque T. This fact can be lead back to the simplified mass transfer at the interfaces between stationary and rotating frames, which directly influences the computation of the runner shaft torque. On the other hand, the head is calculated at the domain boundaries and thus less affected by mathematical limitations of the coupling methods. Nevertheless, the static pressures are accurately observed using both solvers and turbulence models, reaching a maximum deviation of $\Delta p_{rel} < 12\%$ in the runner domain at PL.

Table 8. Relative deviations of integral quantities and static pressure between computed results and measurements using the full model, the $k\omega$-SST/$k\epsilon$ turbulence model, FOAM/commercial code (CC) and the mixing-plane (MP)/frozen rotor (FR) coupling method

OP	Config	Unit	H	T	η	VL01	DT11	DT21	P42	P71	S51
PL	FOAM, SST, MP	%	-1.4	-6.7	-7.0	3.1	0.2	0.5	9.3	6.6	5.4
	FOAM, SST, FR	%	1.1	14.0	5.4	1.9	-1.2	-0.7	10.5	4.9	5.3
	FOAM, kϵ, MP	%	-7.5	-27.3	-18.0	-1.2	-2.4	-1.9	8.4	7.0	3.3
	CC, SST, MP	%	11.9	16.7	3.0	1.7	-1.8	-1.7	8.4	11.7	4.2
	CC, kϵ, MP	%	9.7	10.3	0.4	1.8	-0.9	-0.9	7.8	9.6	4.1
BEP	FOAM, SST, MP	%	0.6	1.9	0.5	1.3	-2.6	-2.1	4.3	3.8	3.2
	FOAM, SST, FR	%	1.5	3.2	0.8	2.2	-2.7	-2.2	4.3	3.8	3.2
	FOAM, kϵ, MP	%	-2.4	-16.1	-13.7	3.4	-2.8	-2.3	4.3	3.6	3.0
	CC, SST, MP	%	12.2	12.5	0.9	5.1	-2.6	-2.2	3.9	1.5	7.9
	CC, kϵ, MP	%	8.4	7.4	-0.3	3.8	-2.6	-2.2	3.8	1.4	7.8
HL	FOAM, SST, MP	%	-1.8	-7.5	-5.6	-0.5	-1.9	-1.0	6.7	5.0	3.8
	FOAM, SST, FR	%	-1.8	-0.5	0.8	-0.5	-1.8	-0.9	7.3	4.9	3.8
	FOAM, kϵ, MP	%	-3.1	-19.4	-15.7	-2.1	-2.1	-1.2	4.6	5.0	3.4
	CC, SST, MP	%	4.5	6.9	2.8	-1.6	-7.3	-6.5	2.5	-3.7	4.5
	CC, kϵ, MP	%	2.2	3.0	1.5	-2.6	-7.3	-6.5	2.4	-3.8	4.5

The computed axial velocities (c_w) reveal an adequate accordance with the measurements at PL (see figure 8), predicting a vortex formation, which mainly blocks the flow in the core region of the draft tube. The observed swirl increases towards the draft tube wall, resulting in a high level of circumferential velocities (c_θ). The computed flow distribution agrees well with the measurements, even the drop towards the wall is predicted by the numerics. The only exception to this is using the combination of FOAM and the $k\epsilon$ turbulence model, which insufficiently computes swirl and near-wall physics. The corresponding results correlate with an almost swirl-free flow distribution in the draft tube core region, reaching a maximum value for both velocity components at the wall. As mentioned before, the observed deviations at the wall can be attributed to the applied standard wall function, which limits the accuracy of the physics in the boundary layer. At BEP, the computed axial- and circumferential velocities agree well with the measurements above a radius ratio of $r/r_0 \sim 0.2$, but continuously deviate towards the hub (see figure 9). Thereby, a hub vortex is computed, leading to an enhanced swirl at a radius ratio of $r/r_0 \sim 0.2$ and to a reduced discharge towards the hub. But since the measurements show an moderate swirl and an increased discharge at the hub, the local effects in this region are not well captured by the numerics. Nevertheless, the observed near-wall physics are in an adequate accordance with the measurements, except of using FOAM with the $k\epsilon$ turbulence model. The computed results at HL are qualitatively similar to those at BEP, since the discharge at the hub is also insufficiently predicted by the numerics (see figure 10). The circumferential component is observed in an accurate manner but using FOAM with the $k\omega$-SST model leads to significant deviations at a radius ratio of $r/r_0 \sim 0.2$.

A comparison between the computed velocities obtained with both solvers and both turbulence models shows a slight preference for commercial code, particularly at BEP and HL. The observed results vary more according to the applied turbulence model using FOAM than commercial code, which can be lead back to the underlying numerics with its implemented damping functions and smoothing algorithms.

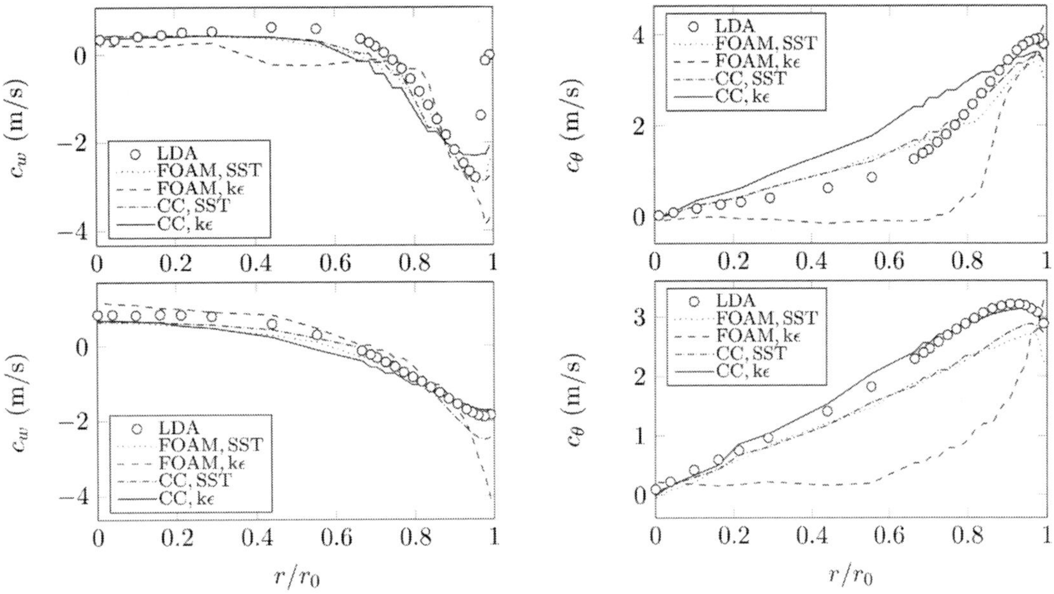

Figure 8. Axial velocity (left) and circumferential velocity (right) on the top line (top) and the bottom line (bottom) at operating point PL using the full model, the kω-SST/kϵ turbulence model, FOAM/commercial code (CC) and the mixing-plane coupling method

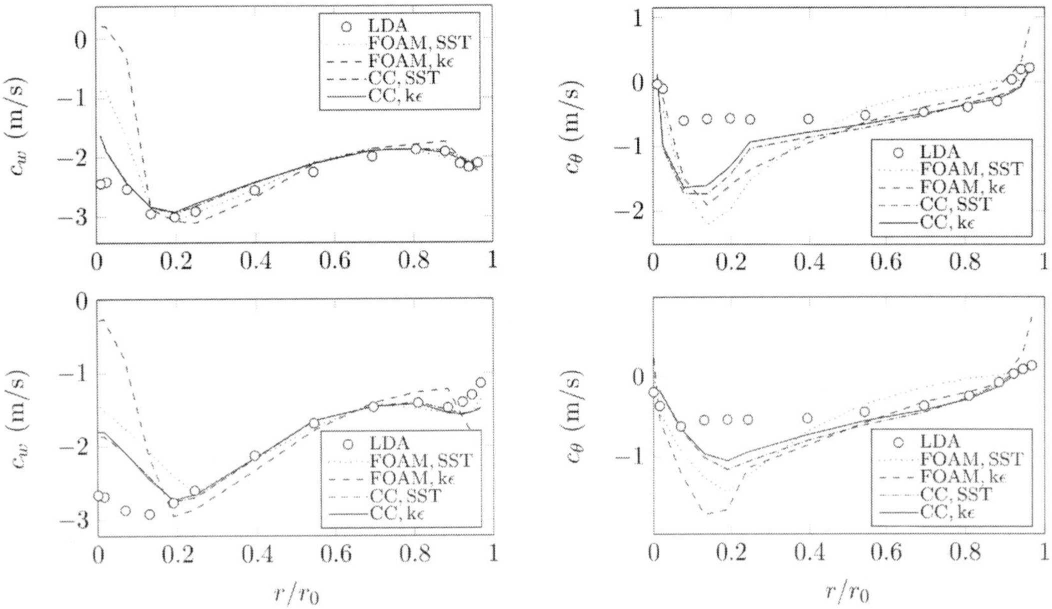

Figure 9. Axial velocity (left) and circumferential velocity (right) on the top line (top) and the bottom line (bottom) at operating point BEP using the full model, the kω-SST/kϵ turbulence model, FOAM/commercial code (CC) and the mixing-plane coupling method

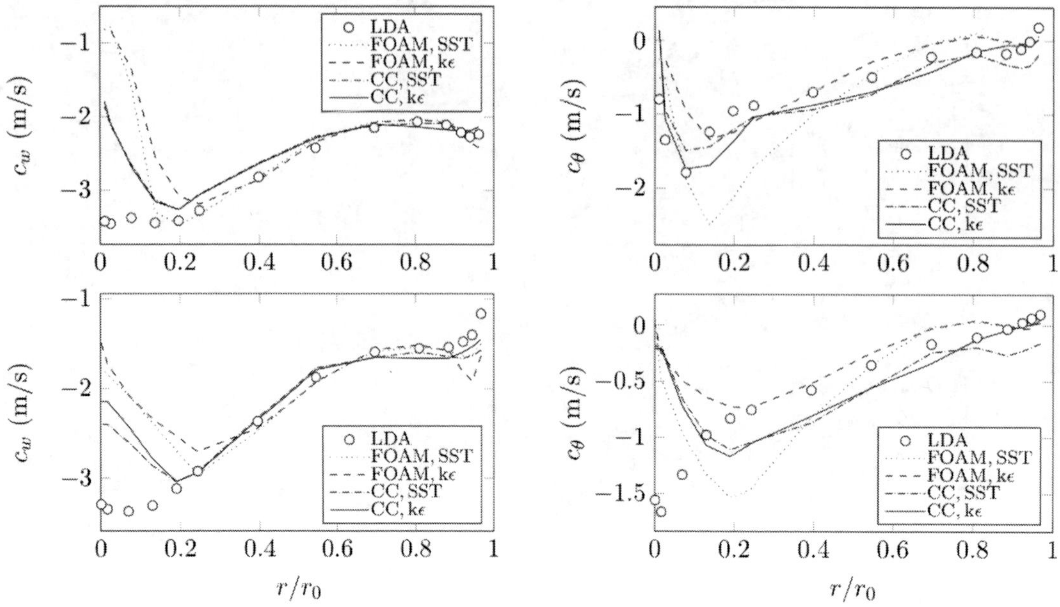

Figure 10. Axial velocity (left) and circumferential velocity (right) on the top line (top) and the bottom line (bottom) at operating point HL using the full model, the kω-SST/$k\epsilon$ turbulence model, FOAM/commercial code (CC) and the mixing-plane coupling method

5. Conclusion

An incompressible turbulent flow in a high head Francis turbine under steady operating conditions is computed using FOAM-extend-3.1. Influences of computational domains (cyclic model, full model), coupling methods (mixing-plane, frozen-rotor), turbulence models (kω-SST, kϵ) and solvers (FOAM, commercial code) are carried out and compared with experimental data of integral quantities, static pressures and velocity components. Therefore, pressure measurements have been done at different locations in the vaneless space, the runner and the draft tube. LDA measurements for the axial- and circumferential velocity components in turn have been performed at several points along two horizontal lines in the draft tube cone. The geometry is spatially discretized and composed both to a cyclic model with a simplified inlet domain and to a full 360° model with an integrated spiral case. The grid refinement is adjusted according to the y^+-criterion with $y^+_{mean} > 30$. A grid independence study is done, resulting in a GCI $< 1\%$ for the chosen computational grid.

At first, numerical flow simulations are performed using the cyclic model with both solvers, coupling methods and the kω-SST turbulence model. Thereby, the frozen-rotor coupling method yields an adequate correspondence to the measurements with deviations of $\Delta\eta_{rel} \sim 5\%$ at part load, $\Delta\eta_{rel} \sim 2\%$ at best efficiency point and $\Delta\eta_{rel} \sim 2\%$ at high load. The corresponding ranges of the static pressures are $\Delta p_{rel} \leq 11.8\%$ at part load, $\Delta p_{rel} \leq 4.3\%$ at best efficiency point and $\Delta p_{rel} \leq 7.2\%$ at high load. On the other hand, using the mixing-plane interface negatively affects the computation of the runner torque at part load and high load, although the head H is well predicted by the numerics. As consequence, the relative deviations of the efficiency increase to a level of $\Delta\eta_{rel} \sim 29\%$ at part load, to $\Delta\eta_{rel} \sim 3\%$ at best efficiency point and to $\Delta\eta_{rel} \sim 8\%$ at high load, whereas the deviations concerning the static pressures are similar to those obtained with the frozen-rotor coupling method. Since those differences are insufficient for predicting such flow parameters, further investigations are done using the full model with different turbulence models. Applying FOAM with the kϵ turbulence model results in enhanced deviations between the computed efficiency and experimental data, reaching values of $\Delta\eta_{rel} = 18\%$ at part load, $\Delta\eta_{rel} \sim 14\%$ at best efficiency point and $\Delta\eta_{rel} \sim 16\%$ at high load.

With focus on the results computed with FOAM, the highest level of accuracy is obtained using a combination of full model, mixing-plane coupling method and kω-SST turbulence model, since the frozen-rotor interface does not achieve a significant improvement in results when the full model is applied. The consideration of the full 360° discretization particularly improves the flow prediction at part load, where the flow field is characterized by an uneven incident flow and by attached flow regimes. The divergences of the computed static pressures to the experimental data are in an acceptable range at part load ($\Delta p_{rel} \leq 9.3\%$), best efficiency point ($\Delta p_{rel} \leq 4.3\%$) and high load ($\Delta p_{rel} \leq 6.7\%$). Therein, the maximum differences are observed in the runner domain, where complex flow phenomena exist and quantities are directly affected by the simplified mass transfer through the interfaces between stationary and rotating frames. The computed efficiencies show relative deviations of $\Delta\eta_{rel} \sim 7\%$ at part load, $\Delta\eta_{rel} \sim 0.5\%$ at best efficiency point and $\Delta\eta_{rel} \sim 5.6\%$ at high load. Comparing the computed results obtained with either solver, FOAM reaches a higher accuracy regarding the integral quantities and similar values regarding the static pressures.

The computed velocity components show an adequate correspondence to the measurements, although local effects near the runner hub are not fully captured at BEP and HL. Especially the flow regime at part load is accurately predicted with slight deviations towards the draft tube wall. Applying the kϵ turbulence model does not yield a clear improving trend with commercial code but even negatively affects the computed results with FOAM. This can be lead back to its underlying numerics, which is not able to capture complex flow structures and attached flow regimes in an accurate manner. For this reason, a high level turbulence model like the Reynolds Stress Model (RSM) should be used instead, decreasing the influence of turbulence modeling and providing improved results. In order to reduce the existing deviations between numerics and experimental data in the the near-wall region, a grid refinement should be applied, resolving the physics in the boundary layer without using a standard wall function. Comparing the computed velocity components obtained with either solver, commercial code yields a slightly higher correspondence to the measurements - especially near the runner hub at BEP and HL, where vortex formations are not well predicted by FOAM. Nevertheless, the observed velocity components using FOAM show an appropriate accordance with the experimental data too.

The numerical results obtained with commercial code are less influenced by varying turbulence models and physical models than those computed with FOAM, which can be lead back to the underlying numerics, the implemented damping functions and smoothing algorithms. Focusing on FOAM, there are still differences concerning accuracy and numerical robustness between mixing-plane and the frozen-rotor coupling method: Using mixing-plane needs more computational time to reach a well converged state and the accuracy of the computed results responds more in a sensitive way to changes in convection-specific interpolation schemes. Nevertheless, FOAM is a reasonable alternative to commercial code that makes it possible to adequately predict integral quantities and even values of local parameters under steady operating conditions at part load, best efficiency point and high load.

6. Nomenclature

Symbol	Definition	Symbol	Definition
BEP	Best efficiency point	H	Head (m)
HL	High load	Q	Volumetric flow rate (m^3/s)

Journal of Physics: Conference Series **579** (2015) 012008 doi:10.1088/1742-6596/579/1/012008

Symbol	Definition	Symbol	Definition
PL	Part load	n	Rotational speed $(1/s)$
OP	Operating point	α	Guide vane angle (deg)
DT	Draft tube domain	β	Incidence angle (deg)
GV	Guide vane domain	η	Efficiency (%)
RN	Runner domain	T	Runner shaft torque (Nm)
SC	Spiral case domain	ρ	Density (kg/m^3)
SV	Stay vane domain	I	Turbulent intensity (-)
LDA	Laser Doppler anemometry	l_t	Turbulent mixing length (m)
FR	Frozen-rotor	c_θ	Circumferential velocity (m/s)
MP	Mixing-plane	c_w	Axial velocity (m/s)
GCI	Grid convergence index	r	Radius (m)
h_i	Grid refinement index	y^+	Dimensionless wall distance (-)
CC	Commercial code	Δp_{rel}	Relative pressure deviation (%)
DT11, DT21, P42, P71, S51, VL01	Pressure sensors	$\Delta\eta_{rel}$	Relative efficiency deviation (%)

7. References

[1] H. Keck, M. Sick: Thirty Years of Numerical Flow Simulation in Hydraulic Turbomachines, Acta Mechanica, 201, pp.211-229 , 2008

[2] OpenFOAM Foundation: OpenFOAM Release History, Retrieved October 23, 2014, from `http://www.openfoam.org/download/history.php`

[3] Sig Turbomachinery, Retrieved October 23, 2014, from `http://openfoamwiki.net/index.php/Sig_Turbomachinery`

[4] O. Bounous: Studies of the ERCOFTAC conical diffuser with openFOAM., Department of Applied Mechanics Chalmers University of technology, Gteborg, Sweden, 2008

[5] O. Petit et al.: The ERCOFTAC Centrifugal Pump OpenFOAM Case-Study, In 3rd IAHR International Meeting of the Workgroup of Cavitation and Dynamic Problems in Hydraulic Machinery and Systems, pages 523-532, Brno, Czech Republic, October 2009

[6] H. Nilsson et al.: The OpenFOAM Turbomachinery working-group, and conclusions from the Turbomachinery session of the third OpenFOAM workshop., IAHR: 24th Symposium on Hydraulic Machinery and Systems, Foz do Iguassu, Brazil. 2008

[7] Sig Turbomachinery and Meetings, Retrieved October 23, 2014, from `http://openfoamwiki.net/index.php/Sig_Turbomachinery_/_Meetings`

[8] M. Beaudouin, H. Jasak: Development of a General Grid Interface for turbomachinery simulations with OpenFOAM, Open Source CFD International Conference, November 2008

[9] H. Jasak, M. Beaudoin: OpenFOAM TURBO TOOLS: FROM GENERAL PURPOSE CFD TO TURBOMACHINERY SIMULATIONS, Joint Fluids Engineering Conference, 2011

[10] OpenFOAM extensions, from `http://sourceforge.net/p/openfoam-extend/foam-extend-3.1/ci/master/tree/ReleaseNotes.txt`, 2014

[11] M. Page et al.: Steady-state Capabilities for Hydroturbines with OpenFOAM, International Journal of Fluid Machinery and Systems, 4(1):160-170, Jan-Mar 2011

[12] M. Karlsson et al.: Numerical estimation of torsional dynamic coefficients of a hydraulic turbine, International Journal of Rotating Machinery, 2009

[13] H. Nilsson: Evaluation of OpenFOAM for CFD of turbulent flow in water turbines., IAHR 2006

[14] H. Nilsson: Experiences with OpenFOAM for water turbine applications. Proceedings of the 1st OpenFOAM International Conference, 26-27 November 2007

[15] H. Nilsson: Some experiences on the accuracy and parallel performance of OpenFOAM for CFD in water turbines." Applied Parallel Computing. State of the Art in Scientific Computing. Springer Berlin Heidelberg, 2007. 168-176.

[16] Ch. Trivedi et. al: Experimental and Numerical Studies for a High Head Francis Turbine at Several Operating Points, Journal of Fluids Engineering, Vol. 135/111102-1 , 2013

[17] R. Moritz: Transient CFD-analysis of a high head Francis turbine., 2014

[18] Francis 99 Workshop, from `http://www.ltu.se/research/subjects/Stromningslara/Konferenser/Francis-99/Test-Case-1.111520`, 2014

[19] L.F. Richardson, J.A. Gaunt: The deferred approach to the limit. Philosophical Transactions of the Royal Society of London. Series A, Containing Papers of a Mathematical or Physical Character, 226, 1927

[20] I. B. Celik et al.: Procedure for Estimation and Reporting of Uncertainty Due to Discretization in CFD Applications, Journal of Fluids Engineering, 078001-1, Vol 130, July 2008

[21] P. J. Roache: Verification and Validation in Computational Science and Engineering, Hermosa publishers, Albuquerque, NM, 1998.

Francis-99 Workshop 1: steady operation of Francis turbines | IOP Publishing
Journal of Physics: Conference Series **579** (2015) 012009 | doi:10.1088/1742-6596/579/1/012009

Predicting the performance of a high head Francis turbine using a fully implicit mixing plane

O Amstutz, B Aakti, E Casartelli, L Mangani, L Hanimann

Lucerne University of Applied Sciences and Arts Engineering & Architecture, Horw Switzerland

E-mail: oliver.amstutz@hslu.ch

Abstract. In the present paper numerical investigations of a complete high head Francis turbine comprehensive of a spiral casing, stay and guide vanes and draft tube have been performed at three operating conditions, namely at part load (PL), best efficiency point (BEP), and high load (HL). The main target of the investigations is to assess the prediction accuracy of a reduced domain of the complete turbine using a novel mixing-plane formulation.

The computational domain is simplified simulating one single passage of the runner, thus assuming rotational periodicity and steady state conditions. The results were compared with experimental data published by the workshop organization.

All CFD simulations were performed at model scale with an in-house adapted, 3D, unstructured, object-oriented finite volume code based on the OpenFOAM-V2.2 framework and designed to solve steady-state incompressible RANS-Equations. The pressure-based solver uses a SIMPLE-C like algorithm and is capable of handling multiple references of frame (MRF). The influence of the turbulence has been considered applying the shear-stress transport model (SST). Full second order upwind scheme for advection discretization has been used for all computations.

1. Introduction

The scope of the first Francis-99 workshop is the assessment of advanced numerical tools for the prediction of steady state conditions in a Francis turbine for design and off-design conditions [1]. The continuous evolution of the numerical techniques and the computer power has lead in the last decade to a large tool-spectrum for turbomachinery investigations. The driver for this trend was mainly the only partly-satisfactory results obtained, especially at off-design conditions. The main reasons for the discrepancies were located firstly in the turbulence models. With the development of more advanced models capable to improve the well-known drawbacks of classical models like $k-\epsilon$ and capture additional flow physics, like curvature correction for the SST model or the v2F model. Secondly they were located in the thought inability of steady state methods to properly predict the inherently unsteady effects present in turbomachines [2], [3].

A detailed analysis of one of the first methods to efficiently investigate the interaction of turbomachinery blade-rows, the mixing plane method [4], has shown that the usual implementation can be improved using a fully implicit approach, compared to the until now used explicit one. The main advantage are (1) an increased stability and robustness, allowing for example to handle backflow at the interface, as well as an increased accuracy, allowing to predict not only globally but also locally reliable distributions of the flow quantities in a time-averaged, i.e. steady-state way over a large operating range [5].

Journal of Physics: Conference Series **579** (2015) 012009 doi:10.1088/1742-6596/579/1/012009

2. CFD - solver technology

2.1. Introduction and concept

The design of modern multi-stage turbomachines is a challenging task for engineers, which can only be tackled with the help of CFD. While in the last few years new approaches based on unsteady methods have been developed for single-passage computations of the rotor-stator interaction, thus promising to capture unsteady effects already in the design procedure though not computing the full annulus [2, 3], the steady-state approach still represents the most used industrial standard [4, 6, 7].

Of the steady-state methods, the mixing-plane approach is the most common and general, assuming, as observed in many real turbomachinery applications, that the flow largely mixes out in the gap between two consecutive blade rows (figure 1). This leads to the assumption that at the interface between the components pitch-wise constant total conditions have to be imposed. Therefore a mixing plane mimics the time averaged impact of rotor-stator interactions over a complete revolution.

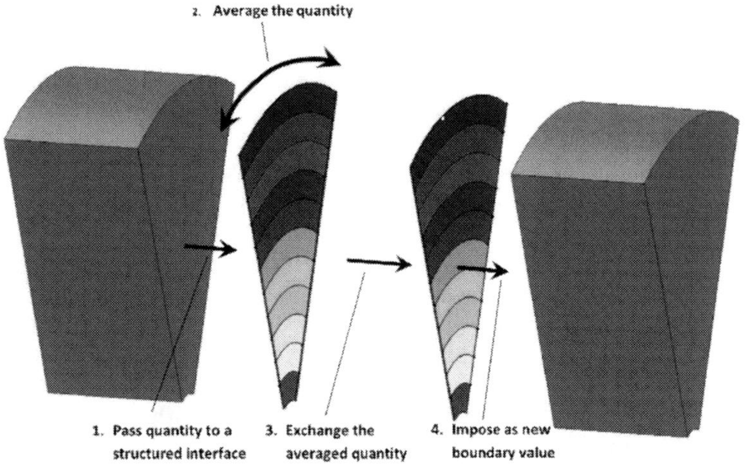

Figure 1: Concept of mixing plane averaging.

Despite the similarity between the mixing process and the mixed-out state, the difference in the way the mixing happens can cause problems. Even if the flow is largely mixed out in reality, the process is more gradual than the instantaneous mixing at the mixing plane. This instantaneous mixing will result in a higher entropy rise and can be seen as a loss of total pressure over the interface. Studies by Fritsch and Giles [8] showed an increase in total loss of about 10% when instantaneous mixing was applied.

2.2. Problems of standard procedure

The usual methodology for the mixing plane implementation is based on an explicit treatment of the interface. This leads to a numerical separation of the problem which has to be coupled with a new boundary-condition set between the two components, constantly exchanging averaged primitive variables. While the approach is simple in the idea, numerical problems can occur by the division of the computational domain in two parts.

A number of authors have investigated the influence of specific averaging methods, agreeing that simple area or mass averaging can not guarantee conservation of mass, momentum and energy. Dzung's [9] consistent averaging is one possible option. Another advanced method using all relevant conservation equations and making the distinction between hypothetically reversible and irreversible quantities was presented in a publication by Kreitmeier [10].

Amongst the difficulty to fulfill conservation equations during the averaging procedure of flow variables (see [9, 10]), one of the most discussed issues are reflections of information at the interface boundaries.

The strong constraint of pitch-wise uniform distribution is accompanied by the problem of reflection of information, as ordinary in- and outlets have. However, for the mixing plane approach these reflections can not be fully assigned to an inappropriate modeling of the farfield at the boundaries of a truncated domain. It is mainly caused by the very basic model-assumption of uniform total conditions in pitch-wise direction. Additionally it has to be mentioned that reflections produce wiggles in the solution, which can cause difficulties in the numerical procedure, deteriorating convergence and thus leading to longer computational time.

As a counter-measure a lot of effort in commercial and proprietary codes was and still is undertaken in the community, for the development of so-called non-reflecting boundary conditions (NRBC), considered a must for a mixing-plane interface in order to work properly (see for example [7], [11], [12]). NRBC are derived from the general unsteady approach, where waves with different frequency have to pass the interface between rotor and stator without being reflected, thus defining appropriate boundary conditions in order to vanish incoming disturbances.

The mixing-plane is thus usually defined in two steps as given in [13]: first, an average state is imposed corresponding to the zero frequency of the wave representing the steady state situation and second, local perturbations are superimposed in order to achieve a non-reflective behavior, based on the Fourier analysis of linearized Euler's equation for the corresponding boundary condition. Out of this approach a series of problems arises in order to achieve both consistent averaging and NRBC behavior and are directly linked to the explicit treatment of the interface. Accordingly, the authors published in [14] a novel approach for the implementation of the mixing-plane, based on a fully implicit treatment of the interface, giving details on the implementation and validating the interface with measurements. The most important fact of the implicit discretization is that it takes directly into account both consistent averaging and reflectivity, as explained in the section "Implicit Mixing Plane".

2.3. New Formulation

A simplified example will be given to point out the main difference between the standard explicit treatment and the new implicit formulation. Figure 2 represents a simplified domain, consisting of four cells and separated by a mixing plane.

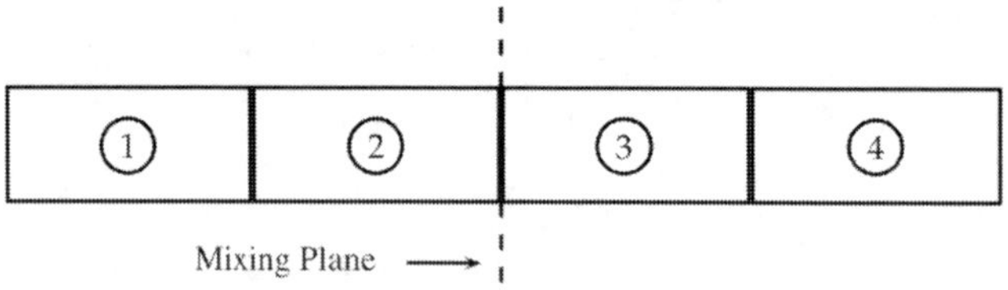

Figure 2: Example domain.

The differences in the coupling of the interface will be pointed out by a numerical discretization of a diffusion operator, given in equation 1.

$$B = \int \nabla \Gamma_{cell} \Phi dV = \int \Gamma_{cell} \Phi dS \cong \sum \Gamma_f \Phi_f S_f \qquad (1)$$

This equation can be rewritten in the general form shown in equation 2.

Journal of Physics: Conference Series **579** (2015) 012009 doi:10.1088/1742-6596/579/1/012009

$$\underline{\underline{A}} \cdot \underline{x} = \underline{b} \tag{2}$$

For further explanations the fluxes across the faces will be discretized with a central difference scheme and the areas of the faces are of uniform size S_f. Furthermore $\Gamma = \Gamma_f * S_f$.

Explicit Algorithm
As mentioned earlier an explicit coupling leads to additional in- and outlets as visualized in figure 3.

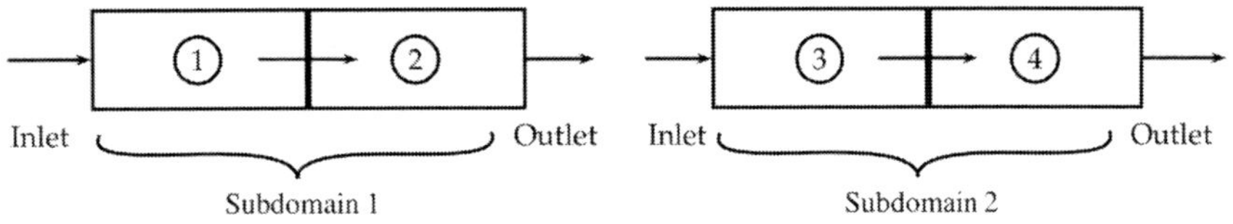

Figure 3: Explicit separation.

The sub-domains are represented by two systems of linearized equations and given in equation 3 and 4.

$$\underbrace{\begin{bmatrix} \frac{\Gamma}{2} & \frac{\Gamma}{2} \\ \frac{\Gamma}{2} & \frac{\Gamma}{2} \end{bmatrix}}_{A_1} * \underbrace{\begin{pmatrix} \Phi_1 \\ \Phi_2 \end{pmatrix}}_{x_1} = \underbrace{\begin{pmatrix} b_1 \\ b_2 \end{pmatrix}}_{b_1} \tag{3}$$

$$\underbrace{\begin{bmatrix} \frac{\Gamma}{2} & \frac{\Gamma}{2} \\ \frac{\Gamma}{2} & \frac{\Gamma}{2} \end{bmatrix}}_{A_2} * \underbrace{\begin{pmatrix} \Phi_3 \\ \Phi_4 \end{pmatrix}}_{x_2} = \underbrace{\begin{pmatrix} b_3 \\ b_4 \end{pmatrix}}_{b_2} \tag{4}$$

The values from the adjacent side of the interface are used as inlet and outlet conditions according to the theory of characteristics. The influence of the adjacent domain fully acts as an explicit contribution to the source. Therefore, the coupling is only carried out before the outer iterations.

Implicit Algorithm
An implicit treatment of a coupled interface can be understood as using two separate matrix of coefficients to build up a system of linearized equations for the complete solution domain as shown in equation 5

$$\underline{b} = \underline{\underline{A}} \cdot \underline{x} = \left(\underline{\underline{A^I}} + \underline{\underline{A^C}}\right) \cdot \underline{x} \tag{5}$$

The interface is no longer build up from an additional in- and outlet (figure 4), instead the fluxes over the interface are integrated into the system of linearized equations.

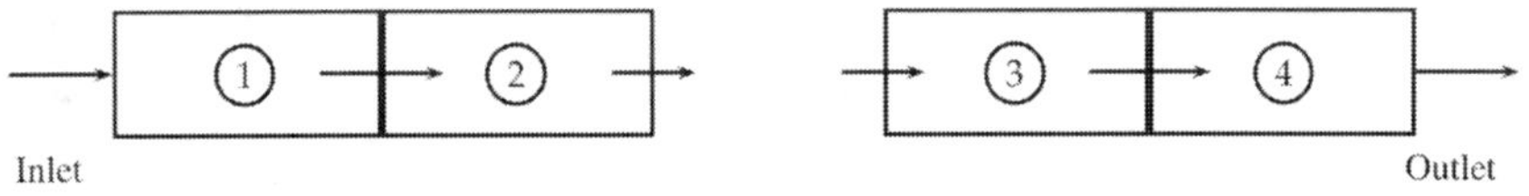

Figure 4: Implicit separation.

The system of linearized equations for the inner coefficients is build up as follows.

$$\underbrace{\begin{bmatrix} \frac{\Gamma}{2} & \frac{\Gamma}{2} & 0 & 0 \\ \frac{\Gamma}{2} & \frac{\Gamma}{2} & 0 & 0 \\ 0 & 0 & \frac{\Gamma}{2} & \frac{\Gamma}{2} \\ 0 & 0 & \frac{\Gamma}{2} & \frac{\Gamma}{2} \end{bmatrix}}_{A^I} * \underbrace{\begin{pmatrix} \Phi_1 \\ \Phi_2 \\ \Phi_3 \\ \Phi_4 \end{pmatrix}}_{x} = \underbrace{\begin{pmatrix} b_1 \\ b_2 \\ b_3 \\ b_4 \end{pmatrix}}_{b} \tag{6}$$

It can be seen, that the system shown in equation 6 can be build up from the equation systems of the explicit approach. However, instead of considering the influence of the coupling in the source term, these fluxes are implicitly discretized. A system of linearized equations for the interface fluxes, respecting the above given example is shown in equation 7.

$$\underbrace{\begin{bmatrix} 0 & 0 & 0 & 0 \\ 0 & \frac{\Gamma}{2} & \frac{\Gamma}{2} & 0 \\ 0 & \frac{\Gamma}{2} & \frac{\Gamma}{2} & 0 \\ 0 & 0 & 0 & 0 \end{bmatrix}}_{A^C} * \underbrace{\begin{pmatrix} \Phi_1 \\ \Phi_2 \\ \Phi_3 \\ \Phi_4 \end{pmatrix}}_{x} = \underbrace{\begin{pmatrix} b_1 \\ b_2 \\ b_3 \\ b_4 \end{pmatrix}}_{b} \tag{7}$$

Adding $A^I + A^C$ will lead to a system of linearized equations for the complete domain.

$$\underbrace{\begin{bmatrix} \frac{\Gamma}{2} & \frac{\Gamma}{2} & 0 & 0 \\ \frac{\Gamma}{2} & \Gamma & \frac{\Gamma}{2} & 0 \\ 0 & \frac{\Gamma}{2} & \Gamma & \frac{\Gamma}{2} \\ 0 & 0 & \frac{\Gamma}{2} & \frac{\Gamma}{2} \end{bmatrix}}_{A} * \underbrace{\begin{pmatrix} \Phi_1 \\ \Phi_2 \\ \Phi_3 \\ \Phi_4 \end{pmatrix}}_{x} = \underbrace{\begin{pmatrix} b_1 \\ b_2 \\ b_3 \\ b_4 \end{pmatrix}}_{b} \tag{8}$$

The coefficients of the linearized equations for the owner cells to their faces are always local while the coefficients from the faces to the coupled cells are averaged coefficients from the adjacent side. With this implicit treatment, the iterative solution procedure constantly considers the complete domain, even during the inner iterations.

Defining the boundary implicitly, i.e. as a boundary which is fully coupled within the domain like for instance periodic boundaries, no additional numerical treatment is needed to avoid numerical artifacts. The key point is that this approach just defines the relation of the boundary interaction directly in the discretization of the domain, without forcing any constraint through an additional BC. This allows to have a fully consistent system based on the accuracy and the physics of the equations solved in the domain. Please note that there are already other implicit boundaries used in CFD, like for instance periodicities. Here no additional non-reflective treatment is needed: there is only an implicit relationship between the two sides. This fact is actually often used in LES and DNS, known as very sensitive regarding the numerics. In this kind of simulations, imposing for example an inlet BC can easily lead to a non-physical flow field, since all the necessary quantities can usually not be given in a consistent way (i.e. divergence free) for all time and length scales resolved.

3. Case setup

3.1. Geometry

In figure 5 the complete model is depicted, with the various components in different colors. As listed in Table 1, the turbine has 15 main and 15 splitter blades, 28 guide and 14 stay vanes. For the three operating points different guide-vane blade-angle are defined: 9.84 degree for the

BEP, 3.91 degree for PL and 12.44 degree for the HL case. Further geometrical details of the model are given in [15], [16].

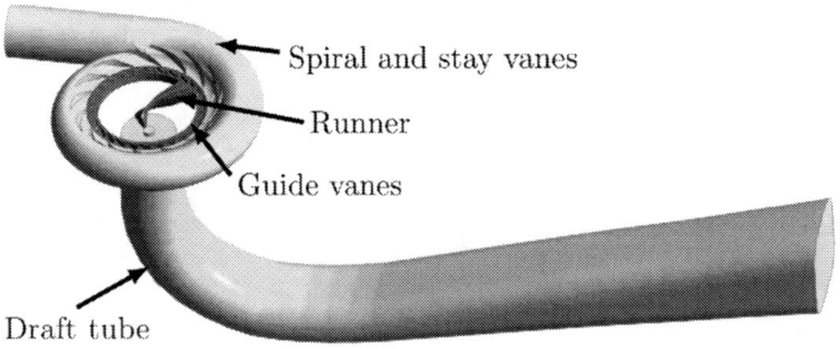

Figure 5: The four simulation domains of the Francis-99 turbine.

Table 1: Main geometry data of the Francis-99 turbine.

Blade No (splitter blades)	15 (+15)	Max diameter runner [mm]	631
GV blade No	28	Max diameter GV [mm]	764
SV blade No	14	Max diameter SV [mm]	980

3.2. Model simplifications

In order to reduce the complexity of the simulations, several model simplifications have been adopted. The clearance gap at the guide vanes are neglected. The model is assumed to behave as a rigid body which is experiencing no deformations. Moreover, the rotor side spaces were neglected so that disc friction as well as leakage flow were not directly available in the simulations. In order to estimate the overall effect of the neglected leakage within the rotor side spaces, an in-house tool [17] has been used to determine the friction torque in the leakage gap. The in-house tool uses interpolation equations which, from an extensive dataset based on detailed parameter studies of rotor side spaces, extract the corresponding correlation values [18]. The calculation is limited to the leakage gap on the shroud side of the runner because on the hub side of the runner there are no pressure relief-holes and therefore the influence is considered as not relevant. At BEP conditions, the friction torque within the leakage gap is 9.53 Nm which corresponds to 1.8% of the measured blade torque. At HL and PL, the friction torque is 1.5% and 8.0%, respectively. These values are used to correct the measured efficiency, thus leading to higher values, which are then in good agreement with the computed efficiency.

3.3. Mesh

In addition to the available geometry and mesh data from the Francis-99 workshop, used in [15], an improved structured mesh, named hereafter HSLU-Mesh, has been created. Particularly, the boundary layer around the blades, splitter blades, guide vanes and stay vanes has been refined. The HSLU-Mesh consists in total of 11 million elements and the average Y^+ value is 22 at BEP. Table 2 summarizes the detailed mesh quality of the individual computation domains.

Table 2: HSLU-Mesh quality.

HSLU-Mesh	Element size	Y^+ (BEP)	Angle	Volume change	Aspect ratio
Global	10'995'677	21.8	2.8	97	1421
Spiral & stay vanes	5'331'788	19.1	3.5	50	1421
Guide vanes BEP	3'780'280	26.9	19.5	56	58
Guide vanes PL	3'780'280	19.4	17.2	97	153
Guide vanes HL	3'780'280	23.6	16.0	55	73
Runner	398'930	15.7	2.8	21	305
Draft tube	1'484'679	21.8	51.8	8	250

3.4. Comparison between LTU-Mesh and HSLU-Mesh

In Table 3 is the mesh data of the provided LTU-Mesh. A main difference between these meshes is the value of the average Y^+ which is 120 and 22 for the LTU-Mesh and the HSLU-Mesh, respectively. The LTU-Mesh consists of 7.5 million elements and is therefore by 32 percent smaller than the HSLU-Mesh in terms of the number of elements. In order to achieve a better mesh quality, the mesh domain of the guide and stay vanes has been separated into two domains.

Table 3: LTU-Mesh quality.

LTU-Mesh	Element size	Y^+ (BEP)	Angle	Volume change	Aspect ratio
Global	7'513'862	119.7	2.8	164	52
Spiral, SV, GV BEP	3'607'016	182.4	12	164	52
Spiral, SV, GV PL	3'441'063	89.6	7.7	81	49
Spiral, SV, GV HL	3'581'240	170.2	7.7	43	46
Runner	335'902	106.0	2.8	54	47
Draft tube	3'570'944	26.3	50.7	3	38

The resulting torque of both meshes is similar, but the LTU-Mesh shows less losses and therefore a overestimation of the net head. Figure 6 visualize the flow around a guide vane. The boundary layers along the walls as well as the wake flow behind the trailing edge is not (well) resolved by the LTU-Mesh being too coarse to capture these phenomena. This could be the main reason for the overestimation of the net head.

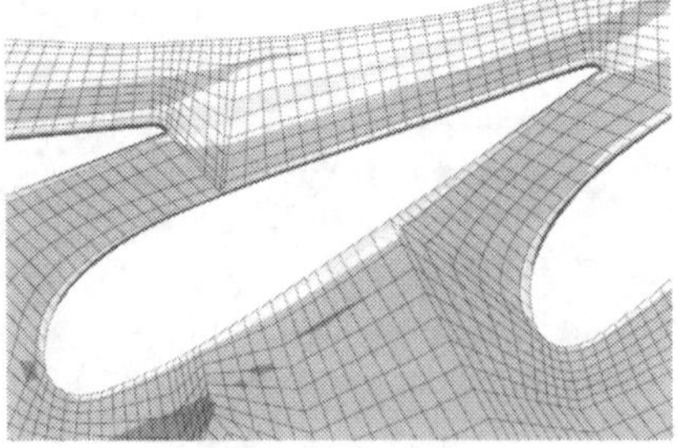

(a) LTU-Mesh

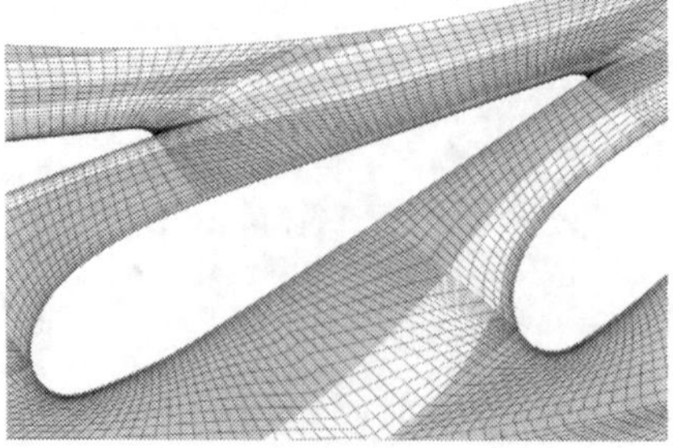

(b) HSLU-Mesh

Figure 6: Guide vanes with the same total pressure contour plot scale.

Journal of Physics: Conference Series **579** (2015) 012009 doi:10.1088/1742-6596/579/1/012009

3.5. Setup
At the inlet of the computational domain the velocity $\mathbf{u}_{\text{inlet}}$, the turbulent kinetic energy (k), and the turbulent frequency (ω) are imposed as a fixed value. The turbulence variables k, ϵ and ω are initialized as $k = 3/2(I|\mathbf{u}_{\text{inlet}}|)^2$, $\epsilon = C_\mu^{0.75} k^{1.5}/\ell$ and $\omega = \epsilon/k/C_\mu$ respectively, where I is the prescribed turbulence intensity of 0.05 respectively 5%, $C_\mu = 0.09$ is a constant turbulence model parameter and ℓ is the turbulent mixing length scale. At the outlet of the computational domain, an average static pressure of $p_{\text{outlet}} = 0$ Pa is used. The walls in the domains are non-rotating, except of course for the runner (hub, blade and shroud). At the walls, a no-slip condition is imposed. The runner domain rotates around the negative z-axis.

Figure 5 shows the individual components of the computational domain. The rotational periodicity of the runner passage is modeled applying a one-to-one interface. Between the guide vanes and the runner as well as between the runner and the draft tube, the implicit HSLU mixing plane interface is applied. Between the stay vanes and the guide vanes an AMI-Interface is placed.

All the simulations were performed using the $k-\omega-SST$ turbulence model which requires the numerical solution of two scalar equations for the turbulent conservation variables in addition to momentum and continuity equations. Although two-equation turbulence models are known to have shortcomings at off-design conditions in the past, even flow instabilities were successfully investigated, as shown in [19, 20], where they were captured with fairly good agreement campared to experimental data. The choice of the SST turbulence model is therefore based on the fact that at present it is regarded as a sort of industrial standard and widely used [21, 22, 23, 24] for this kind of simulation.

All CFD simulations were performed at model-scale with an in-house as well as with a commercial code, solving the steady-state incompressible RANS-Equations in double precision with a full second order upwind scheme for advection discretization.

The in-house code is an adapted, 3D, unstructured, object-oriented finite volume code based on the OpenFOAM-V2.2 framework. The segregated pressure-based solver uses a SIMPLE-C like algorithm and is capable of handling multiple references of frame [25]. The reference commercial finite volume solver is a pressure based coupled solver. All configurations have a mixing plane interface between the stationary and rotating domain.

4. Results
4.1. Integral values
Figure 7 compares the efficiency, the net head, and the torque around the z-axis, at all given operating points. The comparison leads to the following findings:

- The efficiency in the simulations with the HSLU mixing plane is in all three operating points closer to the measurement than that of the commercial code (CC).
- Except for the PL case, the same behavior results for the differential pressure and the net head, respectively.
- The torque of the runner around the z-axis is lower than the simulated torque by the commercial software and is therefore closer to the experiment.

From a comparison with 360° unsteady results [26] it can be stated that no or less influence of unsteadiness exists in BEP and HL operating condition. Since the global results are very similar to the one of the mixing plane. However at part load operating point the rotor-stator interaction seems to be important. The mixing plane underestimates the efficiency because of the neglected unsteady effects. For example in part load condition, the flow in every channel in 360° can be different [26]. The main reason for the better accuracy of the OpenFOAM (OF) simulations compared to the commercial-software results is due, in the opinion of the present authors, to the fully implicit treatment of the mixing plane which imposes a circumferentially uniform distribution of the total pressure immediately downstream of the interface. Denton [4] states that a mixing plane should give a solution which is independent of the relative position of rotor and stator, resulting in uniform pitch-wise enthalpy and entropy for a fixed span-wise height on the downstream side of the domain, in order to produce a better representation of the average effect of the rotor-stator interactions. Defining the boundary implicitly, i.e. as a boundary which is fully coupled within the domain like for instance periodic boundaries, no additional numerical treatment is needed to avoid numerical artifacts. This leads to considerable advantages compared to the state-of-the-art mixing plane: (1) the solution fully satisfies the mixing plane constraint of uniform total conditions in pitch-wise direction, while (2) the continuity of the solution reduces incoming disturbances having therefore a sort of built-in non-reflectiveness. At the same time (3) improved robustness is also demonstrated with the ability to handle backflow at the interface. This leads to a consistent prediction of the blade loading and therefore to more accurate results.

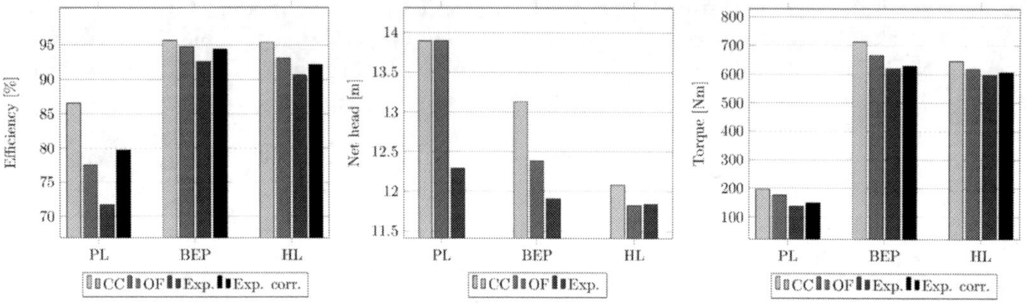

(a) Efficiency (b) Net head (c) Torque in z-axis

Figure 7: Integral values in all three operating points.

Figure 8 shows exemplary the difference between the commercial software and OpenFOAM at the downstream mixing-plane for the BEP conditions. The figures show the total pressure in stationary frame with the same scale, respectively. It can be seen that at the draft tube mixing-plane the HSLU solution is completely axisymmetric, i.e. fully repeating the definition of the mixing plane approach.

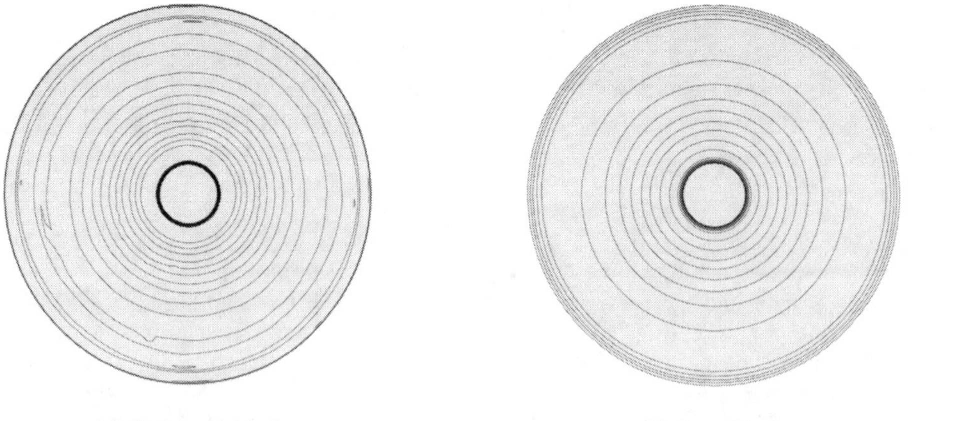

(a) Commercial code (b) OpenFOAM

Figure 8: Total pressure at the mixing plane location between the runner passage and the draft tube. The values of the contour lines are the same in both figures.

Figure 9 shows an axial cut of runner and draft-tube at part load. At this condition the flow through the machine is concentrated along the walls of the draft-tube, while in the center a large back-flow region is present. Here a further strength of the fully implicit approach can be seen: the back flow at the interface can be handled without numerical drawbacks, i.e. not affecting the convergence behavior.

The authors believe that both circumferentially uniform total-pressure and the ability to handle back-flow at the interface are the reasons for the good integral performance of the steady-state one-passage approach compared to experiment and 360° data [26].

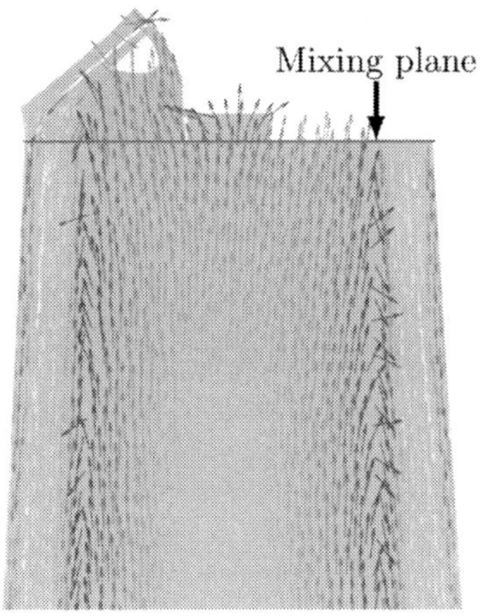

Figure 9: Normalized tangential velocity vectors in stationary frame also colored in velocity. The operating condition is part load.

4.2. LDA Measurements

In the following, the mean axial and tangential velocity profiles along the radius at two measuring sections, as described in figure 10, are compared. In order to ensure a good comparability of the CFD and the Laser Doppler Anemometry (LDA) results, the velocities predicted by CFD have been evaluated exactly at the same position as the LDA measurements.

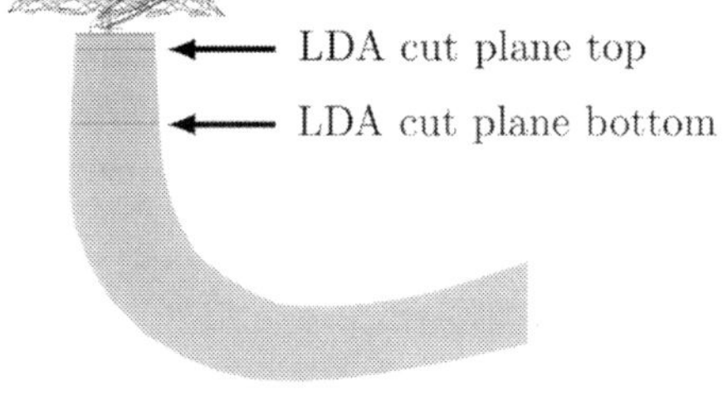

Figure 10: Position of the layers.

The performance of the two CFD codes regarding the prediction-accuracy of the velocity components is similar at off-design conditions for the top and the bottom slice. Therefore, only the results at the top slice are given in figures 11 and 13 for PL and HL.

At PL, both CFD codes predict similar results for the axial and tangential velocity along the radius. The axial velocity is in both codes slightly overestimated compared to the measurement.

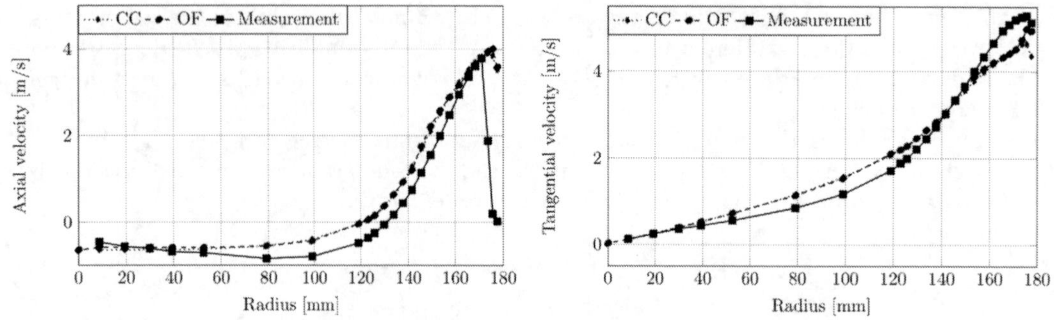

(a) Axial velocity (b) Tangential velocity

Figure 11: Top slice in PL operating point.

At design conditions (figure 12), both CFD codes are fairly close to the experimental results of the axial velocity. OpenFOAM is generally more accurate predicting the tangential velocity. However, both codes clearly overestimates the tangential velocity close to the center. Nevertheless the area involved is small. A discussion of the discrepancy between the numerically predicted tangential velocities and the experimental results is given in the paper by Aakti et al. [26].

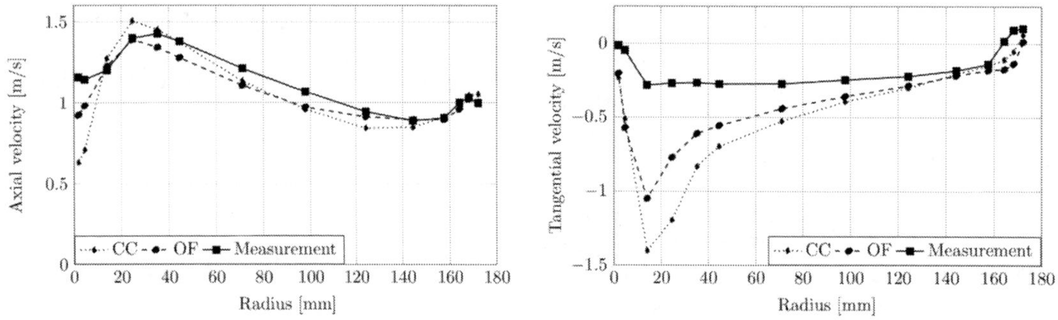

(a) Axial velocity top slice (b) Tangential velocity top slice

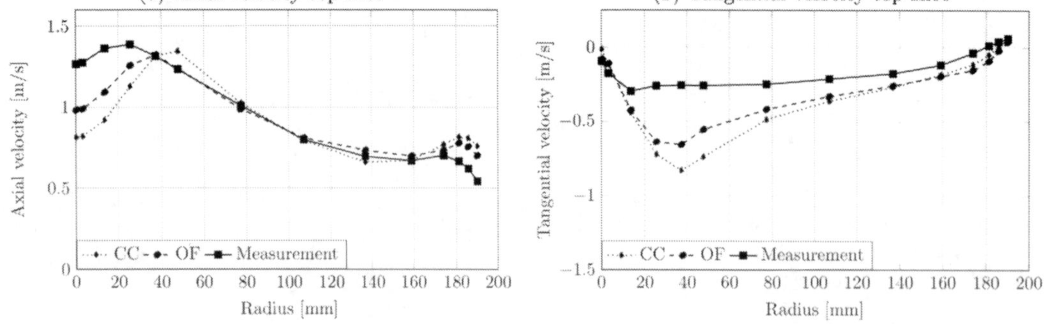

(c) Axial velocity bottom slice (d) Tangential velocity bottom slice

Figure 12: BEP operating point.

At HL, both CFD codes underestimate the axial velocity close to the center, while the commercial code is closer to the experiments. The tangential velocity is overestimated by both CFD codes over the entire radius.

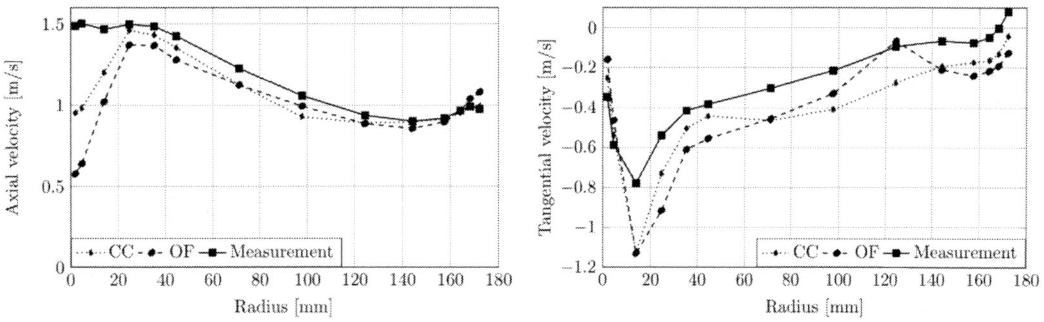

(a) Axial velocity (b) Tangential velocity

Figure 13: Top slice in HL operating point.

5. Conclusion

Steady-state CFD computations of a high-head Francis-turbine model have been performed considering only one runner passage. At the rotor-stator interfaces, a novel fully implicit mixing plane has been applied.

The new mixing-plane formulation has been implemented in an inhouse modified OpenFOAM version. The prediction of the overall performance (efficiency, net head and torque) is fairly accurate compared to experiments and to a reference commercial code. However, both CFD codes overestimated the measured values.

Regarding the local flow distributions at runner outlet, the comparison with the measurements is satisfactory exept for the tangential velocity in the BEP condition, showing that the proposed mixing-plane approach is also able to capture the local flow physics consistently over a wide operating range.

References

[1] NTNU and LTU *Francis-99 workshop* http://www.ltu.se/research/subjects/Stromningslara/Konferenser/Francis-99?l=en accessed august 2014

[2] Biesinger T, Cornelius C, Rube C, Braune A, Campregher R, Godin P G, Schmid G and Zori L 2010 Unsteady CFD methods in a commercial solver for turbomachinery applications *ASME Conference Proceedings* **2010** 44021

[3] Biesinger T, Cornelius C, Nuernberger D and Rube C 2012 Speed line computation of a transonic compressor stage with unsteady cfd methods *ASME Turbo Expo* **2012** 68029

[4] Denton J 1992 The calculation of three-dimensional viscous flow through multistage turbomachines *Journal of Turbomachinery* **114** p 18-26

[5] Hanimann L, Mangani L, Casartelli E, Mokulys T and Mauri S 2013 Development of a novel mixing plane interface using a fully implicit averaging for stage analysis *Journal of Turbomachinery* **136(8)** 081010

[6] Denton J D 2010 Some limitations of turbomachinery CFD *ASME Conference Proceedings* **2010** 44021

[7] Holmes D G 2008 Mixing planes revisited: A steady mixing plane approach designed to combine high levels of conservation and robustness *ASME Conference Proceedings* **2008** 43161

[8] Fritsch G and Giles M 1993 An asymptotic analysis of mixing loss *Journal of Turbomachinery* **117(3)** p 367-374

[9] Dzung L 1971 Konsistente mittelwerte in der theorie der turbomaschinen fuer kompressible medien *Brown-Boveri Mitteilungen* **10-71** p 485-492

[10] Kreitmeier F 1992 Space-averaging 3d flows using strictly formulated balance equations in turbomachinery *ASME COGEN-TURBO* **7** p 397-408

[11] Anker J E, Schrader B, Seybold U, Mayer J F and Casey M V A three-dimensional non-reflecting boundary condition treatment for steady-state flow simulation *44th AIAA Aerospace Sciences Meeting and Exhibit* **44** 1275

[12] Robens S, Jeschke P, Frey C, Kügeler E, Bosco A and Breuer T 2013 Adaptation of giles non-local non-reflecting boundary conditions for a cell-centered solver for turbomachinery applications *ASME Turbo Expo* **GT2013** 94957

Journal of Physics: Conference Series **579** (2015) 012009 doi:10.1088/1742-6596/579/1/012009

[13] Giles M 1988 non-reflecting boundary conditions for the Euler equation *TR* **88** 1
[14] Hanimann L, Mangani L, Casartelli E, Mokulys T and Mauri S 2013 Development of a novel mixing plane interface using a fully implicit averaging for stage analysis in an object-oriented CFD-code *ASME Turbo Expo* **GT2013** 94390
[15] Trivedi C, Cervantes M J, Gandhi B K and Dahlhaug O G 2013 Experimental and numerical studies for a high head francis turbine at several operating points *Journal of Fluids Engineering* **135(11)** 111102
[16] NTNU and LTU *Francis-99 test case* http://www.ltu.se/research/subjects/Stromningslara/Konferenser/Francis-99/Test-Case-1.111520?l=en accessed august 2014
[17] Staubli T, Loetscher B and Cimmino D 2005 Integrale laufrad- simulationen von hydromaschinen Tech. rep. FHZ-HTA
[18] Amoser M 1995 *Stroemungsfelder und Radialkraefte in Labyrinthdichtungen hydraulischer Stroemungsmaschinen* Ph.D. thesis ETH Zuerich
[19] Widmer C, Staubli T and Ledergerber N 2011 Unstable characteristics and rotating stall in turbine brake operation of pump-turbines *Journal of Fluids Engineering* **133** 041101
[20] Yan J P, Seidel U and Koutnik J 2012 Numerical simulation of hydrodynamics in a pump-turbine at off-design operating conditions in turbine mode *26th IAHR Symposium on Hydraulic Machinery and Systems* **15** 032041
[21] Feng J J, Li W F, Wu H, Lu J L, Liao W L and Luo X Q 2014 Investigation on pressure fluctuation in a francis turbine with improvement measures *27th IAHR Symposium on Hydraulic Machinery and Systems* **2014** 2.2.1
[22] Tamura Y, Tani K and Okamoto N 2014 Experimental and numerical investigation of unsteady behavior of cavitating vortices in draft tube of low specific speed francis turbine *27th IAHR Symposium on Hydraulic Machinery and Systems* **2014** 2.2.6
[23] Heo M W, Kim K Y, Ma S B, Yoo I S, Choi W C, Kim J H and Choi Y S 2014 Parametric performance evaluation of a hydraulic centrifugal pump *27th IAHR Symposium on Hydraulic Machinery and Systems* **2014** 3.1.6
[24] Guo P C, Guo D M, Sun L G and Luo X Q 2014 Numerical investigation of different volutes in a low specific speed centrifugal pump *27th IAHR Symposium on Hydraulic Machinery and Systems* **2014** 3.1.7
[25] Mangani L, Casartelli E and Mauri S 2011 Assessment of various turbulence models in a high pressure ratio centrifugal compressor with an object oriented CFD code *ASME Conference Proceedings* **2011** 54679
[26] Aakti B, Amstutz O, Casartelli E, Romanelli G and Mangani L 2014 On the performance of a high head francis turbine at design and off-design conditions *Francis-99 workshop* **1**

Francis-99 Workshop 1: steady operation of Francis turbines | IOP Publishing
Journal of Physics: Conference Series **579** (2015) 012010 | doi:10.1088/1742-6596/579/1/012010

On the performance of a high head Francis turbine at design and off-design conditions

B Aakti, O Amstutz, E Casartelli, G Romanelli, L Mangani

Lucerne University of Applied Sciences and Arts Engineering & Architecture, Horw Switzerland

E-mail: ernesto.casartelli@hslu.ch

Abstract. In the present paper, fully 360 degrees transient and steady-state simulations of a Francis turbine were performed at three operating conditions , namely at part load (PL), best efficiency point (BEP), and high load (HL), using different numerical approaches for the pressure-velocity coupling. The simulation domain includes the spiral casing with stay and guide vanes, the runner and the draft tube. The main target of the investigations is the numerical prediction of the overall performance of the high head Francis turbine model as well as local and integral quantities of the complete machine in different operating conditions. All results were compared with experimental data published by the workshop organization. All CFD simulations were performed at model scale with a new in-house, 3D, unstructured, object-oriented finite volume code within the framework of the open source OpenFOAM library. The novel fully coupled pressure-based solver is designed to solve the incompressible RANS-Equations and is capable of handling multiple references of frame (MRF). The obtained results show that the overall performance is well captured by the simulations. Regarding the local flow distributions within the inlet section of the draft-tube, the axial velocity is better estimated than the circumferential component.

1. Introduction

The scope of the first Francis-99 workshop is the assessment of advanced numerical tools for the prediction of steady-state operating points in a Francis turbine for design and off-design conditions [1]. The continuous evolution of the numerical techniques and the computer power has lead in the last decade to a large tool-spectrum for turbomachinery investigations. The driver for this trend was mainly the only partly-satisfactory results obtained, especially at off-design conditions. The main reasons for the discrepancies were located firstly in the turbulence models, with the development of more advanced models capable to improve the drawbacks of classical models like $k-\epsilon$ and capture additional flow physics, like curvature correction for the SST model or the v2F model and secondly in the thought inability of steady-state methods to properly predict the inherently unsteady effects present in turbomachines [2], [3].

In the present paper a comparison of steady and unsteady computations for the 360 degrees model are presented within the framework of OpenFOAM, performing the simulations with two different computational technologies, the segregated and the coupled one. For steady-state computations, the coupled technology has clear and well known advantages [4], leading to a speed-up of factors, typically 3 to 10, depending on the mesh size.

For transient cases, with a typical time step corresponding to mesh motion of 1 degree or less, the CFL can be considerably larger than 1. In these cases the use of a fully implicit approach like in the coupled solver is beneficial. In these cases a speed up of additional 50% is obtained, thus underlining the power of a coupled approach also for unsteady cases.

2. CFD - solver technology

2.1. Governing equation

The Unsteady Reynolds Averaged Navier-Stokes (URANS) equations governing the dynamics of an incompressible, viscous flow can be written within an Arbitrary Lagrangian Eulerian (ALE) framework in differential form as follows:

$$\nabla \cdot \mathbf{u} = 0 \tag{1}$$

$$\frac{\partial \mathbf{u}}{\partial t} + \nabla \cdot (\mathbf{u}_\mathrm{r}\mathbf{u}) = -\frac{1}{\rho}\nabla p + \nabla \cdot (\nu_\mathrm{eff}\nabla \mathbf{u}) \tag{2}$$

where $\mathbf{u}$ is the absolute velocity, p the pressure and ρ the density of the fluid. The convecting flux is written in terms of the relative velocity $\mathbf{u}_\mathrm{r} = \mathbf{u} - \mathbf{u}_\mathrm{ALE}$, where $\mathbf{u}_\mathrm{ALE}$ is the velocity associated to the dynamic motion (either rigid and/or deformable) of the physical domain $V = V(t)$ and it is evaluated enforcing the so-called Geometric Conservation Law (GCL). For instance, for a purely rigid rotation with constant angular velocity $\mathbf{\Omega}$, it corresponds to $\mathbf{u}_\mathrm{ALE} = \mathbf{\Omega} \times \mathbf{r}$. Finally the effective kinematic viscosity is the sum of the laminar and turbulent kinematic viscosities ($\nu_\mathrm{eff} = \nu + \nu_\mathrm{t}$).

2.2. Numerical discretization

Speed and robustness are among the most important requirements for any software that has to be used for the time-accurate numerical analysis of complex highly unsteady phenomena. When programming a CFD code the best way of combining both requirements is to couple the governing equations implicitly, since resolving efficiently the pressure-velocity coupling is essential for the performance of any CFD code. However up until today the SIMPLE family of algorithms [5], which couples the governing equations only by means of sub-looping, solving sequentially each governing equation, still remains the predominant methodology used in the CFD community. Therein a segregated approach in resolving the pressure velocity coupling is followed. Compared to block coupled implicit algorithms, segregated algorithms lack scalability with mesh size, robustness and calculation speed, which is inherent due in part to the under-relaxation needed to stabilize the algorithm. This is even more critical for unsteady simulations, as a set of inner non-linear iterations must be accounted for in order to drive the solver to the desired accuracy level in time.

In order to overcome these shortcomings Mangani et al. [4] developed a block coupled incompressible solver using the open-source CFD library OpenFOAM® as programming platform. Therein the algebraic equations resulting from the Navier-Stokes equations are solved simultaneously with the following system of equations, where 'C' are the cell values while 'NB' the neighbor contributions:

$$\begin{bmatrix} a_C^{uu} & a_C^{uv} & a_C^{uw} & a_C^{up} \\ a_C^{vu} & a_C^{vv} & a_C^{vw} & a_C^{vp} \\ a_C^{wu} & a_C^{wv} & a_C^{ww} & a_C^{wp} \\ a_C^{pu} & a_C^{pv} & a_C^{pw} & a_C^{pp} \end{bmatrix} \cdot \begin{bmatrix} u_C \\ v_C \\ w_C \\ p_C \end{bmatrix} + \sum_{NB} \begin{bmatrix} a_{NB}^{uu} & a_{NB}^{uv} & a_{NB}^{uw} & a_{NB}^{up} \\ a_{NB}^{vu} & a_{NB}^{vv} & a_{NB}^{vw} & a_{NB}^{vp} \\ a_{NB}^{wu} & a_{NB}^{wv} & a_{NB}^{ww} & a_{NB}^{wp} \\ a_{NB}^{pu} & a_{NB}^{pv} & a_{NB}^{pw} & a_{NB}^{pp} \end{bmatrix} \cdot \begin{bmatrix} u_{NB} \\ v_{NB} \\ w_{NB} \\ p_{NB} \end{bmatrix} = \begin{bmatrix} b_C^u \\ b_C^v \\ b_C^w \\ b_C^p \end{bmatrix} \tag{3}$$

To further enhance computational performance an algebraic multi-grid solver has been implemented and used for the solution of the block-coupled system of linearized equations presented above.

While segregated algorithms operate using many under-relaxed sub-loops to account for inter-equation coupling, the inter-variable coupling is much stronger and less sub-loops are needed. Moreover no under-relaxation is needed. The solution of such a discretized system of equations therefore results to be numerically much more stable than that of segregated algorithms and also turns out to be significantly faster in terms of calculation time, which has been demonstrated by Mangani et al. [4]. The solver described in [4] has been used for the unsteady simulations carried out in this work and compared with the results of a reference commercial solver. More details regarding the modeling choices, the numerical set-up and the initial and boundary conditions are provided below.

2.3. Turbulence modeling

All the simulations were performed using a $k-\omega$ SST turbulence model which requires the numerical solution of two scalar equations for the turbulent conservation variables (not coupled) in addition to momentum and continuity equations. Although two-equation turbulence models are known to have shortcomings at off-design conditions, even flow instabilities can be investigated, as shown in [6, 7], where they are captured and fairly agree with experimental data. The choice of the SST turbulence model is based on the fact that it is a sort of industrial standard and widely used [8, 9, 10, 11]. Moreover, the automatic wall functions are applicable using the SST.

2.4. Domain discretization

The simulation domain ranges from the spiral case inlet, the stay vanes, the guide vanes and runner to the draft tube as shown in figure 1. For the three operating points, the guide vane opening angles are defined as 9.84, 3.91 and 12.44 degrees at BEP, PL and HL, respectively. Further geometrical details of the model are given in [12], [13].

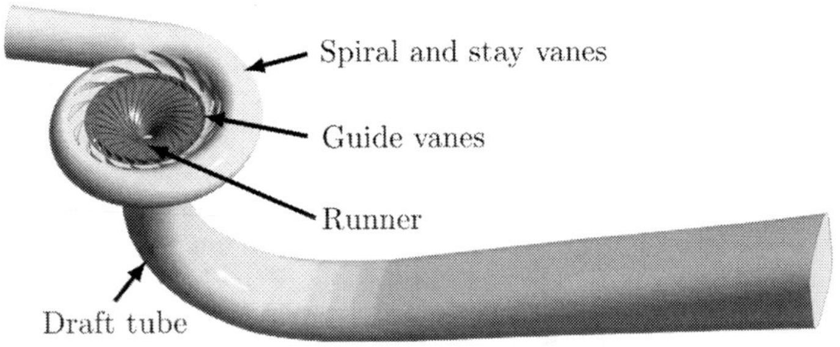

Figure 1: The four simulation domains of the Francis-99 turbine.

The global structured mesh encompassed approximately 16.5 million hexahedral distributed in 4 cell zones (spiral/stay-vanes, guide-vanes, rotor and draft tube). The runner domain has been discretized appling ANSYS ICEM v14.5, where as the other domains have been created with Pointwise v17.2R1. Rotating and stationary domains were matched using transient rotor stator Arbitrary Mesh Interfaces (AMI). The geometry of the mesh is taken by the provided data from the Francis-99 workshop, but the mesh is recreated. No further mesh study was performed. The quality of the mesh corresponds the usual industrial standard. Table 1 summarizes the detailed mesh quality of the individual computation domains.

Table 1: Mesh Quality.

HSLU-Mesh	Element size	Y^+ (BEP)	Angle	Volume change	Aspect ratio
Global	16'580'697	21.8	2.8	97	1421
Spiral & stay vanes	5'331'788	19.1	3.5	50	1421
Guide vanes BEP	3'780'280	26.9	19.5	56	58
Guide vanes PL	3'780'280	19.4	17.2	97	153
Guide vanes HL	3'780'280	23.6	16.0	55	73
Runner	5'983'950	15.7	2.8	21	305
Draft tube	1'484'679	21.8	51.8	8	250

The rotor side spaces were neglected so that disc friction as well as leakage flow were not directly available in the simulations. In order to estimate the overall effect of the neglected leakage within the rotor side spaces, an in-house tool [14] has been used to determine the friction torque in the leakage gap. The in-house tool uses interpolation equations which, from an extensive dataset based on detailed parameter studies of rotor side spaces, extract the corresponding correlation values [15]. The calculation is limited to the leakage gap on the shroud side of the runner because on the hub side of the runner there are no pressure relief-holes and therefore the influence is considered as not relevant. At BEP conditions, the friction torque within the leakage gap is 9.53 Nm which corresponds to 1.8% of the measured blade torque. At HL and PL, the friction torque is 1.5% and 8.0%, respectively. These values are used to correct the measured efficiency, thus leading to higher values, which are then in good agreement with the computed efficiency.

2.5. Numerical schemes

All the simulations were performed with fully second order upwind space discretization operators. The second order upwind operator uses a special gradient extrapolation[4]. Moreover ad-hoc switching to spatial first order was not necessary at the beginning of the simulations for stability issues as instead usually required by the standard simpleFoam solver [16]. A fully second-order time discretization is achieved by means of inner non-linear iterations (for the present simulations 4-6 sub-iterations were sufficient to achieve a drop of 5 orders of magnitude of the inner residuals), using a time step which is corresponding to a resolution of the rotor motion of 0.25 deg. The small timestep is chosen according to the large number of main- and splitterblades. For example a resolution of the rotor motion of 1 deg. would lead to only 12 steps per blade. Starting from the steady-state solution, the time step has been decreased stepwise (waiting each time until a for periodic oscillation was reached) from a rotor motion corresponding to 5 deg, 1 deg, 0.5 deg to finally 0.25 deg. The convergence of the frozen rotor calculations is at least 10^{-4} RMS respective 10^{-2} maximum residuals. The convergence of the transient simulation is for RMS 10^{-5} and for the maximum residuals 10^{-3}.

2.6. Initial and boundary conditions

All unsteady simulations have been initialized with a steady-state Multiple Reference Frames (MRF) solution. At the inlet of the computational domain the velocity $\mathbf{u}_{\text{inlet}}$, the turbulent kinetic energy (k), and the turbulent frequency (ω) are imposed as a fixed value. The turbulence variables k, ϵ and ω are initialized as $k = 3/2(I|\mathbf{u}_{\text{inlet}}|)^2$, $\epsilon = C_\mu^{0.75}k^{1.5}/\ell$ and $\omega = \epsilon/k/C_\mu$ respectively, where I is the prescribed turbulence intensity of 0.05 respectively 5%, $C_\mu = 0.09$ is a constant turbulence model parameter and ℓ is the turbulent mixing length scale. At the outlet of the computational domain, an average static pressure of $p_{\text{outlet}} = 0$ Pa is used. The walls in the domains are non-rotating, except for the runner (hub, blade and shroud). At the walls, a no-slip condition is imposed. The runner domain rotates around the negative z-axis. The only difference in the numerical setup between the coupled and the segregated is the discretization of the convection term. Nevertheless, both are second order [17].

2.7. Comparison of computational time

In the frozen rotor case (steady-state), the coupled solver is approximately 4 to 5 times faster than the segregated solver, depending on the operating point. For transient simulations, where small time steps are used, the segregated solver is usually a more adequate tool to use. Nevertheless the coupled solver is still faster by a factor of 1.5 than the segregated one, mainly because of the big mesh size.

3. Results

3.1. Integral values

In order to predict the performance of the Francis turbine, time accurate transient simulations as well as steady-state (frozen rotor) simulations were carried out. Figure 2 summarizes the results of the transient as well as the steady-state results. In each sub-figure, a comparison of the Commercial Code (CC), OpenFOAM coupled solver (OF coupled) , OpenFOAM segregated solver (OF segregated) and the measurements provided by the Francis-99 workshop [1] is given.

In all simulations (see figure 2a), the efficiency has been overestimated. The main reason for this overestimation is because the rotor side spaces are neglected in the simulation and therefore all the associated losses are not considered. Assuming that the efficiency difference is proportional to the friction torque, the measured efficiency has been corrected by the estimated friction torque (Exp. corr.). For the sake of clarity, the measured efficiency is corrected and not the results of the simulations. The effect of the leakage flow at the rotor side spaces is not used in the correction of the efficiency.

At BEP operating conditions, all transient simulations predict the corrected measured efficiency well. At off-design conditions, the segregated OpenFOAM solver estimates the efficiency of the turbine closest to the measurements. At PL, the coupled OpenFOAM solver is furthest away from the measurements and at HL, the reference commercial solver.

Figure 2b gives an overview of the performance of the steady-state simulations using the frozen rotor approach. Overall, the steady-state frozen rotor results matches the measurements well. This could be an impact of the evenly distributed flow in the guide vane channels. At all operating conditions, the OpenFOAM segregated solver estimates the efficiencies closest to the measurements followed by the coupled OpenFOAM solver. While at BEP operating conditions all CFD codes perform approximately the same, at off-design conditions, the presented OpenFOAM solvers predict the integral values of the turbine closer to the experiments than the reference commercial solver. The difference in the results between the segregated and the coupled solver should be due to the different second order convection term scheme.

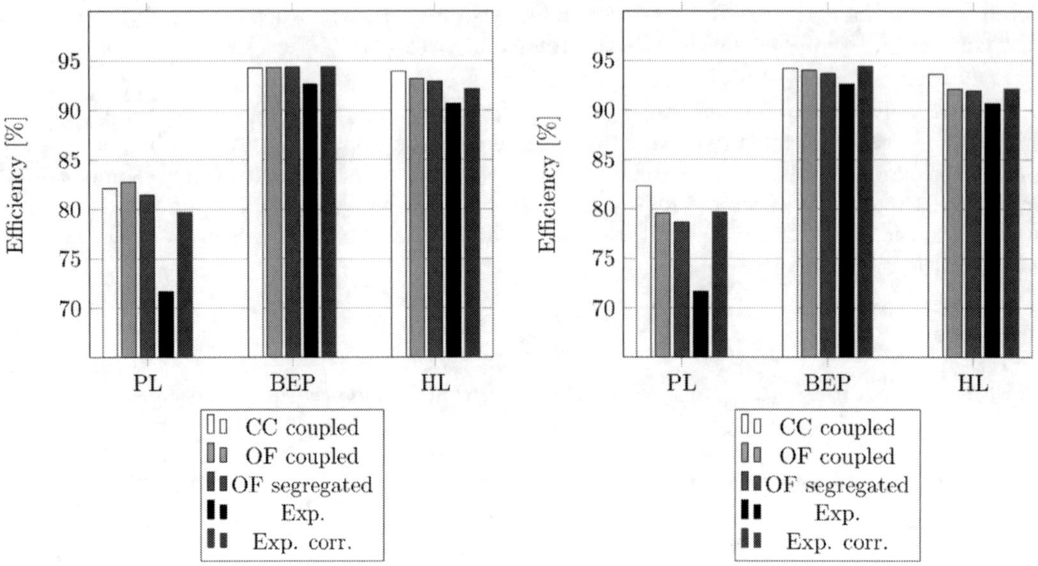

(a) Efficiency transient (b) Efficiency frozen rotor

Figure 2: Efficiency in all operating points.

According to figure 3, the vortex rope in all operating conditions but especially in part load is surprisingly straight. The straight vortex rope could be a explanation for the good frozen rotor results compared to the measurements as well as the transient simulation results.

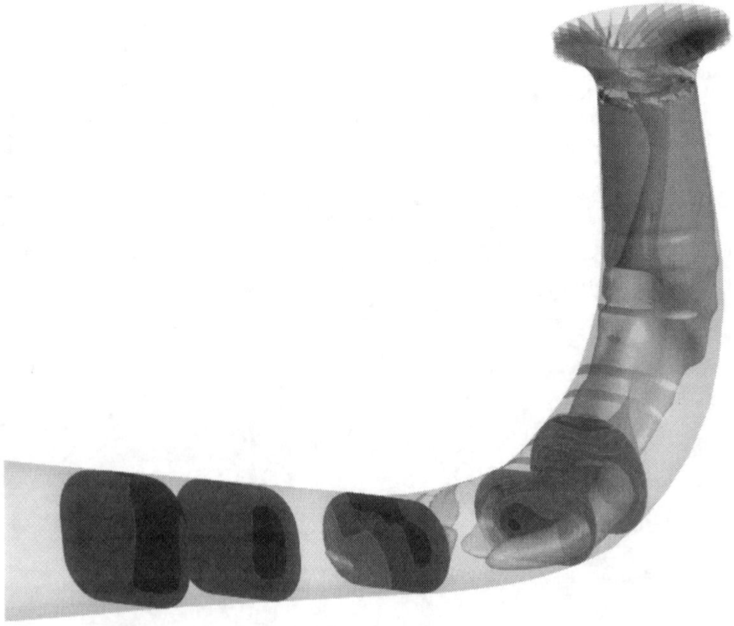

Figure 3: Vortex rope in PL operating point. On the contour planes is the velocity plotted and the vortex rope is displayed with the swirling strength.

3.2. Local data

The Francis-99 workshop provided pressure measurement data at different probe point locations. There are two points within the simulation domain, as shown in figure 4.

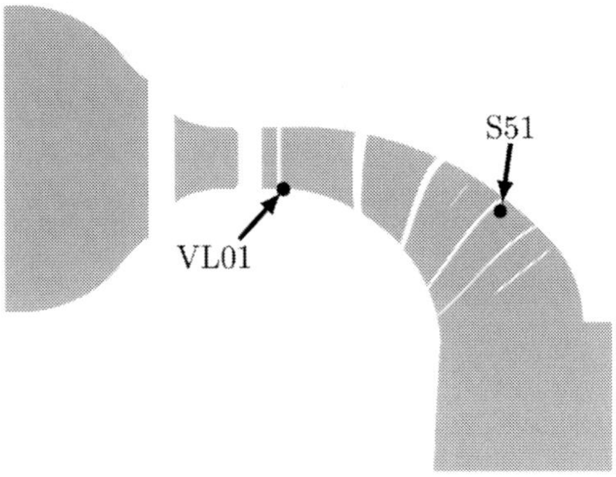

Figure 4: Probe point locations.

In order to visualize the dominant frequencies, fast Fourier transformations (FFT) have been performed. The y-axis represents the normalized pressure amplitude. The measurement provided by the workshop showed broad band noise after the FFT, therefore only the CFD data has been analyzed. The frequency of the main and splitter blades at the given operating conditions are 167.7 Hz at BEP, 184.8 Hz at HL and 203.1 Hz at PL. As shown in figures 5 and 6, the blade frequency is dominant. All three CFD codes are showing the same characteristic frequencies.

In S51 is the amplitude in the part load condition dominant. Due to the fact that at this operating point the flow is largely separated at the entrance of the runner, i.e. just before the measuring point.

At the probe point VL01, which is located very close or even between the guide vanes depending on the operating point, the frequency of the main blades (84 Hz / 92 Hz / 102 Hz) is clearly visible. Due to this position the amplitude of the pressure signal is relatively low.

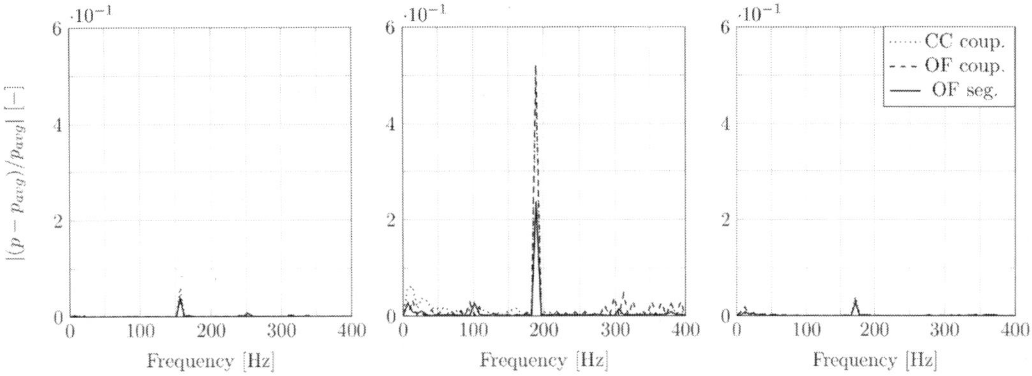

(a) BEP (b) PL (c) HL

Figure 5: FFT signal of S51.

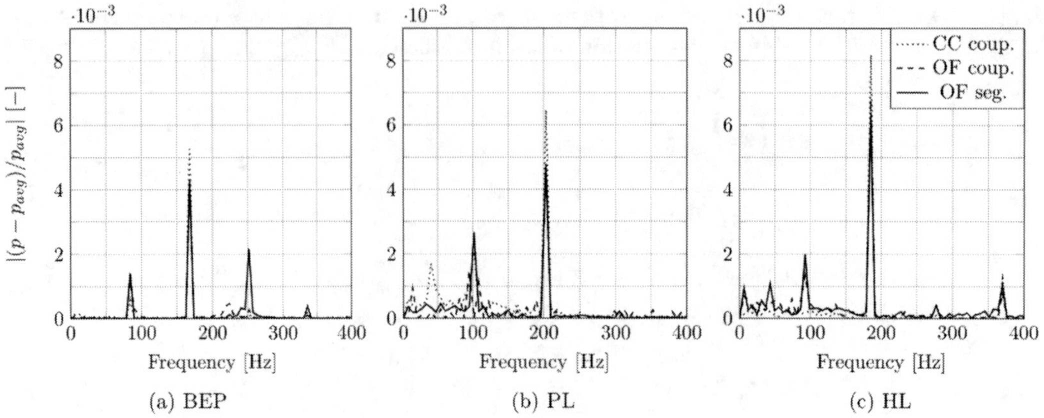

Figure 6: FFT signal of VL01.

3.3. LDA Measurements

In the following, the mean axial and tangential velocity profiles along the radius at two measuring sections (LDA cut planes), as described in figure 7, are compared. All presented velocities have been normalized using the mean axial velocity at the draft tube inlet, which is upstream of the top plane. In order to ensure a good comparability of the CFD and the Laser Doppler Anemometry (LDA) results, the velocities predicted by CFD have been evaluated exactly at the same position as the LDA measurements. The results of the transient simulation have been averaged over one runner passage.

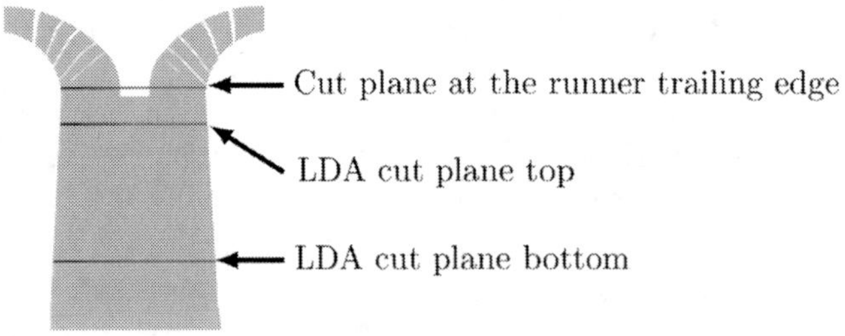

Figure 7: Position of the evaluation planes.

The resulting velocity profiles of the transient simulations are shown in figures 8 to 14. The discrepance between measurements and the different frozen rotor results are similar to the transient ones. Therefore these results are not presented. In general the axial velocity distribution is captured better than the tangential component. The reference commercial solver usually predicted the axial velocity better compared to the OpenFOAM solvers. On the other hand the tangential velocity has been highly overestimated by all three solvers. The OpenFOAM coupled solver is at BEP conditions and at HL condition the furthest away form the measurement, whereas the reference commercial solver is closest to the experimental results. A discussion of the discrepancy between the numerically predicted tangential velocities and the experimental results is given in section 3.4. At PL operating conditions all simulation-results are similar. The difference to the measurement is small compared to that observed at BEP and HL.

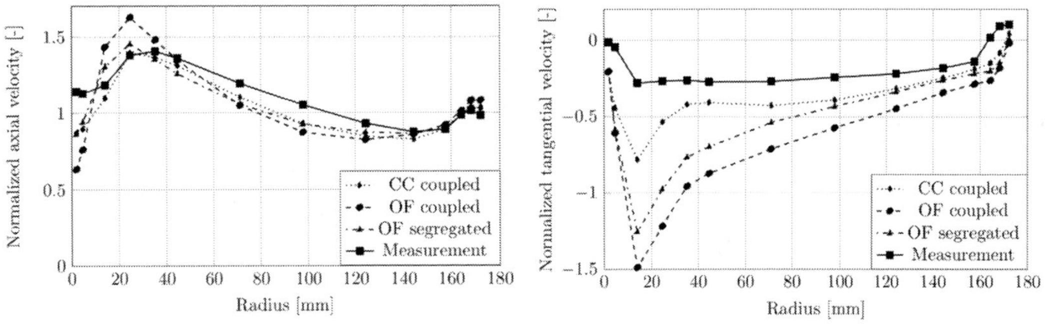

(a) Axial velocity (b) Tangential velocity

Figure 8: Normalized axial and tangential velocity at top slice in transient BEP operation.

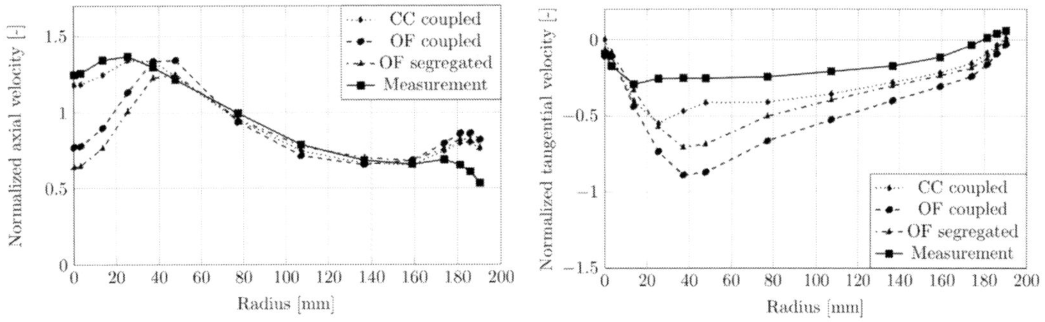

(a) Axial velocity (b) Tangential velocity

Figure 9: Normalized axial and tangential velocity at bottom slice in transient BEP operation.

3.4. *Open issues*

An open issue is the discrepancy of the tangential velocity evaluated by the LDA measurements and CFD simulations, especially at BEP conditions. All the numerically predicted tangential-velocities are far away from the experimental results. At the top slice, the experimental results show that the tangential velocity is approximately constant along the radius. As there are no flow guiding devices between the runner blade trailing edge and the evaluation planes, the flow in-between should more or less follow the free vortex law, where the product of the tangential velocity and the radius is a constant. Therefore, the distribution of the tangential velocity predicted by CFD seems qualitatively correct. Questionable is the development of the tangential velocity of the measurement from the top to the bottom plane. The tangential velocity peaks in the top plane at a radius of approximately 16 mm, while further down stream at the bottom plane is the maximum velocity close the zero radius. Never the less the area involved is rather small. In order to analyze the described discrepancy, an additional cut plane has been placed close to the runner trailing edge, as shown in figure 7. Figure 10 compares the numerical results of the coupled OpenFOAM solver applying the frozen rotor approach to the experimental results at the three evaluation planes. The tangential velocity evaluated at the plane closest to the runner trailing edge is approximately identical to that at the other evaluation planes.

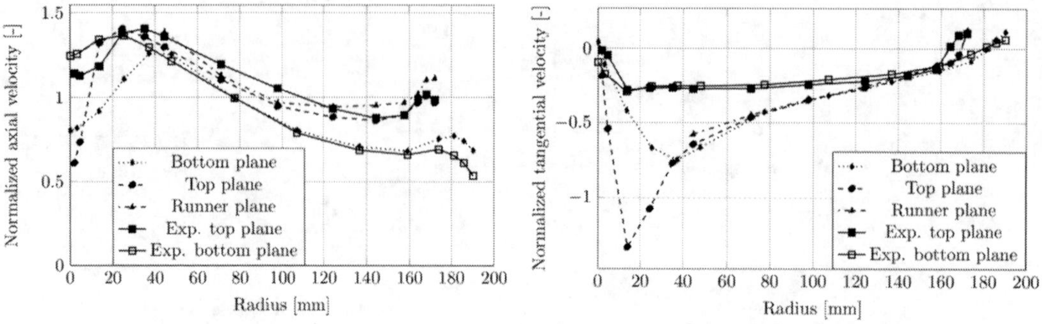

(a) Axial velocity (b) Tangential velocity

Figure 10: Normalized axial and tangential velocity at different cut plane positions at BEP operating conditions applying the steady-state coupled OpenFOAM solver.

Another open issue is that the integration of the measured axial velocity at PL conditions (figure 13a and 14a) does not well agree with the given volume flow rate. Below a radius of approximately 130 mm the measured axial velocity is negative and larger compared to the simulation and above it is smaller, which results in a overall lower volume flow rate compared to the simulation and therefore also the given data by the Francis-99 workshop [13].

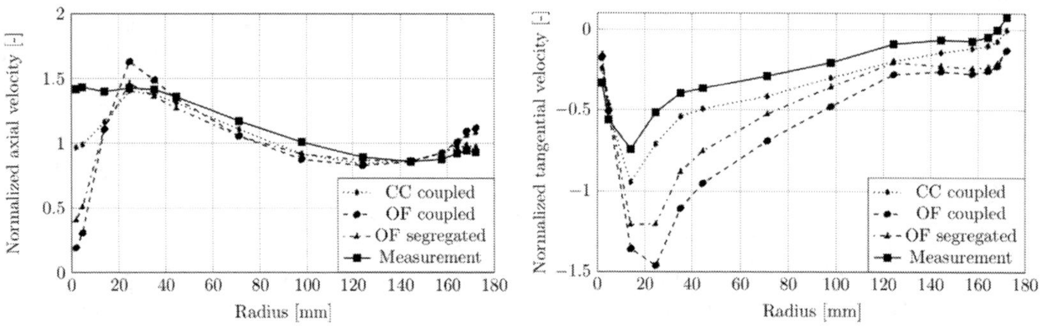

(a) Axial velocity (b) Tangential velocity

Figure 11: Normalized axial and tangential velocity at top slice in transient HL operation.

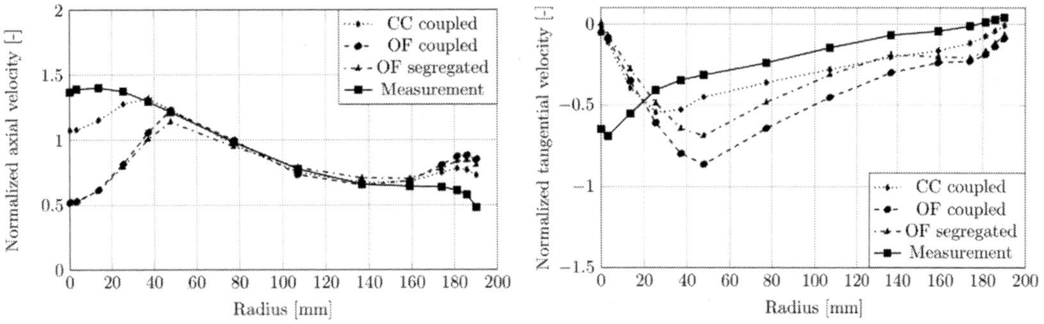

(a) Axial velocity (b) Tangential velocity

Figure 12: Normalized axial and tangential velocity at bottom slice in transient HL operation.

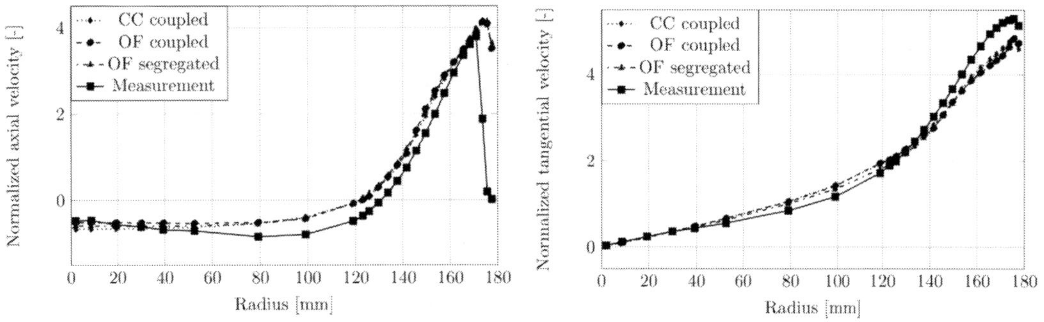

(a) Axial velocity (b) Tangential velocity

Figure 13: Normalized axial and tangential velocity at top slice in transient PL operation.

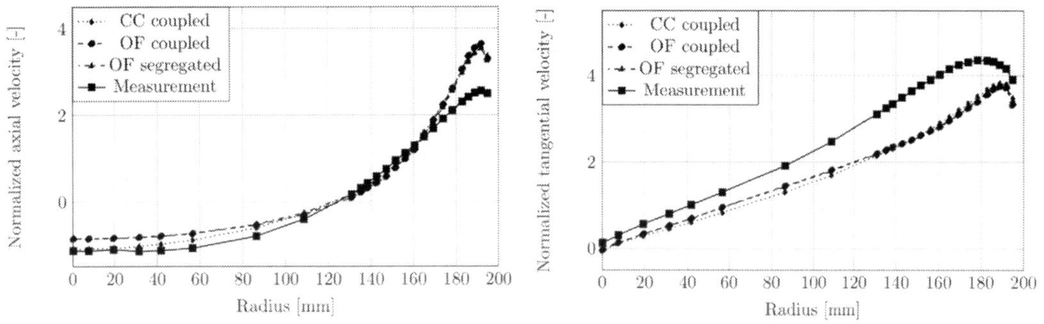

(a) Axial velocity (b) Tangential velocity

Figure 14: Normalized axial and tangential velocity at bottom slice in transient PL operation.

4. Conclusion

Steady-state as well as transient CFD computations of a 360 degrees high-head Francis-turbine model have been performed. Using a coupled approach for the pressure-velocity-coupling, a speed up of the steady-state and transient computational times of 4 to 5 and 1.5, respectively,

Journal of Physics: Conference Series **579** (2015) 012010 doi:10.1088/1742-6596/579/1/012010

has been achieved. The developed coupled solver based on the framework of OpenFOAM, allows to predict the overall efficiency accurately compared to measurements and a reference commercial code. In general, the CFD codes overestimated the efficiency, because the simulation domain does not consider the rotor side spaces and the hereby associated losses. The steady-state frozen rotor results are in a surprisingly good agreement with the measurements, a reason for this could be the straight vortex rope especially in PL condition. Regarding the time dependent signals, the FFT shows a good comparison of the data with all three solvers in all operating points. The dominant frequencies are the blade passings and their multiples. The comparisons with the LDA-measurements are showing that the numerical prediction of the tangential velocity downstream of the runner trailing does not match with the measured velocity. The distribution of the measured velocities are questionable.

References

[1] NTNU and LTU *Francis-99 workshop* http://www.ltu.se/research/subjects/Stromningslara/Konferenser/Francis-99?l=en accessed august 2014

[2] Biesinger T, Cornelius C, Rube C, Braune A, Campregher R, Godin P G, Schmid G and Zori L 2010 Unsteady CFD methods in a commercial solver for turbomachinery applications *ASME Conference Proceedings* **2010** 44021

[3] Biesinger T, Cornelius C, Nuernberger D and Rube C 2012 Speed line computation of a transonic compressor stage with unsteady cfd methods *ASME Turbo Expo* **2012** 68029

[4] Mangani L, Buchmayr M and Darwish M 2014 Development of a novel fully coupled solver in openfoam: Steady state incompressible turbulent flows *Numerical Heat Transfer, Part B: Fundamentals* **66** p 1-20

[5] Patankar S 1980 *Numerical Heat Transfer and Fluid Flow* (Hemisphere Publishing, Washigton)

[6] Widmer C, Staubli T and Ledergerber N 2011 Unstable characteristics and rotating stall in turbine brake operation of pump-turbines *Journal of Fluids Engineering* **133** 041101

[7] Yan J P, Seidel U and Koutnik J 2012 Numerical simulation of hydrodynamics in a pump-turbine at off-design operating conditions in turbine mode *26th IAHR Symposium on Hydraulic Machinery and Systems* **15** 032041

[8] Feng J J, Li W F, Wu H, Lu J L, Liao W L and Luo X Q 2014 Investigation on pressure fluctuation in a francis turbine with improvement measures *27th IAHR Symposium on Hydraulic Machinery and Systems* **2014** 2.2.1

[9] Tamura Y, Tani K and Okamoto N 2014 Experimental and numerical investigation of unsteady behavior of cavitating vortices in draft tube of low specific speed francis turbine *27th IAHR Symposium on Hydraulic Machinery and Systems* **2014** 2.2.6

[10] Heo M W, Kim K Y, Ma S B, Yoo I S, Choi W C, Kim J H and Choi Y S 2014 Parametric performance evaluation of a hydraulic centrifugal pump *27th IAHR Symposium on Hydraulic Machinery and Systems* **2014** 3.1.6

[11] Guo P C, Guo D M, Sun L G and Luo X Q 2014 Numerical investigation of different volutes in a low specific speed centrifugal pump *27th IAHR Symposium on Hydraulic Machinery and Systems* **2014** 3.1.7

[12] Trivedi C, Cervantes M J, Gandhi B K and Dahlhaug O G 2013 Experimental and numerical studies for a high head francis turbine at several operating points *Journal of Fluids Engineering* **135(11)** 111102

[13] NTNU and LTU *Francis-99 test case* http://www.ltu.se/research/subjects/Stromningslara/Konferenser/Francis-99/Test-Case-1.111520?l=en accessed august 2014

[14] Staubli T, Loetscher B and Cimmino D 2005 Integrale laufrad- simulationen von hydromaschinen Tech. rep. FHZ-HTA

[15] Amoser M 1995 *Stroemungsfelder und Radialkraefte in Labyrinthdichtungen hydraulischer Stroemungsmaschinen* Ph.D. thesis ETH Zuerich

[16] OpenCFD 2014 *OpenFOAM User Guide* OpenCFD Limited

[17] Mangani L, Casartelli E and Mauri S 2011 Assessment of various turbulence models in a high pressure ratio centrifugal compressor with an object oriented CFD code *ASME Conference Proceedings* **2011** 54679

Journal of Physics: Conference Series **579** (2015) 012011 doi:10.1088/1742-6596/579/1/012011

Steady and unsteady numerical simulations of the flow in the Tokke Francis turbine model, at three operating conditions

Lucien Stoessel[1], **Håkan Nilsson**[2]

[1] Master's student, Laboratory for Hydraulic Machines, École polytechnique fédérale de Lausanne, CH-1015 Lausanne, Switzerland

[2] Professor, Division of Fluid Dynamics, Department of Applied Mechanics, Chalmers University of Technology, SE-41296 Gothenburg, Sweden

E-mail: [1] `lucien.stoessel@alumni.epfl.ch`, [2] `hakan.nilsson@chalmers.se`

Abstract. This work investigates the flow in the scale model of the high-head Tokke Francis turbine at part load, best efficiency point and high load, as a contribution to the first Francis-99 workshop. The work is based on the FOAM-extend CFD software, which is a recent fork of the OpenFOAM CFD software that contains new features for simulations in rotating machinery. Steady-state mixing plane RANS simulations are conducted, with an inlet before the guide vanes and an outlet after the draft tube. Different variants of the k-ε and k-ω turbulence models are used and a linear explicit algebraic Reynolds stress model is implemented. Sliding grid URANS simulations, using a general grid interface coupling, are performed including the entire turbine geometry, from the inlet to the spiral casing to the outlet of the draft tube. For the unsteady simulations, the k-ω SSTF model is implemented and used in addition to the standard k-ε model. Both the steady and unsteady simulations give good predictions of the pressure distribution in the turbine compared to the experimental results. The velocity profiles at the runner outlet are well predicted at off-design conditions. A strong swirl is however obtained at best efficiency point, which is not observed in the experiments. While the steady-state simulations strongly overestimate the efficiency, the unsteady simulations give good predictions at best efficiency point (error of 1.16%) with larger errors at part load (10.67%) and high load (2.72%). Through the use of Fourier decomposition, the pressure fluctuations in the turbine are analysed, and the main rotor-stator interaction frequencies are predicted correctly at all operating conditions.

1. Introduction

Water turbines are often used to regulate the electric grid - to balance electricity demand and supply - since they are able to change operating point within a short time frame [1]. This requires the turbines to frequently run at off-design conditions and during transients between different operating conditions. The turbines are traditionally designed for steady operation at a single operating point, which is characterised by a stable flow with minimal swirl at the turbine outlet. In contrast, off-design operation lead to more complex flow through the turbine and the creation of a swirl at the runner outlet [2]. Such unsteady flow features can lead to fluid-structure interaction phenomena and oscillating forces on the runner [3, 4]. It is therefore of importance to investigate the flow in water turbines both at best efficency point and at off-design points. The turbines should be designed to give a high efficiency and to avoid unstable conditions. Validated numerical results can be used to extend the knowledge of the flow in the turbines, and to improve turbine design.

The present work uses the FOAM-extend fork of the OpenFOAM CFD software to study the flow in the Tokke high-head Francis turbine model (Francis-99). FOAM-extend was referred to as OpenFOAM-dev and OpenFOAM-ext in previous work. The FOAM-extend software and its tools for turbomachinery simulations have been validated and applied on hydro-power simulation cases by Nilsson [5], Petit [6] and Javadi [7]. Features for steady-state turbomachinery simulations were developed by Beaudoin *et al.* [8], and tested by Page *et al.* [9]. Petit *et al.* [10] set up an open case study that proved the validity of FOAM-extend for unsteady sliding grid simulations of swirling flows. Zhang and Zhang [11] used OpenFOAM on high-head Francis turbines, and included a model for cavitation. Recent work on Francis turbines at off-design conditions using proprietary software was performed by Nicolet *et al.* [4] for part load and by Shingai *et al.* [2] for high load operation. Susan-Resiga *et al.* [12] also included efforts to reduce the computational effort by the use of symmetry assumptions. The use of different turbulence models for RANS simulations of Francis turbines was investigated and compared to experiments at best efficiency point by Maruzewski *et al.* [13], while Wu *et al.* [14] validated their turbulence model in various operating conditions.

The present work examines the flow in the Tokke high-head Francis turbine model at best efficiency point, part load and high load using several turbulence models and using two rotor-stator-interaction approaches. Steady-state simulations are performed using a single guide vane and runner blade passage, and a mixing plane coupling between the stationary and rotating parts of the domain. Unsteady simulations are performed using all the blade passages and a sliding general grid interface (GGI). The steady-state simulations include the guide vane passage, the runner, and the draft tube. The unsteady simulations also include the spiral casing and the stay vanes. The pressure and velocity distributions are compared with the experimental and numerical results by Trivedi *et al.* [15], and additional experimental data that was distributed before the first Francis-99 workshop (www.francis-99.org).

2. Case description and numerical details

This section gives a brief description of the Tokke Francis turbine model case, and the numerical set-up used in the present work.

2.1. Turbine description and operating points

The Tokke turbine is a high-head Francis turbine with 15 full-length runner blades and 15 splitter blades, with a distributor consisting of 14 stay-vanes and 28 guide-vanes. The prototype has a runner outlet diameter $D = 1.779m$, a rated head of $H = 377m$ and a rated output power of $110MW$. The model turbine, studied in the present work, has a runner outlet diameter $D = 0.349m$ and a rated head of $H = 12m$. A cut view of the entire turbine geometry with all the components is shown in Figure 1. The turbine is limited by the inlet section I at the inlet of the spiral casing, and the outlet section $\bar{I}$ at the draft tube outlet. Three different operating points are examined; part load, best efficiency point and high load. A summary of the operating conditions is shown in Table 1. Due to experimental flow instabilities, the data for the best efficiency and high load conditions were given at two slightly different conditions, by the Francis-99 workshop organizers. The present steady-state simulations are therefore conducted at the three initial operating points, as well as the two additional operating points at best efficiency and high load. Due to the large computational cost, the unsteady simulations are only conducted at the initial conditions that were suggested at the time of those simulations. The velocity distributions of the unsteady results are thus compared with the experimental results at slightly different conditions.

Journal of Physics: Conference Series **579** (2015) 012011 doi:10.1088/1742-6596/579/1/012011

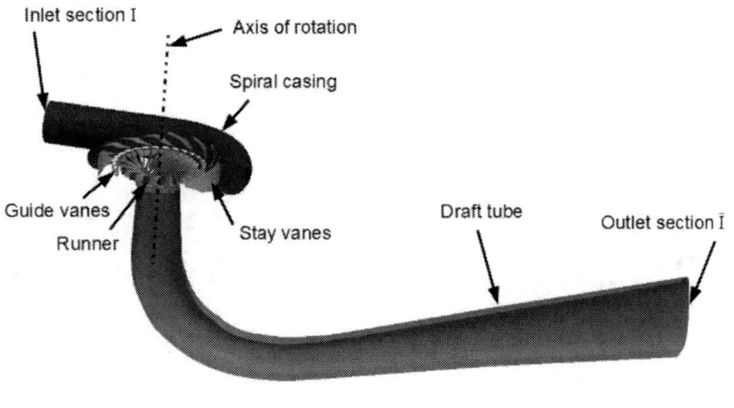

Figure 1: Cut view of the entire turbine geometry.

Parameter	Part load	BEP 1	BEP 2	High load 1	High load 2
H $[m]$	12.29	11.91	12.77	11.84	12.61
Q $[m^3/s]$	0.071	0.203	0.21	0.221	0.23
n $[rpm]$	406.2	335.4	344.4	369.6	380.4
$\eta_{h,F99} = \omega T/(\rho Q \Delta p_0)[-]$	71.69	92.61	92.4	90.66	91.0

Table 1: Physical parameters for all operating points. BEP 2 and high load 2 correspond to the velocity measurements close to the original BEP 1 and high load 1 conditions, and are therefore used for the comparison of the velocity profiles.

It should be noted that in Table 1, the definition of the hydraulic efficiency neglects the potential energy. Here Δp_0 is the difference in total pressure between the inlet section I and the outlet section $\bar{I}$, given by

$$\Delta p_0 = p_{0,I} - p_{0,\bar{I}} = p_I - p_{\bar{I}} + \frac{\rho Q^2}{2} \cdot (\frac{1}{A_I^2} - \frac{1}{A_{\bar{I}}^2}). \qquad (1)$$

This expression is not consistent with the definition in the IEC standard [16], which includes the gravitational term.

2.2. *Mesh and boundary conditions*

Original block-structured hexahedral meshes were supplied by the workshop organizers. Due to some severe imperfections, they were slightly improved in the present work. A detailed description of the mesh modifications is given in the Master's thesis by Stoessel. The meshes were also adapted for the different types of simulations. The steady-state mixing plane simulations are conducted for one guide vane passage and one blade channel, and the complete draft tube, as seen in Figure 2. The inlet boundary condition is set at the inlet of the guide vane passage. The inlet radial velocity is given by the flow rate and the inlet cross-sectional area. The inlet tangential velocity is given by a ratio of $\frac{U_R}{U_\theta} = \frac{2}{3}$ between the radial and tangential components, approximating the angle of the stay vanes. The inlet turbulence is specified by a turbulence intensity of 10%, and a ratio between the turbulent and laminar viscosities of $\nu_t/\nu = 10$. At the draft tube outlet the static pressure is set to zero, and a zero gradient boundary condition is applied for the other variables. A mixing plane interface is used at the runner inlet and outlet, yielding a circumferential average of the flow field [9] at those interfaces. A refined mesh is used

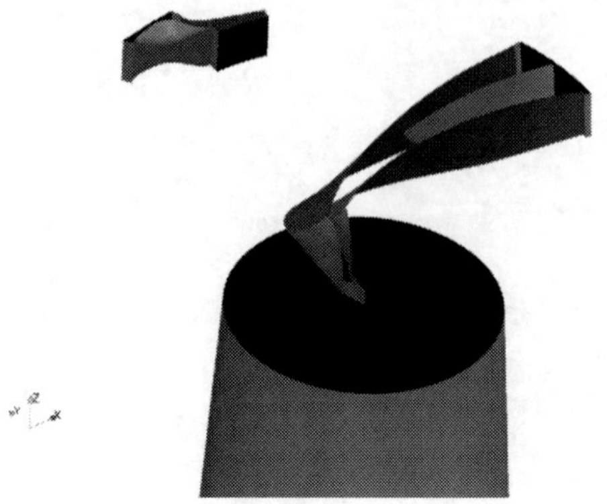

Figure 2: Computational domain for the steady-state simulations.

Figure 3: New shape of the interface between runner and draft tube, for the unsteady simulations. Originally flat, now conical.

for the best efficiency point, with improvements in the region below the hub and between the guide vanes. The off-design meshes are kept similar as the distributed meshes, except for the most important improvements. The resulting meshes contain 3,947,947 cells at best efficiency point, 3,931,653 at high load and 3,929,101 at part load.

The entire turbine geometry is used in the unsteady simulations, as seen in Figure 1. The boundary conditions are set as for the steady-state cases except that the velocity is specified normal to the inlet boundary. The mesh topology at the runner outlet is modified and the flat interface between the runner outlet and the draft tube inlet is replaced by a conical surface, as seen in Figure 3. This improves the mesh quality in the region below the hub. A general grid interface (GGI) [17] is used for the coupling between the rotating and stationary parts, allowing the use of a sliding grid method with non-conformal meshes. The meshes consist of 12,214,685 cells for the best efficiency point, 12,357,362 for the high load point, and 12,709,257 for the part load point. The supplied spiral casing and stay vane mesh has severe quality issues that affect the pressure convergence and should be solved in the future by re-meshing this region.

2.3. Numerical method

The software used for the simulations is FOAM-extend-3.0-Turbo, which is a modified version of FOAM-extend-3.0. It includes recent turbomachinery-specific developments, such as an improved implementation of the mixing plane interface. The code is based on the finite volume method, and the pressure-velocity coupling is handled using the SIMPLE algorithm in all the present simulations. The convection terms are discretised with a second-order linear upwind scheme.

For the steady-state simulations, a solver that uses multiple rotating frames of references (MRF) is chosen to mimic the behaviour of a rotating runner. A mixing plane is applied between stationary and rotating parts of the domain. Different schemes are available in FOAM-extend-3.0-Turbo to link the flow variables on either side of the mixing plane interface. For the best efficiency point simulations, the fluxAveragingAdjustMassFlow scheme is used for the velocity, where the values are adjusted to guarantee a conserved mass flow. A scheme based on a zero gradient boundary condition, preserving a corrrect mean value, called zeroGradientAreaAveragingMix is used for the pressure. At off-design, the default areaAveraging scheme is used for all variables. All variables are solved with a stabilised bi-conjugate gradient method and a diagonal incomplete LU-decomposition for preconditioning.

The unsteady simulations are performed with a solver that allows a physical rotation of the rotating part of the domain, and a coupling between the stationary and rotating parts of the domain with a GGI interface. The pressure is solved with a conjugate gradient method and diagonal incomplete Cholesky preconditioning, while the other variables are solved using a bi-conjugate gradient method with diagonal incomplete LU-decomposition for preconditioning. The time step corresponds to 0.25° of runner rotation, yielding maximum CFL numbers of about 4.5, 2.85 and 2.27 for best efficiency, part load and high load, respectively.

2.4. Turbulence models

The turbulence models used for steady-state simulations are the standard k-ε, RNG k-ε, realizable k-ε and k-ω SST. In addition, the linear explicit algebraic Reynolds stress model (EARSM) proposed by Wallin [18, paper 6, 196-199] was implemented and tested. For the unsteady simulations, the standard k-ε and the k-ω SSTF model developed by Gyllenram and Nilsson [19] are compared. All of the turbulence models use standard high-Reynolds wall functions.

3. Results

The results are here evaluated, based on the hydraulic efficiency, the static pressure and velocity distributions, and the pressure and torque fluctuations.

3.1. Hydraulic efficiency

Figure 4 compares the hydraulic efficiency of both the steady and unsteady simulations with the experimental and numerical results by Trivedi *et al.* [15]. While both of the present unsteady simulations give good agreement with the experimental data at best efficiency point, yielding an error of 1.16%, the difference to the experimental data increases to 2.72% at high load and 10.67% at part load. The behaviour of the unsteady numerical results is very similar to that of the numerical results of Trivedi *et al.* [15]. The steady-state simulations, on the other hand, give a strong over-prediction of the efficiency. In addition, the steady-state peak efficiency does not occur at the same flow rate as in the experimental data. This may be due to the definition of the inlet boundary conditions at the guide vane inlet, and that the hydraulic efficiencies of the steady-state cases are directly evaluated using those conditions. An example of other possible reasons for this effect is the mixing plane. The present work uses a newly implemented mixing plane, and this is the first comparison with experimental data. There may still be some issues with the implementation, but there may as well be some fundamental issues with the mixing plane concept. A study of the exact cause of this effect is left for future work. It should be recalled that the unsteady simulations have the inlet before the spiral casing, where the flow is more well-known with less streamwise gradients.

3.2. Head and torque

Depending on the type of calculations and the turbulence models that are used, different values are obtained for the resulting hydrostatic head and the torque at the shaft (Tables 2 and 3). For simplicity, only the results obtained with the standard k-ε are shown. For steady-state simulations (Table 2), the variations of the torque with other turbulence models are within 4.4% at part load, 0.7% at best efficiency point, and 0.9% at high load. For the head, the values are within 11.8% at part load, 1.1% at best efficiency point and 1.1% at high load. The time-averaged values of the unsteady simulations (Table 3) present differences to the standard $k-\varepsilon$ results of 0.28% at part load, 4% at best efficiency point and 1.9% at high load for the torque. The values of the time-averaged head have deviations of 0.07% at part load, 9.5% at best efficiency point and 8.7% at high load.

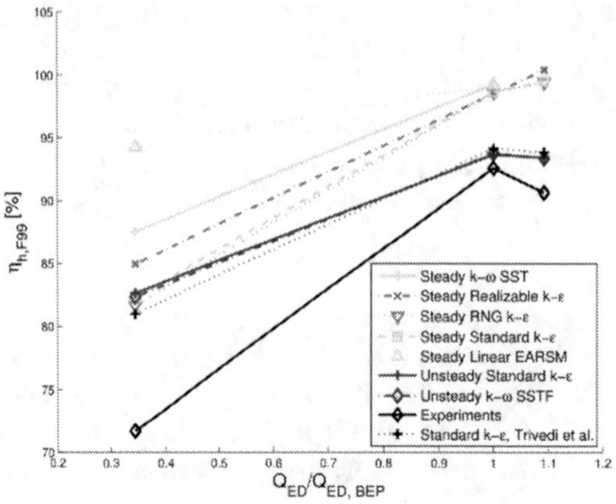

Figure 4: Hydraulic efficiency as a function of the relative unit flow rate, $Q_{ED}/Q_{ED,BEP} = (Q\sqrt{H_{BEP}})/(Q_{BEP}\sqrt{H})$.

Parameter	Part load	BEP	High load
H_{F99} $[m]$	12.47	13.148	11.01
T $[Nm]$	168.15	735.47	613.54

Table 2: Head and torque results of the steady-state standard k-ε calculations for the three original operating points.

Parameter	Part load	BEP	High load
H_{F99} $[m]$	13.78	13.87	11.99
T $[Nm]$	186.34	736.07	627.06

Table 3: Time-averaged head and torque results of the unsteady standard k-ε calculations for the three original operating points.

3.3. Static pressure distribution

Figure 5 shows the locations of the experimental static pressure probes, and 6 shows the corresponding static pressure distributions. The numerical static pressure level at the outlet of the draft tube is adjusted to that of the experiment. Those are $p_{\bar{I}} = 101.562kPa$ at best efficiency point, $p_{\bar{I}} = 99.536kPa$ at part load, and $p_{\bar{I}} = 95.977kPa$ at high load. The mixing plane simulations show good agreement with the unsteady simulations by Trivedi *et al.*, regardless of the turbulence model. The unsteady simulations give a higher static pressure on the blade suction side, while the static pressure in the vaneless space between the guide vanes and runner blades is closer to the experimental data. All the simulations predict a decrease in pressure between the $P71$ probe and the draft tube, while the experiment shows a slight increase. This indicates that certain phenomena occurring at the runner outlet are not represented the same way in the experiment and in the numerical simulations. Figure 7 gives an overview of the static pressure distribution in the symmetry plane of the draft tube. The static pressure probes, $DT11$ and $DT21$, are located on opposite walls just below the upper velocity measurement section. A significant reduction in static pressure is observed at the centerline, and its predicted width must be very important for the static pressure at the walls in that region.

Figure 8 shows that all numerical pressure distributions at part load agree to a high degree, and follow a similar trend as the experimental data. Figure 9 shows that there is no local region

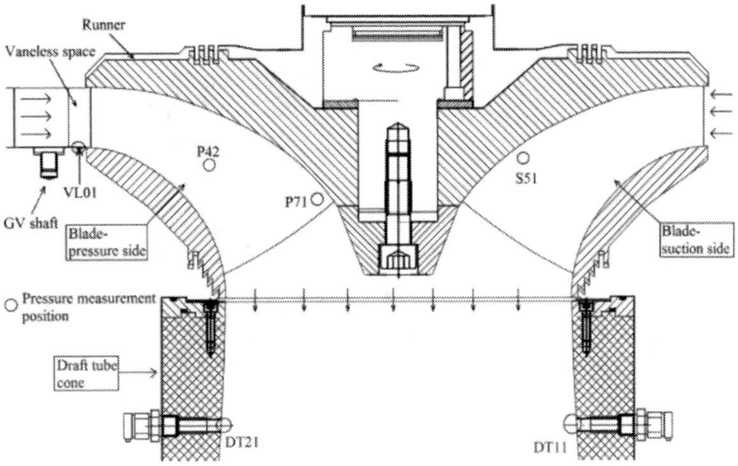

Figure 5: Locations of the static pressure probes in the guide vane cascade, runner and draft tube (illustration taken from the Francis-99 workshop website with kind permission of Prof. M.Cervantes).

of reduced pressure at the centerline in this case, and the all the numerical results do predict a similar static pressure increase between $P71$ and the draft tube probes as the experimental data.

Figure 10 shows that all numerical pressure distributions at high load are quite similar, and correspond to the experimental data except for the two probes in the draft tube. Figure 11 shows that there is again a local reduction of the static pressure at the centerline which may be a cause of the discrepancy.

The static pressure level of the numerical results is set to the same level as that of the experiment at the outlet of the draft tube. A numerical problem between probe $P71$ and the draft tube probes will thus change the level of the upstram probes accordingly. For the best efficiency and part load points, the numerically predicted static pressure at the draft tube probes correspond quite well with the experimental value, indicating that the draft tube performance is predicted reasonably well. Shifting the upstream points to lower values would in most cases improve the results. This indicates that additional attention should be paid to the numerical methods and the computational mesh in that region. On the other hand, the high load results would get worse with such an operation, although there is also a strong indication of problems in that region. The curves indicate that also the performance of the draft tube may not be predicted as well for the high load condition.

3.4. Velocity profiles

The velocity distributions in the draft tube cone were measured experimentally along two lines, located at $z = -0.2434m$ and at $z = -0.5614m$ respectively, shown as black lines in Figures 7, 9 and 11. The experimental data presented in this section corresponds to the original operating condition at part load, but to the two additional operating conditions at best efficiency and high load. The steady-state results are taken from the same conditions as the experimental data, while the unsteady results are taken from the original conditions.

Figure 12 shows the velocity distributions for the best efficiency point. All numerical results show a strong swirl near the axis of rotation which is not present in the experiments. This is related to the local reduction in the static pressure at the centerline, as discussed for Figure 7. The velocity distributions at the second measurement line shows that the swirl is

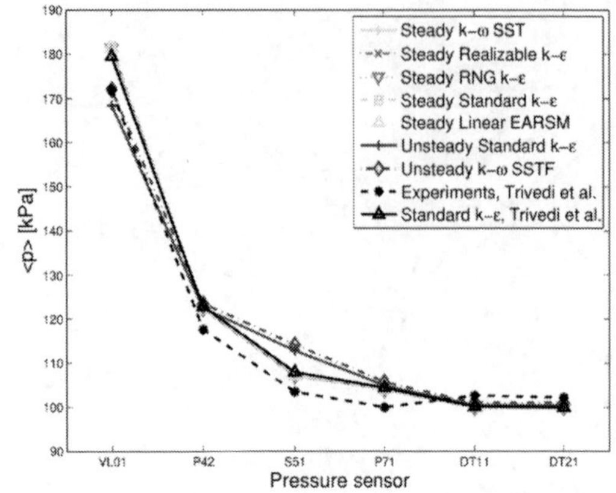

Figure 6: Static pressure distribution at best efficiency point.

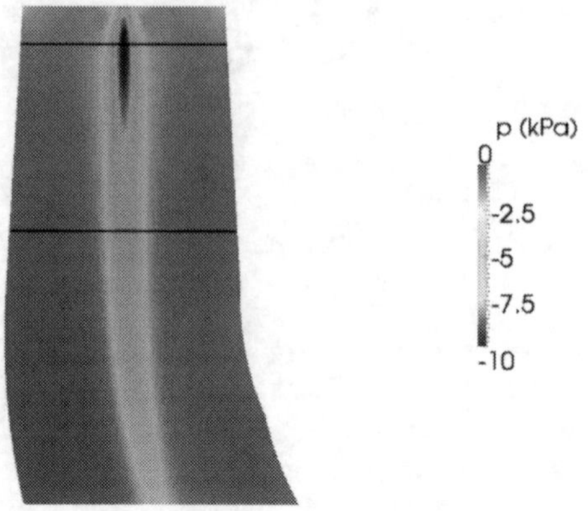

Figure 7: Static pressure distribution at the symmetry plane of the draft tube cone for a steady-state simulations at best efficiency point. The black lines indicate the location of the velocity measurement sections.

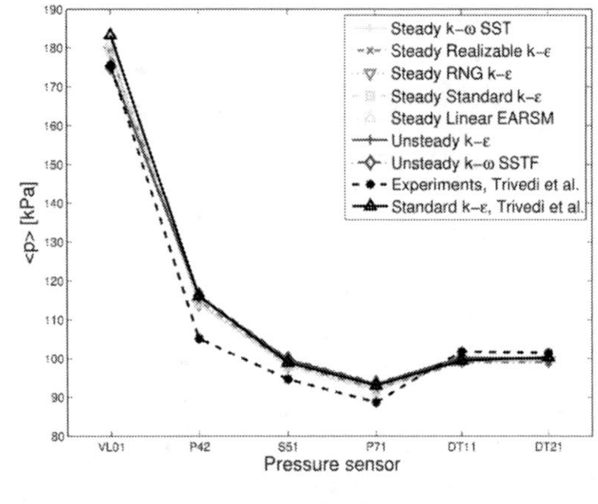

Figure 8: Static pressure distribution at part load.

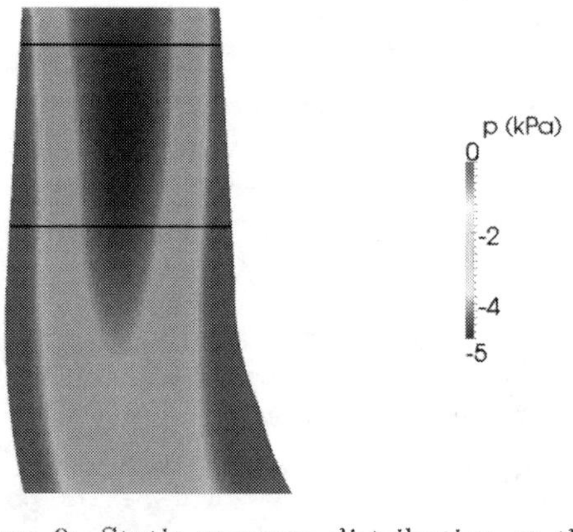

Figure 9: Static pressure distribution at the symmetry plane of the draft tube cone for a steady-state simulations at part load. The black lines indicate the location of the velocity measurement sections.

still over-predicted by the steady-state simulations, while the unsteady profile approaches the experimental data. The numerical axial velocity distributions agree well with the experimental data over most of the draft tube width, except for the region near the rotation axis. At the first line, only the unsteady standard k-ε simulation predicts the recirculation zone below the hub correctly, while it is under-predicted by the steady simulations and over-predicted by the unsteady k-ω SSTF results. All simulations strongly over-predict this wake region at the second line, especially the steady EARSM and the unsteady simulations, and only the steady realizable k-ε simulation gives a correct width of the recirculation zone at this location.

Figure 13 shows that there is a strong swirl near the draft tube walls at the part load condition.

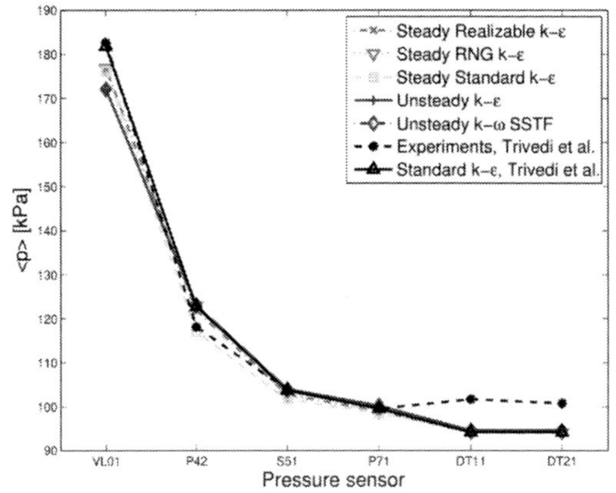

Figure 10: Static pressure distribution at high load.

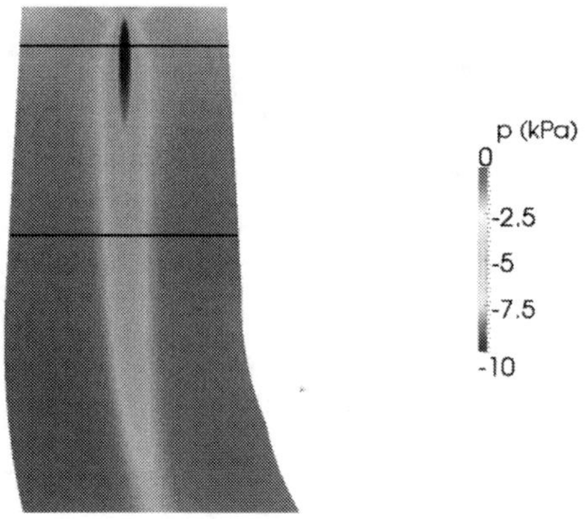

Figure 11: Static pressure distribution at the symmetry plane of the draft tube cone for a steady-state simulations at high load. The black lines indicate the location of the velocity measurement sections.

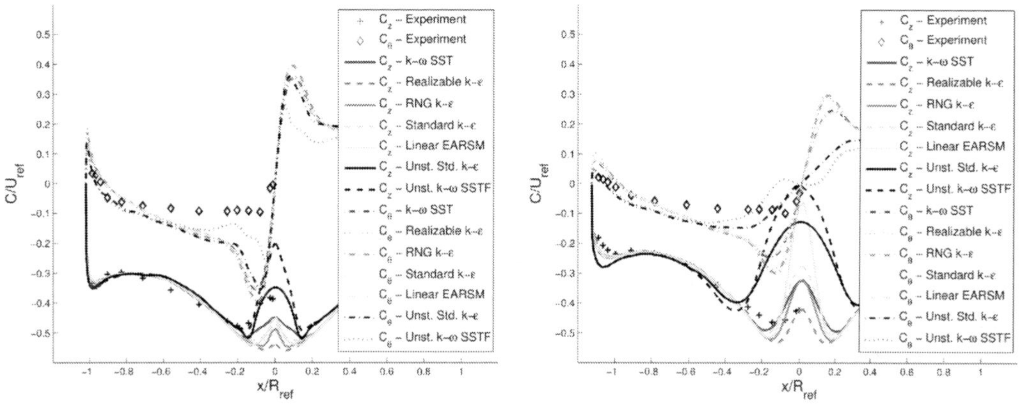

(a) Line 1.

(b) Line 2.

Figure 12: Dimensionless velocity distributions in the draft tube cone at the best efficiency point. C_z: axial. C_θ: tangential. Note that the unsteady simulations were conducted at a slightly different operating condition than that of the experiment.

Most of the flow is passing in this region, while there is a stagnation with some back-flow in the rest of the cross-section. This recirculation region is clearly visible on the pressure contours in Figure 9, as a large low pressure region that extends over a large part of the draft tube width. These phenomena are well captured both by the steady and unsteady simulations. The steady k-ω SST gives the best prediction at the first line. The best agreement at the second line is given by the steady-state standard and RNG k-ε simulations, and those models give reasonable results also at the first line.

Figure 14 shows that the axial velocity distributions correspond well to the measurements in a large portion of the draft tube at high load. However, all numerical results predict a

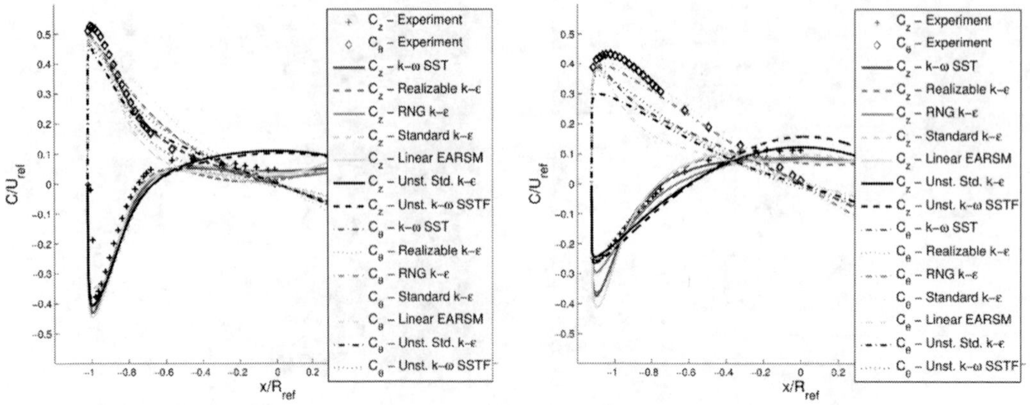

(a) Line 1. (b) Line 2.

Figure 13: Dimensionless velocity distributions in the draft tube cone at part load. C_z: axial. C_θ: tangential.

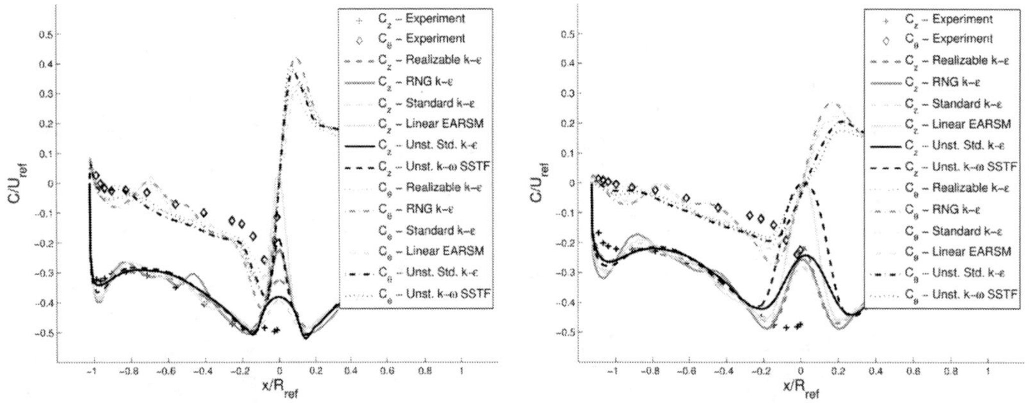

(a) Line 1. (b) Line 2.

Figure 14: Dimensionless velocity profiles in the draft tube cone at high load. C_z: axial. C_θ: tangential. Note that the unsteady simulations were conducted at a slightly different operating condition than that of the experiment.

much stronger on-axis recirculation than in the experiment. The predicted swirl behaves as the experimental, but is overestimated by all models. This swirl is then attenuated until the second measurement lines, but is still larger than the experimental one. The RNG k-ε and linear EARSM results show some unexpected variations in the radial direction, which is probably due to numerical instability.

3.5. Pressure and torque fluctuations

The pressure and torque fluctuations of the unsteady simulations are analysed using discrete fast Fourier transform. Only the results for standard k-ε are shown here. The frequency spectra of the pressure fluctuations at the best efficiency point are shown in Figure 15. The fluctuations in the vaneless space contain a frequency of $14.9n$. This is also found in the experimental results,

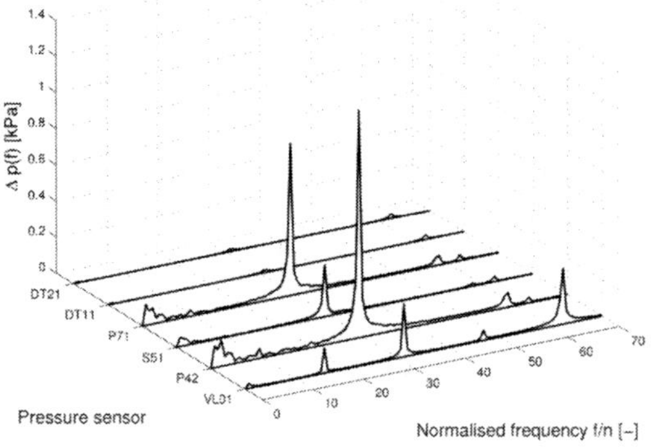

Figure 15: Frequency spectra of the pressure fluctuations at best efficiency point, normalised by the runner rotation frequency. Unsteady k-ε.

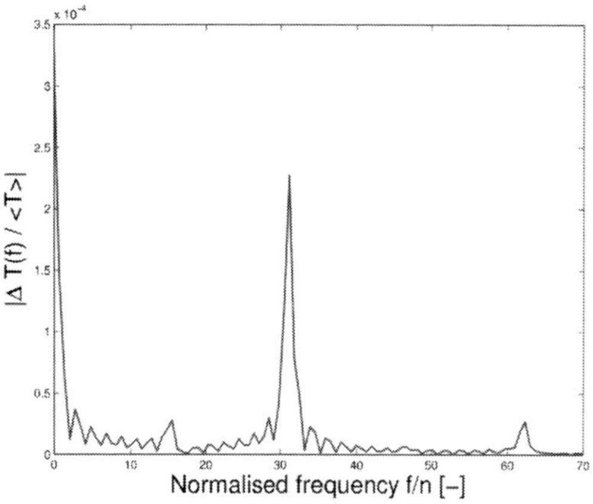

Figure 16: Frequency spectrum of the axial torque at best efficiency point, normalised by the runner rotation frequency. Unsteady k-ε.

and corresponds to the full blade passing frequency. It shows that the splitter blades do not give exactly the same upstream imprint on the pressure distribution as the full blades. A harmonic corresponding to the total number of runner blades is found at $31.2n$. The corresponding experimental peak is at $29.6n$. The pressure on the blade surface shows a strong oscillation with a frequency of $29.1n$. The corresponding experimental peak is at $27.7n$. This frequency corresponds to the blades passing the 28 guide vanes. A series of peaks on the blade surfaces located between $17.8n$ and $18n$ which is measured experimentally is not found in the simulations. On the other hand, the frequency located at $62.3n$, in the vane-less space and in the draft tube, is found in the numerical results but is not present in the measurements. Figure 16 shows that axial torque oscillations occur with the main peak at $31.2n$, corresponding to the total numer of runner blades, and a small peak at $15.6n$, corresponding to the full blade passage. An additional peak is observed at $62.3n$, corresponding to the high-frequency pressure fluctuations predicted at all sensor locations, which are not observed experimentally.

The frequency spectrum of the pressure fluctuations at part load is presented in Figure 17. In the vaneless space, the passing frequency of the full blades is observed at $15.4n$, corresponding to the experimentally observed $15n$. The frequency corresponding to the total number of blades is found at $30.77n$, which corresponds to the experimentally observed $30n$. On the blade surface, the guide vane passing frequency predicted at $28.7n$, corresponding to the experimentally observed $28n$. The experimental data shows a second peak at $29.55n$, which is not observed numerically. The experimental data also show a strong peak at $44.3n$. A tendency of such a peak is predicted numerically, but shifted to $46.1n$. The draft tube probes show the blade passing frequency at $15.4n$, experimentally observed at 14.8. The draft tube frequency corresponding to the total number of runner blades is predicted at $30.8n$, which corresponds to two experimental peaks, at $29.55n$ and $30n$. The numerically predicted frequency of $46.1n$ is experimentally measured at $44.3n$. A low frequency oscillation in the draft tube is observed at $0.27n$ in the experiments, while the numerical results show one at $0.5n$. However, the peak is very weak and no clear single frequency is present. This is consistent with the observation made by Dörfler *et al.* [20, pp. 40-41], that the stable vortex rope is replaced by random pulsations at very low part load. Figure 18 again shows that axial torque oscillations occur at $30.71n$. There is as well a harmonic at $92.13n$.

Figure 19 shows the frequency spectrum of the simulated pressure fluctuations at high load.

Journal of Physics: Conference Series **579** (2015) 012011 doi:10.1088/1742-6596/579/1/012011

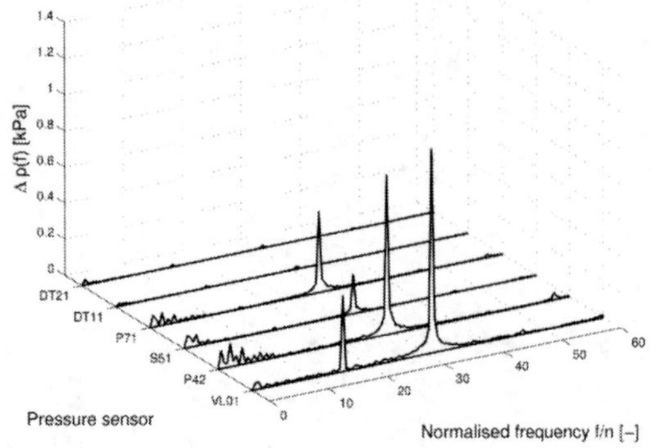

Figure 17: Frequency spectrum of the pressure fluctuations at part load, normalised by the runner rotation frequency. Unsteady k-ε.

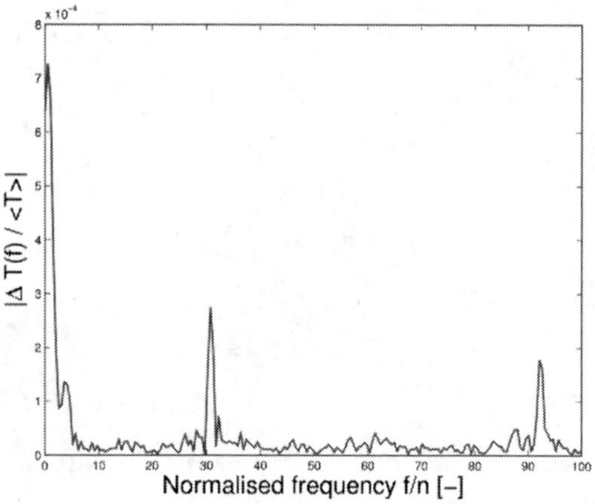

Figure 18: Frequency spectrum of the axial torque at part load, normalised by the runner rotation frequency. Unsteady k-ε.

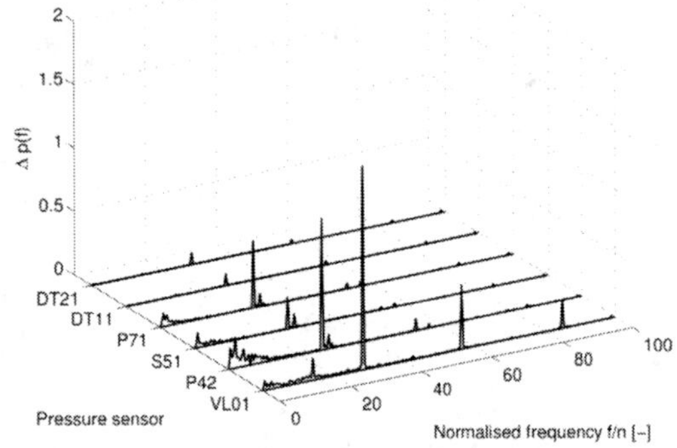

Figure 19: Frequency spectrum of the pressure fluctuations at high load, normalised by the runner rotation frequency. Unsteady k-ε.

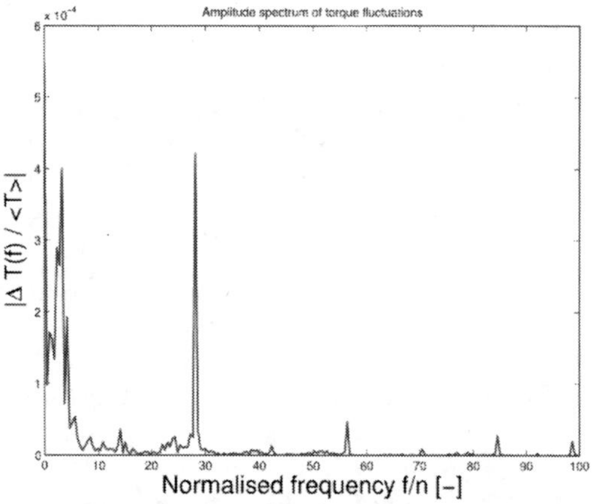

Figure 20: Frequency spectrum of the axial torque at high load, normalised by the runner rotation frequency. Unsteady k-ε.

In the vaneless space, the frequencies corresponding to the full blade passage and the total runner blade passage are shifted from their expected values of $15n$ and $30n$ to $14.1n$ and $28.2n$, respectively. These frequencies are not observed experimentally. The frequency of the total number of runner blades is present in the draft tube at $28.2n$. On the blade surface, the guide vane passing frequency is at $26.3n$, while two experimental peaks are observed at $28n$ and $30n$. The experimental frequencies at $48.7n$ are not present in the simulations. On the other hand, the peak at $56.4n$ in the vane-less space is not present in the experimental results. Figure 20 shows that the axial torque fluctuations have a main peak at $28.3n$, which is the same as the most significant pressure fluctuation. A minor peak is observed at $14.1n$, which is also observed in the pressure data.

4. Conclusions

Steady-state and unsteady numerical simulations are performed at three operating conditions of the Tokke high-head Francis turbine model, using several turbulence models. The steady-state simulations use a mixing plane, and the unsteady simulations use a sliding GGI interface, for the coupling of the stationary and rotating parts of the domain. The steady-state simulations use a multiple frames of reference approach, and the unsteady simulations use a rotating mesh, to represent the runner rotation. A single blade passage of the guide vanes and runner blades is used in the steady simulations, together with a cyclic GGI, while the unsteady simulations include all the blades. The steady simulations have an inlet just before the guide vanes, while the unsteady simulations have the inlet before the spiral casing. All simulations have the outlet after the draft tube. Most of the numerical predictions show reasonable agreement with the experimental data, with exceptions as summarized below.

The hydraulic efficiency predictions of the unsteady simulations closely correspond to those of previously published simulations by other authors using a proprietary CFD code. The error compared with the experimental data is 1.16% at the design point, 2.27% at high load, and 10.67% at part load. The similar behaviour of the present and previous numerical results suggest that there is a systematic error in the case description, the mesh, or the common methods. All of those were similar in the present and the previously published study. The steady-state results do however yield much higher errors in the prediction of the hydraulic efficiency. The losses in the spiral casing are not included in the steady-state simulations, and the choice of inlet conditions just before the guide vanes should be revised to investigate their effects on the predicted hydraulic efficiency. This is the main deficiency of the steady-state results. If this deficiency is removed, the fast steady-state results are as accurate as the time-consuming unsteady results as long as it is the mean flow features and engineering quantities that are of interest.

The overall behaviour of the numerically predicted static pressure distributions is very similar to the experimental ones. A main source of difference is related the region between the trailing edges of the runner blades and the first measurement point in the draft tube. The effect of the mesh resolution and mesh quality in that region should be further investigated. The presently used turbulence models are quite similar in nature, and they may not be able to correctly capture the effects of turbulence in that region. At high load, the draft tube pressure recovery is predicted different than the experimental one by all simulations, including the previously published simulation.

The velocity distributions are predicted reasonably similar to the experimental ones at part load and high load. At those conditions it is mainly the width and strength of the recirculation region below the hub that differs between models and compared to the experimental distributions. Even furher convergence of the steady-state results, and longer simulation times for the unsteady results, should be considered to verify that the flow is indeed fully developed in that region. At the best efficiency condition also the tangential velocity distribution shows an unexpected and unexplainable large discrepancy compared to the experimental one. This behaviour is the same for all the turbulence models used in the present work, and it is one of the top priorities for further investigations.

A Fourier analysis shows that the unsteady simulations are capable of resolving the most important pressure and torque fluctuations. However, there is a slight frequency shift compared to the experimental results and the expected values. The accuracy of the frequency spectrum could be improved by choosing a smaller time step and increasing the observation period.

All turbulence models used in the present work are part of the FOAM-extend CFD code except for the linear EARSM model, which is implemented in the present work. It shows similar behaviour as the other models. A main difference is that it gives a narrower and stronger recirculation region below the hub. The implementation should be further verified before any major conclusions can be drawn, and the full EARSM model should as well be investigated.

Acknowledgements
The authors would like to thank the Swedish national infrastructure for computing (SNIC) and the Chalmers centre for computational science and engineering (C^3SE) for providing support and the computational resources for this work. Professorn Nilsson is financed by the Swedish Hydro Power Center (SVC). SVC was established by the Swedish Energy Agency, Elforsk, and Svenska Kraftnät together with Luleå University of Technology, the Royal Institute of Technology, Chalmers University of Technology, and Uppsala University. Our thanks go also to Ardalan Javadi, who contributed with his experience. Furthermore, the efforts made by Martin Beaudoin and Prof. Hrvoje Jasak for the continuous improvement of FOAM-extend for turbomachinery applications as well as the fruitful discussions are much appreciated.

References
[1] Nicolet C 2007 *Hydroacoustic modelling and numerical simulation of unsteady operation of hydroelectric systems* Ph.D. thesis École polytechnique federale de Lausanne
[2] Shingai K, Okamoto N, Tamura Y and Tani K 2014 *J. Fluids Eng.* **136** 071105
[3] Keck H and Sick M 2008 *Acta Mech.* **201** 211 – 229
[4] Nicolet C, Zobeiri A, Maruzewski P and Avellan F 2011 *I. J. Fluid Mach. and Syst.* **4** 179 – 190
[5] Nilsson H 2006 *23th IAHR Symp. on Hydr. Mach. and Syst.*
[6] Petit O 2012 *Towards full predictions of the unsteady incompressible flow in rotating machines, using OpenFOAM* Phd thesis Department of Fluid Dynamics, Chalmers University of Technology Göteborg, Sweden
[7] Javadi A 2014 *Time-accurate Turbulence Modeling of Swirling Flow for Hydropower Application* Licentiate thesis Department of Fluid Dynamics, Chalmers University of Technology Göteborg, Sweden
[8] Beaudoin M, Nilsson H, Page M, Magnan R and Jasak H 2014 *27th IAHR Symp. on Hydr. Mach. and Syst.* (Montreal, Canada)
[9] Page M, Beaudoin M and Giroux A M 2011 *I. J. Fluid Mach. and Syst.* **4** 161–171
[10] Petit O, Bosioc A I, Nilsson H, Muntean S and Susan-Resiga R F 2010 *25th IAHR Symp. on Hydr. Mach. and Syst.* **12** 012056
[11] Zhang H and Zhang L 2012 *Procedia Eng.* **31** 156 – 165
[12] Susan-Resiga R, Muntean S, Stein P and Avellan F 2009 *I. J. Fluid Mach. and Syst.* **2** 295 – 302
[13] Maruzewski P, Hayashi H, Munch C, Yamaishi K, Hashii T, Mombelli H P, Sugow Y and Avellan F 2010 *IOP Conf. Ser.: Earth Environ. Sci.* **12** 1–9
[14] Wu Y, Liu S, Wu X, Dou H, Zhang L and Tao X 2010 *IOP Conf. Ser.: Earth Environ. Sci.* **12** 012004
[15] Trivedi C, Cervantes M J, Gandhi B and Dahlhaug O G 2013 *J. Fluids Eng.* **135** 111102
[16] IEC TC/SC 4 1999 *IEC 60193 Hydraulic turbine, storage pumps and pump-turbines - Model acceptance tests* (Geneva, Switzerland: International Electrotechnical Commission)
[17] Beaudoin M and Jasak H 2008 *Open Source CFD International Conference 2008* (Berlin, Germany)
[18] Wallin S 2000 *Engineering turbulence modelling for CFD with a focus on explicit algebraic Reynolds stress models* Phd thesis Royal Institute of Technology (KTH) Stockholm, Sweden
[19] Gyllenram W and Nilsson H 2008 *J. Fluids Eng.* **130** 051401
[20] Dörfler P, Sick M and Coutu A 2013 *Flow-Induced Pulsation and Vibration in Hydroelectric Machinery* (London: Springer)

Journal of Physics: Conference Series **579** (2015) 012012 doi:10.1088/1742-6596/579/1/012012

Performance study for Francis-99 by using different turbulence models

Zhao Yaping[1], **Liao Weili, Ruan Hui, Luo Xingqi**
Institute of Water Resources and Hydro-Electric Engineering, Xi'an University of Technology, Xi'an, 710048, China

E-mail: zhaoyp0168@hotmail.com

Abstract. The three-dimensional numerical investigation for turbine-99 at the best efficiency operation point, part load operation point and full load operation point was conducted by using the different turbulence models. By comparing the results of numerical simulation and experimental results, it was found that: there is a certain deviation between the numerical simulation results obtained by different turbulence models and experimental values, and the deviation increase with the reduction of output. Compared to other turbulence model, the result obtained by the standard k-ε turbulent model has a relatively small difference with the experimental results. The main causes for the big difference between the numerical simulation and model test include two aspects: (1) the mesh generation and boundary conditions setting lead to differences between the research object and the actual model, (2) it is difficult to accurately simulate the unstable flow such as impact, flow separation and vortex in the turbine. Therefore, in the future actual flow pattern simulation, besides the reasonable choice of turbulence model, based on the actual flow characteristics, the boundary conditions and the simulation results will be amended to reduce the deviation between the numerical simulation and experimental results as much as possible.

1. Introduction

An increasing number of hydraulic workers focus on accurately simulating the flow characteristics of turbine and estimating its energy characteristics based on CFD numerical computation. Dragica[1] used three turbulent models to predict non- cavitating and cavitating vortex rope in a Francis turbine draft tube, and concluded that SAS-SST, RSM and LES models are suitable for vortex rope prediction and when cavitation was not modeled there was no significant difference in the accuracy of the results obtained by these three turbulent models. S Kurosawa[2] etal simulated the efficiency characteristic of Francis turbine by RSM and LES turbulent models, and obtained that it is necessary to predict the efficiency characteristic by using the unsteady calculation basically. Wu Yulin [3-4] applied SST *k-ω* turbulence model to simulate cavitating flows around a model marine propeller in both uniform flow and wake flow, and proved that SST k-ω turbulence model was more accurate in the simulation of cavitating flow in turbomachineries.

Firstly, several numerical simulations for Francis-99 model were conducted under three operating conditions by means of different turbulence models and boundary conditions. Then, the differences

[1] To whom any correspondence should be addressed.

between the experimental and numerical heads, torques and efficiencies were compared, the uncertainty of turbine efficiency calculation has summarized.

All the model test results for Francis-99 shown in this paper were provided by the web-site of Luleå tekniska universitet - Forskning och utbildning.

2. Numerical simulation method

The computational region of Francis -99 consisted of three components: stationary domain1 (casing, 14 stay vanes, 28 guide vanes); rotating domain (runner with 15blades and 15 splitters); stationary domain-2 (draft tube). The mesh of Francis -99 was provided by workshop, the total number of mesh was 13096727, in which, the stationary domain1 is about 3600000; the rotating domain is about 5170000 and the stationary domain2 is about 3630000. The numerical simulations of Francis turbine at three operating points have achieved, table1 shows the parameters of calculated operating points. Table 2 shows the solution parameters used for performing the numerical simulations.

Table 1. Parameters of calculated operating points

Parameter	Symbol	Part load	BEP	Full load
Net head[m]	H	12.29	11.91	11.84
Flow rate[m^3/s]	Q	0.071	0.203	0.221
Runner angular speed[rev/sec]	n	6.77	5.59	6.16
Guide vane angle[°]	α	3.91	9.84	12.44
Density[kg/m^3]	ρ	999.23	999.19	999.20
Kinematic viscosity[m^2/s]	ν	9.57e-7	9.57e-7	9.57e-7

Table 2. Boundary physics and solution parameters used in the numerical simulations

Parameters	Description
Analysis type	Steady State
Interfaces	Frozen Rotor; discretization type-GGI
Fluid	Water properties updated with actual density, viscosity
Boundary conditons	Inlet: mass flow inlet Outlet: opening-type with static pressure, outlet-type with average static pressure Reference pressure: 0 kPa
Discretization and solution controls	Advection scheme: high resolution Turbulence numeric: high resolution
Turbulence models	Standard k-ε, k-ω SST
Convergence control	rms of pressure, mass-momentum, and turbulent parameters≤10E-5
Total run	Run-1: three operating points total, standard k-ε and opening-type with average static pressure Run-2:three operating points total, k-ω SST and opening-type with average static pressure Run-3:three operating points total, standard k-ε and outlet-type with average static pressure

3. Results analysis of numerical simulation

The basic equation of turbine can be described as:

$$H\eta = \omega\left(C_1 - C_2\right)/2\pi g \tag{1}$$

Where, H is the head, η is the hydraulic efficiency, ω is the angular speed, C_1 is the circulation of inlet edge , C_2 is the circulation of outlet edge, g is the acceleration of gravity.

The energy parameters of turbine can be represented by the circulation change amount of inlet and outlet. Figures 1-3 show the average circulation distribution from inlet edge to outlet edge of runner with different turbulent model at three operating points, the circulation difference of runner and rate-of-change of circulation estimated from the standard k-ε turbulence model were smaller than that from the SST model at the same operating conditions.

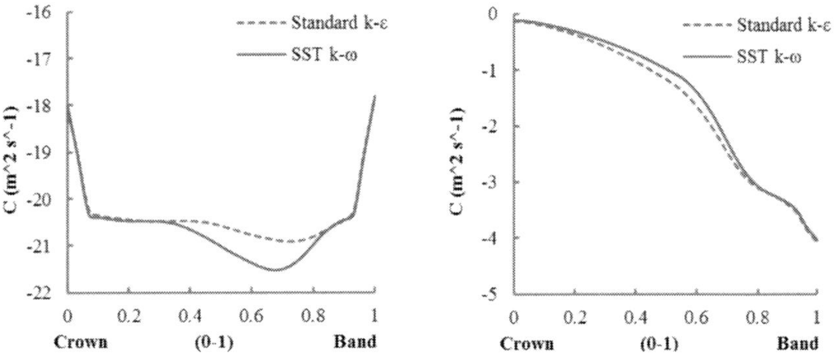

(a) Inlet of runner (b) Outlet of runner

Figure 1. Comparison of the standard k-ε and SST turbulent model circulation at part load operating condition

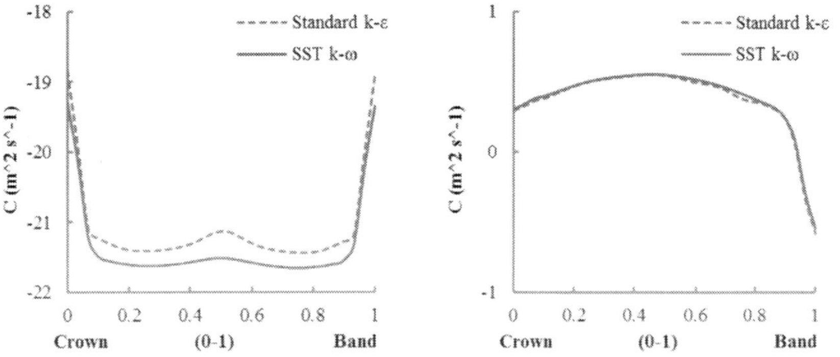

(a) Inlet of runner (b) Outlet of runner

Figure 2. Comparison of the standard k-ε and SST turbulent model circulation at BEP operating condition

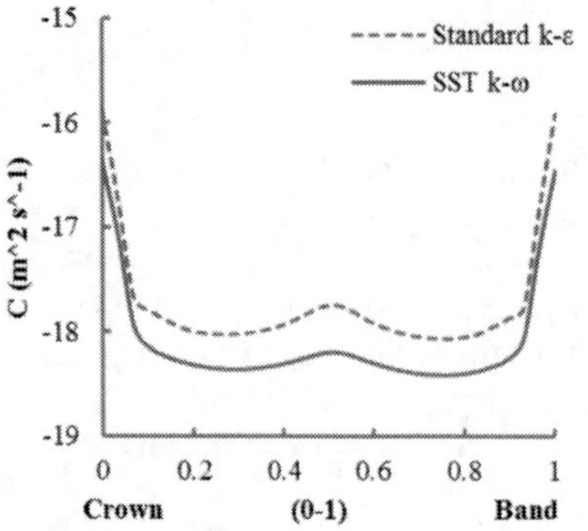

(a) Inlet of runner

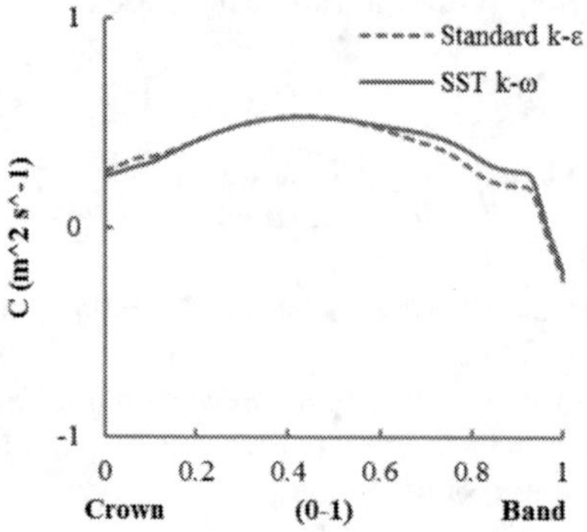

(b) Outlet of runner

Figure 3. Comparison of the standard k-ε and SST turbulent model circulation at full load operating condition

Figure 4 shows the comparison of torque obtained by the model test and numerical simulation at three operating points. The torque calculated by SST turbulence model and opening-type outlet condition has the maximum deviation compared with experimental torque. The numerical torque is 56.68N m (part load), 81.38 N m (BEP), and 62.22N m (full load) higher than the experimental torque, the ratio with experimental torque was 41.21%, 13.13% and 10.4%, respectively. The numerical torques calculated by standard *k-ε* turbulence model with opening boundary conditions and outlet boundary conditions were similar, slightly less than that computed by SST turbulence model and opening-type outlet condition. Under the condition of standard k-ε turbulence model and opening-type outlet condition, the torque difference between the experimental and numerical was 47.56N m (part load), 71.11 N m (BEP) and 45.032 N m (full load), the ratio with experiment torque was 34.58%, 11.47% and 7.53%.

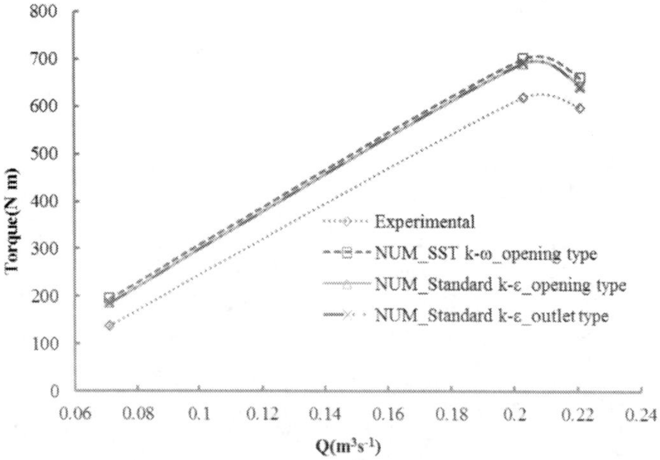

Figure 4. Comparison of the experimental and numerical torque of the turbine at three operating points

Hydraulic efficiency of the turbine can be calculated as:

$$\eta = \frac{T \cdot \omega}{\gamma Q H} \tag{2}$$

Where T is torque, ω is the angular speed, γ is the bulk density of water, Q is the volume flow rate, H is the head.

It can be seen that hydraulic efficiency accuracy of numerical simulation is bound up with the torque, especially in the part load operating conditions, the internal flow of turbine is very complex, unsteady flow such as impact, flow separation and vortex is the main reason caused turbine efficiency reduction.

The comparison of the experimental and numerical hydraulic efficiency at three operating points is shown in Figure 5. Under the opening-type outlet condition, the maximum difference between the experimental and numerical efficiencies was observed at a low discharge (Q=0.071 m^3s^{-1}, α=3.91 degree) operating condition. The numerical hydraulic efficiency was 11.06% (k-ε) and 12.42% (SST) higher than the experimental efficiency. The lowest difference between the experimental and numeric results was 0.24% (k-ε) and 1.29% (SST) at the BEP (Q=0.203 m^3s^{-1}, α=9.84 degree). The difference between the experimental and numerical efficiencies at the high discharge operating point (Q=0.221m^3s^{-1}, α=12.44deg) was 2.71% and 3.67% for k-ε and SST, respectively. The hydraulic efficiency estimated from the standard k-ε turbulence model showed a smaller deviation than that from the SST model.

Under the standard k-ε turbulent model, the simulation with outlet type and the standard k-ε model produced a hydraulic efficiency that was different from the experimental efficiency by 11.17%, 0.74% and 3.06% for low discharge, BEP and high discharge respectively. The hydraulic efficiency estimated from the outlet type model showed a higher deviation than that from the opening type model.

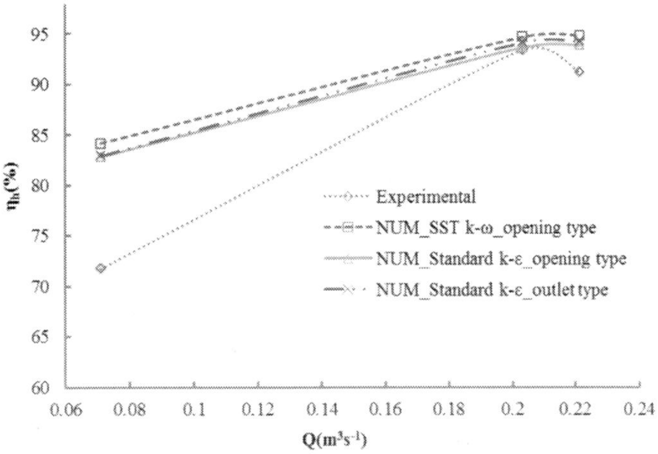

Figure 5. Comparison of the experimental and numerical hydraulic efficiency of the turbine at three operating points

Figure 6 shows the head comparison of the experiment and numerical simulation. The maximum difference occurs at the low discharge operating conditions, the difference between the experimental and numerical head was 1.35m (k-ε, outlet), 1.37m (k-ε, opening) and 1.81m (SST, opening). The ratio of head difference and experimental head was 11%, 11.14% and 14.72%, respectively. At BEP operating point, the numerical head was 1.1m (k-ε, outlet), 1.13m (k-ε, opening) and 1.19m (SST, opening) higher than the experimental head, the ratio of head error and experimental head were 9%, 9.48% and 9.99%. The smallest head difference appears in the high discharge operating point, the difference between numerical simulation and experimental head was 0.29m (k-ε, outlet), 0.41m (k-ε, opening), 0.61m (SST, opening), the ratio of head difference and experimental head were 2%, 3.46% and 5.15% ,respectively. It can be seen that head difference of the standard k-ε turbulence model and outlet boundary conditions simulation was the least relatively.

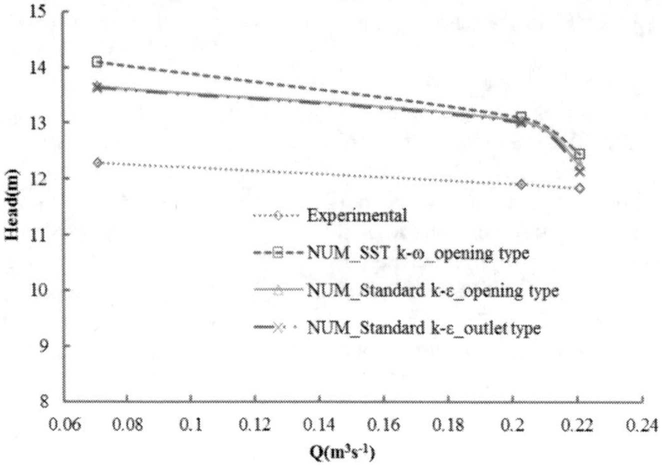

Figure 6. Comparison of the experimental and numerical head of the turbine at three operating points

4. Conclusions

The numerical simulations with the standard k-ε turbulence model, the SST turbulence model, the opening and outlet boundary conditions respectively have been fulfilled, and the results of numerical simulation are compared with experimental data, the following conclusions are obtained:

1) The larger circulation deviation of SST turbulence model simulation lead to the higher hydraulic efficiency than the standard *k-ε* turbulence model simulation. And the differences between the experimental and numerical efficiency, head and torque simulated by standard *k-ε* turbulence model are smaller than that simulated by SST turbulence model.

2) The minimum difference operating point between the experimental and numerical efficiencies was the BEP operating condition, high discharge operating point seconded, and low discharge operating condition is the largest.

3) The unstable flow such as impact, flow separation and vortex may cause the torque calculation inaccuracy in the numerical simulation, which lead to the numerical efficiency greater than the experimental data.

So, besides the reasonable choice of turbulence model, how to accurately amending the boundary conditions and the simulation results will be an urgent problem to reduce the deviation between the numerical simulation and experimental results.

References:

[1] Dragica Jost-Andrej Lipej. Numerical prediction of non-cavitating and cavitating vortex rope in a Francis turbine draft tube. Journal of Mechanical Engineering 57(2011)6, 445-456.

[2] S Kurosawa, S M Lim and Y Enomoto. Virtual model test for a Francis turbine. IOP Conf. Series: Earth and Environmental Science 12(2010) 012063.

[3] B.Ji,X.W.Luo,X.X.Peng,Y.L.Wu and H.Y.Xu. Unsteady numerical simulation of cavitating turbulent flow around a highly skewed model marine propeller, Journal of Fluid Engineering, 133(1)(2011)011102-1-8.

[4] Yulin Wu, Jintao Liu, Yuekun Sun, Shuhong Liu and Zhigang Zuo. Numerical analysis of flow in a Francis turbine on an equal critical cavitation coefficient line. Journal of Mechanical Science and Technology 27(6)(2013)1635-1641.

Application of the non-linear harmonic method to study the rotor-stator interaction in Francis-99 test case

J.D. Buron[1], S. Houde[1], R. Lestriez[2] and C. Deschênes[1]

[1] Laval University, 1065 Avenue de la Médecine, Québec, QC, G1V 0A6, Canada
[2] NumIberica, C/ Ferraz 49, 1°D, 28007 Madrid, Spain

E-mail: jean-david.buron.1@ulaval.ca

Abstract. Steady state and non-linear harmonic (NLH) flow simulations were performed within the framework of the Francis-99 project in order to assess the capacity of the NLH method to capture the main pressure fluctuations associated with the rotor-stator interactions between the distributor and the runner of the turbine. This paper focusses on the methodology developed to obtain harmonic solutions and presents preliminary results from the simulations using the flow solver NUMECA FineTURBO on intermediate grid level meshes. Comparisons of the first simulations to experimental data reveal good agreement concerning the predicted pressure amplitudes notably at high load operating condition.

1. Introduction

The process of energy extraction in hydraulic turbines requiring the presence of rotating and non-rotating hydraulic surfaces leads to pressure fluctuations that translate in unsteady hydrodynamic forces on the different parts of the machines. Those fluctuations originate from different flow phenomena like the part load vortex, the inter-blade channel vortex or the rotor-stator interaction. This last phenomenon, present for all operating conditions at fixed frequencies based on the runner rotation speed, is highly dependent on the flow rate and the geometry of the turbines. It will mainly affect the runner blades and the guide vanes through the interaction of the non-uniform pressure field associated with the flow in both rows of blade.

The prediction of the unsteady loading stemming from rotor-stator interaction is of the foremost importance to perform structural analysis that will provide life span and maintenance estimates at the design stages. Nowadays, Computational Fluid Dynamics (CFD) has deepened the general understanding of the rotor-stator interaction as well as giving what might be the only tool to evaluate it. The typical method consists of performing simulations of the complete interactions between the components in the time-domain. Although this approach can be quite precise, it does require significant computing resources and more importantly leads to long simulation time.

The nonlinear harmonic decomposition method (NLH) implemented within the CFD package FineTURBO of the NUMECA Corporation was aimed specifically at lowering significantly those computing costs by focusing only on the main frequencies associated with the rotor-stator interaction. NLH simulations have already been performed successfully on different cases using the FineTURBO solver by [5, 9, 10], where the reduction in computational costs was also highlighted. The Francis-99 test case [6], with its high head Francis turbine, is well suited for the application of the method since the rotor-stator interaction causes the dominant fluctuations within the runner. This paper focusses on preliminary NLH simulations results for two operating conditions validated through experimental measurements performed at the Waterpower laboratory, NTNU, Norway [2]. The paper presents a brief overview of the NLH method, the methodology and an analysis of the first results.

2. Non-linear harmonic method

The non-linear harmonic method (NLH), introduced by He and Ning [4], is based on the unsteady Reynolds Averaged Navier-Stokes (URANS) formulation. For an incompressible fluid, the equations are:

$$\begin{gathered}\frac{\partial U_i}{\partial x_i} = 0 \\ \frac{\partial U_i}{\partial t} + U_j \frac{\partial U_i}{\partial x_j} = -\frac{1}{\rho}\frac{\partial P}{\partial x_i} + \frac{\partial}{\partial x_j}\left(\nu \frac{\partial U_i}{\partial x_j} - \tau_{ij}\right)\end{gathered} \tag{1}$$

where U_i are the mean velocity components ($u_i = U_i + u_i'$), P the mean pressure, ν the kinematic viscosity and $\tau_{ij} = \overline{u_i' u_j'}$ the Reynolds stresses. A simple derivation of the harmonic method is done by decomposing these equations into a time-averaged flow and unsteady perturbations. The decomposition of the conservative variables into a mean value $\bar{\phi}$ and N periodic perturbations ϕ' with complex amplitude $\tilde{\phi}$ has the form:

$$\phi(\vec{x}, t) = \bar{\phi}(\vec{x}) + \sum_{k=1}^{N} \phi_k'(\vec{x}, t) \tag{2}$$

$$\phi_k'(\vec{x}, t) = \tilde{\phi}_k(\vec{x}) e^{i\omega_k t} \tag{3}$$

Substituting equation (2) in the unsteady equations (1) and time-averaging them yields the mean flow equations. Considering only one perturbation and adding a pseudo time-dependence, this gives:

$$\begin{gathered}\frac{\partial \bar{U}_i}{\partial x_i} = 0 \\ \frac{\partial \bar{U}_i}{\partial t} + \bar{U}_j \frac{\partial \bar{U}_i}{\partial x_j} + \frac{\partial \overline{U_i' U_j'}}{\partial x_j} = -\frac{1}{\rho}\frac{\partial \bar{P}}{\partial x_i} + \frac{\partial}{\partial x_j}\left(\nu \frac{\partial \bar{U}_i}{\partial x_j} - \tau_{ij}\right)\end{gathered} \tag{4}$$

Because of the non-linearity of the URANS equations, this operation brings new terms that must be modeled, the deterministic stresses $\overline{U_i' U_j'}$, which are similar to the Reynolds stresses appearing when time-averaging the Navier-Stokes equations. These terms are responsible for the coupling between the perturbations and the time-mean flow, hence the nonlinearity of the method. Subtracting the time-mean flow equations (4) from the URANS equations (1) and keeping only the first order terms yields the transport equations required to solve the unsteady perturbations:

$$\begin{gathered}\frac{\partial U_i'}{\partial x_i} = 0 \\ \frac{\partial U_i'}{\partial t} = -\frac{1}{\rho}\frac{\partial P'}{\partial x_i} + \nu \frac{\partial^2 U_i'}{\partial x_j^2}\end{gathered} \tag{5}$$

The complex perturbations (3) can then be substituted in the transport equations to obtain the non-linear harmonic method equations. Equation (4) is equivalent to a RANS formulation that is solved to obtain the mean flow field and equation (5) is solved simultaneously for each of the complex amplitudes $\tilde{\phi}_k$, which are space dependent only. The final solution can be constructed in time by computing the perturbations at a given time using equation (3) and adding them to the time-mean flow. This operation allows to obtain an approximation of the unsteady flow field. It should be noted that interactions between the perturbations themselves can only be achieved through their respective interaction with the mean flow by the deterministic stresses. More details about this formulation and its implementation can be found in [4, 9].

The resolution of the perturbations equations in the frequency domain allows the usage of periodic computational domains with mixing plane interfaces, thus lowering the required memory and simulation time. Periodic boundary conditions are implemented using a simple phase shift of the complex amplitudes, which are solved for the frequencies of the perturbations coming from adjacent computational domains with different rotational speeds.

3. Francis-99 test case

3.1. Geometry

The Francis-99 test case is a model turbine installed and investigated at the Water Power Laboratory at the Norwegian University of Science and Technology (NTNU) in Norway. It is a 1:5.1 model of a high head Francis turbine in operation at the Tokke power plant containing a spiral casing, 14 stay vanes, 28 guide vanes, and a runner with 15 full length and 15 splitter blades. The model is illustrated in Figure 1. More details about the test case and its geometry can be found in [2, 6]. For the needs of this project, the complete turbine geometry was reconstructed with ProEngineer according to the model drawings and compared with the provided geometry. The process resulted in a slight decrease, below 1%, of the distributor height and adjustment of the runner meridional profile.

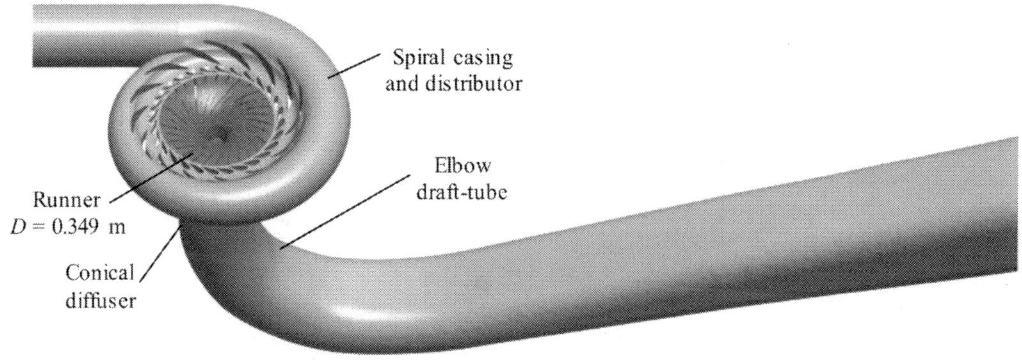

Figure 1. Tokke model turbine geometry.

3.2. Operating points

For the first Francis-99 workshop, three operating conditions are studied with experimental measurements data provided at part load, best efficiency and high load. These operating conditions are summarized in Table 1.

Table 1. Experimental operating points.

	Part load	BEP	High load
Net head [m]	12.29	11.91	11.84
Guide vanes angle α [°]	3.91	9.84	12.44
Flow rate [m^3/s]	0.071	0.203	0.221
Runner frequency [Hz]	6.77	5.59	6.16
Efficiency [%]	71.69	92.61	90.66

3.3. Pressure measurements

Pressure measurements are provided at six locations illustrated in Figure 2: one in the vaneless space between the guide vanes and the runner blades (VL01), two on a blade pressure side (P42 and P71), one on the suction side of the adjacent blade (S51) and two in the draft tube cone (DT11 and DT21). Again, more details concerning the pressure measurement campaign can be found in [2].

4. Numerical methodology

The simulations are conducted with both standard RANS method and nonlinear harmonic method using the same computational domains for the distributor and the runner. The numerical methodology and simulation parameters used are described in this section.

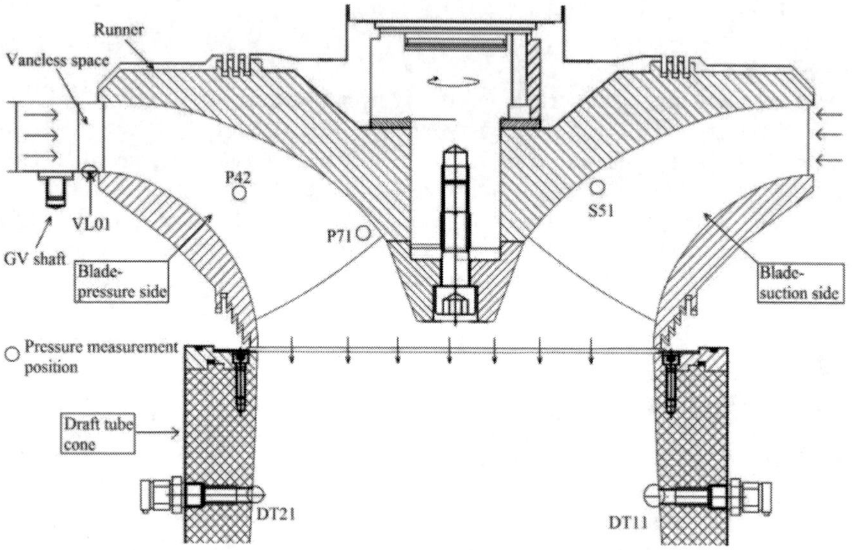

Figure 2. Pressure sensors locations [3].

4.1. Computational domains and boundary conditions

The computational domains of the distributor and runner have a periodicity of 25.7° and 24°, respectively, and are illustrated in Figure 3*a*. The value of the periodicity angle is important since it is used to compute the perturbation frequency and harmonics by the NLH solver. The axisymmetric geometry of the distributor is obtained by the revolution of a representative spiral casing section. The inlet is located at the radius where the spiral casing height is maximal and the distributor domain includes one stay vane and two guide vanes. The runner domain includes one full length and one splitter blade, as well as the conical part of the draft tube. The inclusion of the conical diffuser in the periodic runner domain implies that one cannot expect to capture different perturbations than the NLH frequencies coming from the distributor domain.

At the inlet, the mass flow rate, flow angle and turbulent viscosity are imposed. The flow angle is set to a constant value of $\alpha = 5°$ relative to the boundary and a turbulent viscosity $\nu_t = 1x10^{-4}$ m^2/s is imposed, both values coming from spiral casing simulations. It was verified with steady simulations that the results were not highly sensitive to these parameters. At the outlet of the conical diffuser, an average static pressure is imposed. Modelling of the rotor-stator interface between the distributor and runner is made with a mixing plane approach for the steady simulations while a non-reflecting mixing plane approach is used for the NLH simulations.

4.2. Flow solver and numerical parameters

The software used for the computations is NUMECA FineTURBO v9.0-3, a structured multigrid solver specialized in turbomachinery cases. The Spalart-Almaras turbulence model is used with an extended wall function and a finite volume second order central spatial discretization is used. A multi-stage Runge-Kutta explicit scheme is used to advance the solution in time with local time stepping and a CFL number equal to 3. Finally, nonlinear harmonic simulations are set to compute one perturbation per domain, i.e. the perturbation coming from the domain separated by the rotor-stator interface. Simulations are conducted using two ($n = 2$) or four ($n = 4$) harmonics to represent the perturbation, with harmonic number two being related to the relative blade passing frequency.

4.3. Meshes

The meshes are made using NUMECA Autogrid5 9.0-3 for the distributor and runner and NUMECA IGG 9.0-3 for the diffuser. These meshers produce structured multigrid meshes readily functional with the FineTURBO solver. The simulations are conducted on two different grid levels of the meshes: an intermediate grid level denoted *111*, and a fine grid level denoted *000* which represents the finer grid of a given mesh. Examples of these two grid levels are shown in Figure 3*b* and 3*c*, respectively.

Grid orthogonality, aspect ratios and expansion ratios are checked to assure a good quality according to NUMECA recommendations, and grid distribution is made to be identical on both sides of the mixing plane interface in the three spatial directions. In addition, all meshes are checked to produce $y+$ values above ~30 on the finer grid level so the turbulence model with wall functions can be used.

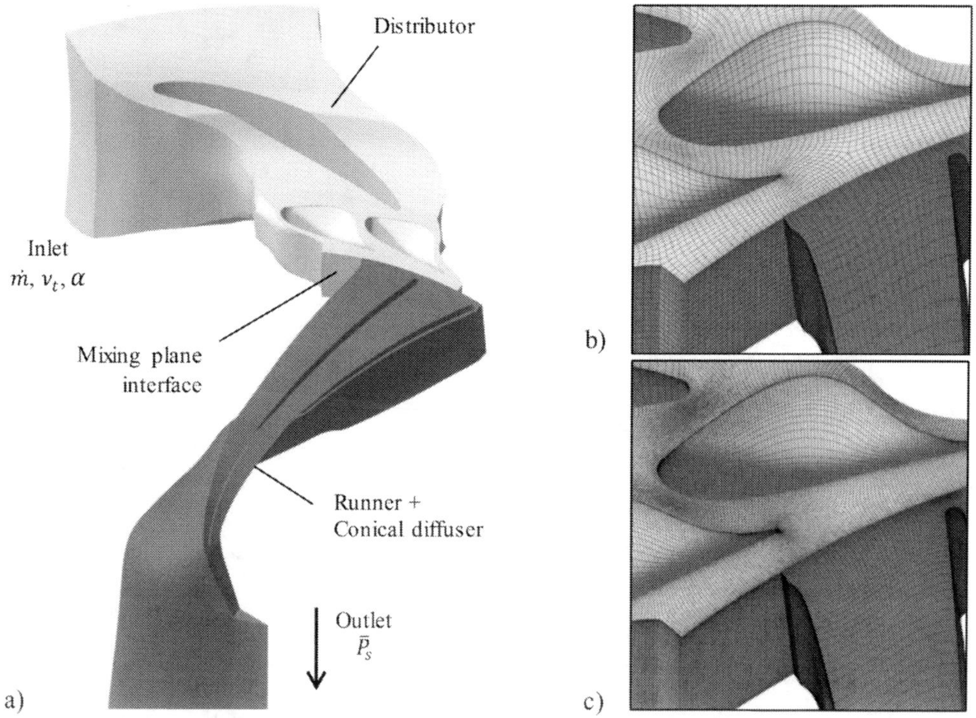

Figure 3. a) Computational domains and boundary conditions; b) Intermediate grid level of the mesh (*111*); c) Fine grid level of the same mesh (*000*).

In order to obtain solutions that are not dependent on the mesh refinement level, steady state simulations were conducted with three different meshes (coarse, medium and fine) at the two grid levels for each of them. Analysis of the results showed that pressure values are weakly affected by the grid level or mesh used. For a given grid level, differences range from 0.1% in the runner to 1.5% in the vaneless space between the three mesh refinements. For a given mesh, the differences between the grid levels *111* and *000* are around 1.5% in the vaneless space, and 0.5% in the runner.

Figure 4 shows the head loss in the different components of the simulations using the three meshes and the two grid levels, indicating that the medium mesh, whose grid levels are illustrated in Figure 3, is adequate to obtain good results. Computation of the grid convergence index (GCI) for this mesh on each component yields errors of 1.7%, 0.6% and 0.4% for the distributor, runner and conical diffuser respectively [1]. Since this refinement level also produces acceptable computation times, the medium mesh is used for the simulations presented in this paper. Its statistics and quality on the finer grid level are detailed in Table 2.

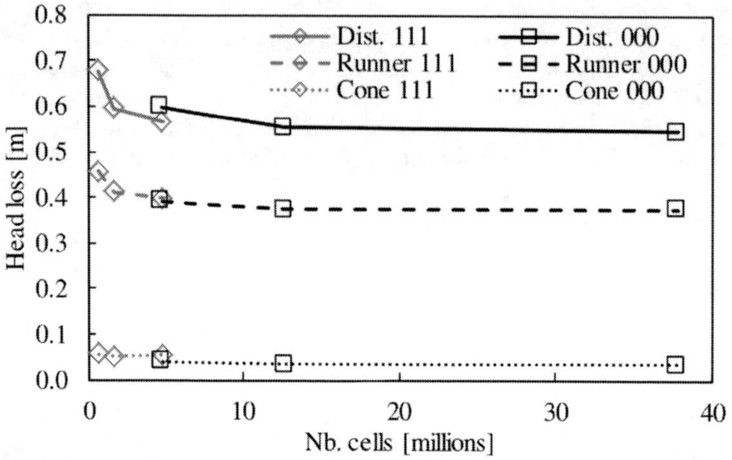

Figure 4. Head losses in the distributor, runner and conical diffuser for the three meshes at the two different grid levels.

Table 2. Medium mesh statistics and quality on the finer grid level (*000*).

	Nb. cells	Min. angle [°]	Max. aspect ratio	Max. exp. ratio	Max. y+
Distributor	5 107 200	34.9	20.2	2	80.7
Runner	6 300 992	33.7	66.7	1.77	87.6
Diffuser	2 078 272	33.6	141	1.34	58.2

5. Results

In this section, preliminary results are shown from steady and NLH simulations at the best efficiency point and high load. For both operating points, steady simulations were made at grid levels *111* and *000*. Non-linear harmonic simulations are presented on grid level *111* only with 2 and 4 harmonics. The residuals, the real and imaginary parts of the pressures at each measurement points for each harmonic and other performance quantities were monitored during the simulations. The simulations were considered converged when all the performance and harmonic values were stabilized.

5.1. Mean pressure values

Comparison of the steady and harmonic simulations on both grid levels showed negligible differences in the averaged pressure predictions for all average pressures at VL01, P42, P71, S51 and DT when compared to the fine grid steady simulations, with differences ranging from ~0.0% in the conical diffuser up to ~0.5% in the vaneless space for the simulation on intermediate grid level at high load. The mean flow stemming from the harmonic computations is thus in accordance with the results from the steady computations with regard to the mean pressure field prediction.

The averaged numerical pressure values from the steady simulation on the fine grid for the best efficiency condition are compared to the average of the experimental pressure signals in Figure 5. Globally, the simulations are showing good agreement with the experimental values with a maximum difference of ~5.5% in the vaneless space. The simulations do appear to overestimate slightly the average pressure in the distributor and runner. It should be noted that the computed data is corrected according to the atmospheric pressure and water level in the test rig downstream tank. The draft tube experimental value shown corresponds to the average of sensors DT11 and DT21.

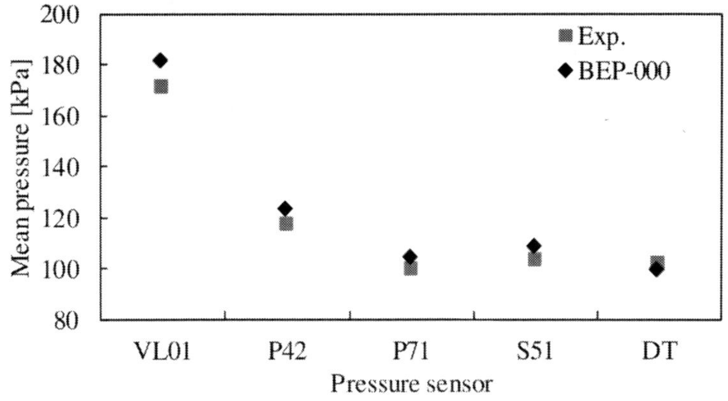

Figure 5. Mean experimental and numerical pressure values at BEP.

The computed values at high load are shown and compared to the averaged experimental pressures in Figure 6. In the same way as for the BEP, the simulations overestimate the pressure in the distributor and runner, and underestimate it in the draft tube. Here, the maximum difference of ~7.5% arises at the P42 sensor on the runner blade pressure side.

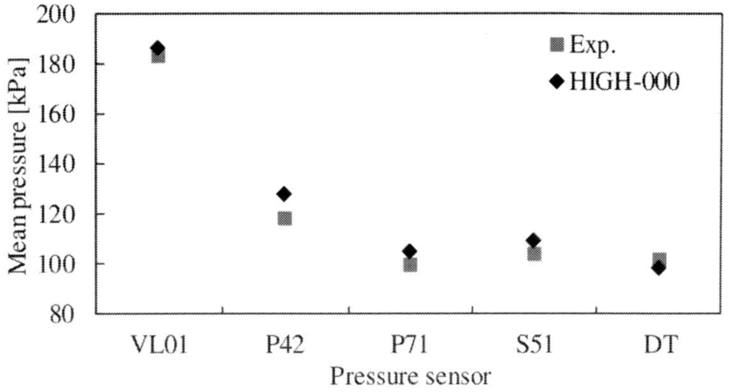

Figure 6. Mean experimental and numerical pressure values at high load.

5.2. *BEP pressure amplitudes and signals*

Pressure frequency spectra are computed with the complex amplitudes from the intermediate grid harmonic simulations and compared to the Fourier spectral analysis of the experimental values at the NLH frequencies. The FFT were performed on the experimental pressure signals using a Hanning window to prevent side lobe leakage.

The spectra for the best efficiency condition are illustrated in Figure 7 and show that the simulations generally overestimate the amplitude related to the blade passing frequency (F/F_{runner} = 28 or 30) in the vaneless space and underestimate the other fluctuation harmonics. Caution should be exercised when comparing amplitudes below 0.05 kPa for the VL01 sensor and 0.2 kPa in the runner since they fall in the measurements noise range. The computation with $n = 4$ allows a slight prediction improvement over the $n = 2$ simulations, with a decrease of ~7.2% of the relative blade passing frequency amplitude in the distributor and ~1.2% for each of the sensors on the runner blade. The relative amplitudes between the third and fourth harmonics also appear to match.

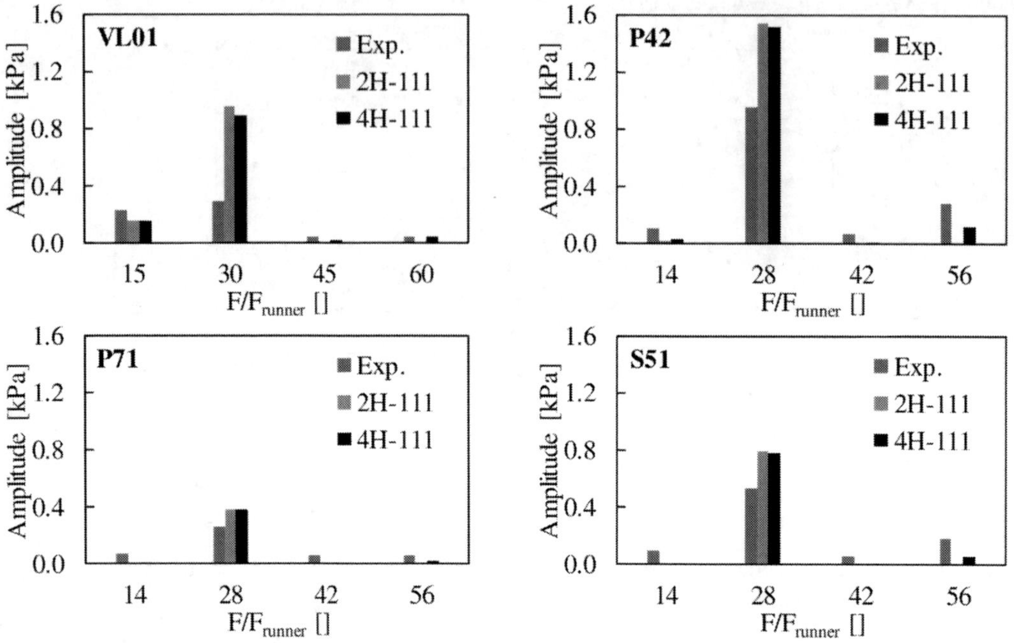

Figure 7. Experimental and numerical frequency spectra for the main perturbation harmonics at best efficiency point.

The reconstructed signals from the NLH simulations are compared to the experimental ones for the VL01 and P42 sensors in Figure 8 and Figure 9, respectively. The experimental pressures are averaged over ten full runner rotations and a bandstop filter was used to remove the electrical grid frequency (50 Hz) and its harmonics. The numerical VL01 signal shows similarities with the experimental data regarding the global amplitude and frequencies. The mean difference lies in the over prediction of the second harmonic relatively to the three others and in the other frequencies present experimentally.

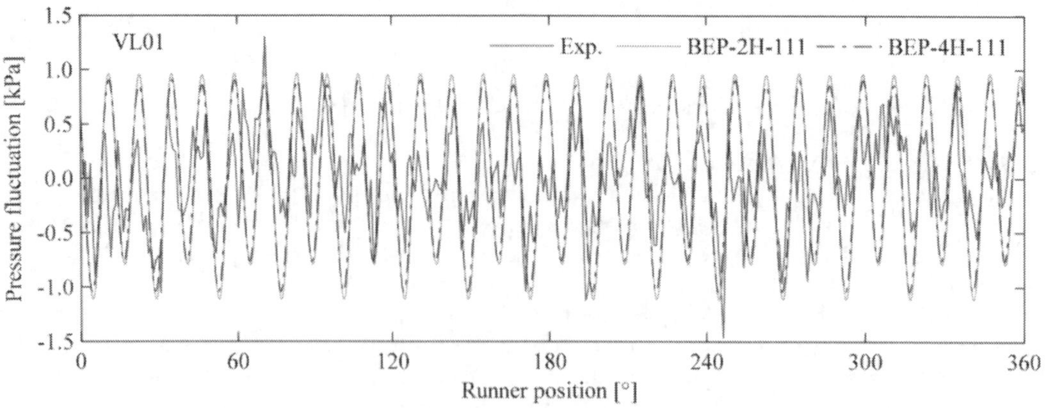

Figure 8. Experimental and NLH pressure signals for the VL01 sensor at best efficiency point.

Similar observations are made when comparing the numerical and experimental signals for the P42 sensor, where the mean amplitude of the signal seems to be respected even though the fluctuations from the simulations are still slightly lower than the measured values. The important energy content from the turbulent fluctuations over a wide frequency range, absent from the simulations, can explain this behaviour.

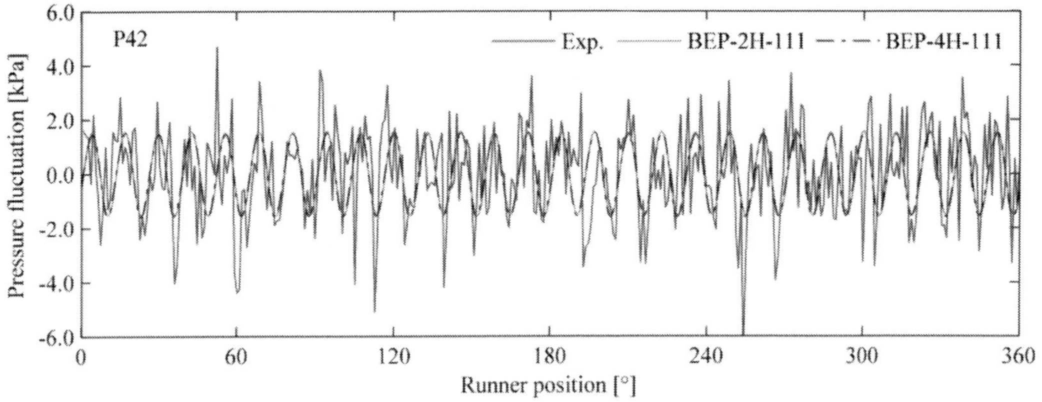

Figure 9. Experimental and NLH pressure signals for the P42 sensor at best efficiency point.

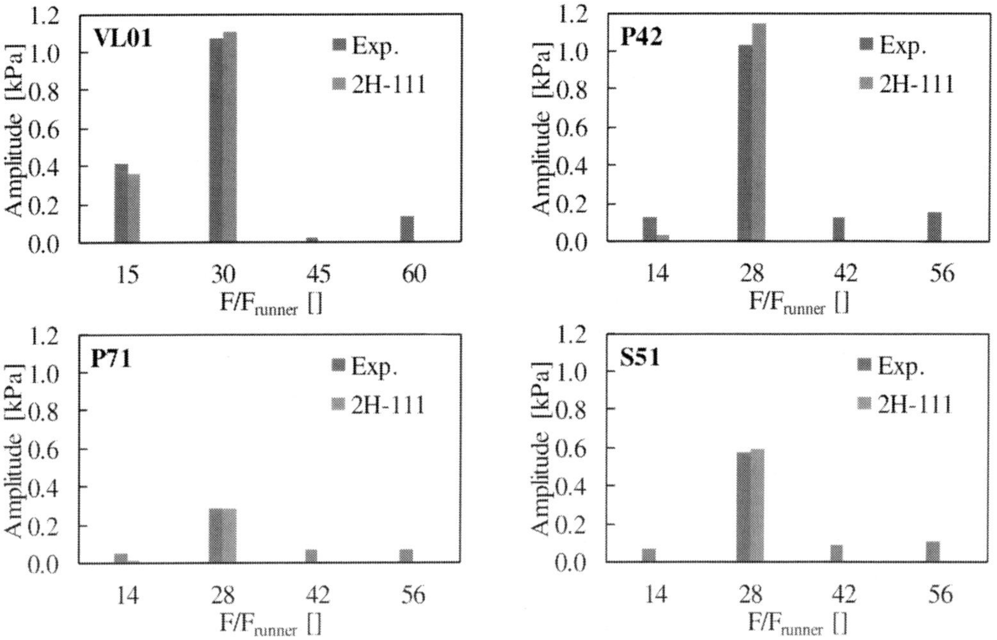

Figure 10. Experimental and numerical frequency spectra for the main perturbation harmonics at high load.

5.3. High load pressure amplitudes and signals

The results from the two harmonics simulation at high load, illustrated in Figure 10, show a better agreement with the experimental values for the blade passing frequency, with differences between ~0.9% for the P71 sensor and ~10.5% for the P42 sensor. The first harmonic for the VL01 sensor is

Journal of Physics: Conference Series **579** (2015) 012013 doi:10.1088/1742-6596/579/1/012013

also predicted with a better accuracy than the BEP case. Once again, the general tendency of the computations is to overestimate the second harmonics and underestimate the first one.

Comparison of the reconstructed signal at VL01 with the experimental measurements also shows a very good agreement. The amplitudes are, in fact, well predicted as well as the relative phase between the two simulated harmonics. This comparison also provides confidence in the NLH base assumption to represent the unsteady flow field with a superposition of periodic sinusoidal values.

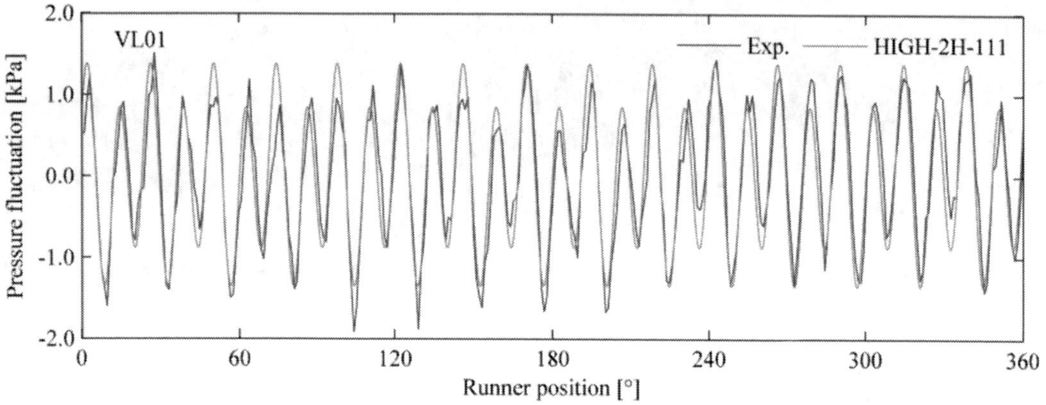

Figure 11. Experimental and NLH pressure signals for the VL01 sensor at high load.

5.4. Pressure visualisations

The distribution of the second harmonic amplitude on the blade for simulations with $n = 2$ is illustrated with contours in Figure 12. At the best efficiency point, the maximum amplitudes are obtained near the leading edge of the blade on the suction side, as it can be seen on the top-right picture. At high load, the leading edge is characterized by a diminution of the amplitude. In both cases, the fluctuations tend to be higher on the pressure side than on the suction side, and decrease linearly until they become almost non-existent near the blade trailing edge.

Colour contours are shown at midspan in the distributor and runner for the second harmonic in Figure 13. Again, the maximum amplitudes, which are higher than 4 kPa, are obtained for the best efficiency point and are located in the runner domain near the rotor-stator interface. The amplitude distributions for both operating points follow a similar pattern, but the gradients at BEP are higher particularly between the guide vanes. At high load, the lower gradients allow the perturbation to be more spread out in the upstream part of the distributor.

5.5. Computation times

The computations shown in this paper were conducted on a cluster and were partitioned over 32 processors. The average computation times per iteration were checked for each simulation type on different grid levels and are summarized in Table 3. For the finer grid level, NLH simulations with $n = 2$ require about four times as much time as the steady simulations. The simulation of four harmonics requires about 35% more computational time as the simulation of two harmonics.

Table 3. Average computation times per iteration in seconds.

Grid level	Nb. cells	Steady	NLH $n = 2$	NLH $n = 4$
111	1 733 664	0.10	2.40	3.24
000	13 486 464	2.54	10.4	n/a

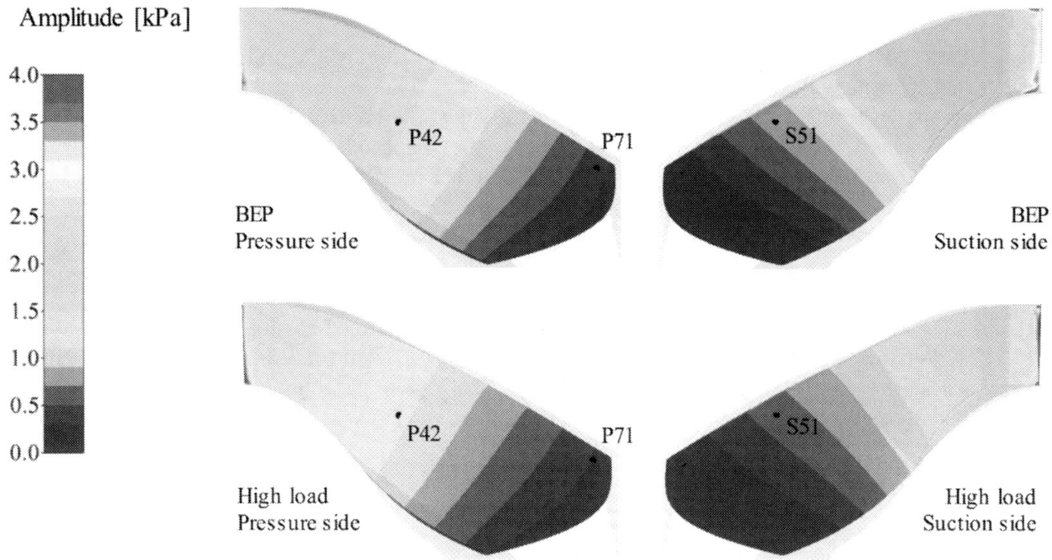

Figure 12. Pressure amplitude of the second harmonic on the runner blade. *Top*: BEP; *Bottom*: High load; *Left*: Pressure side; *Right*: Suction side.

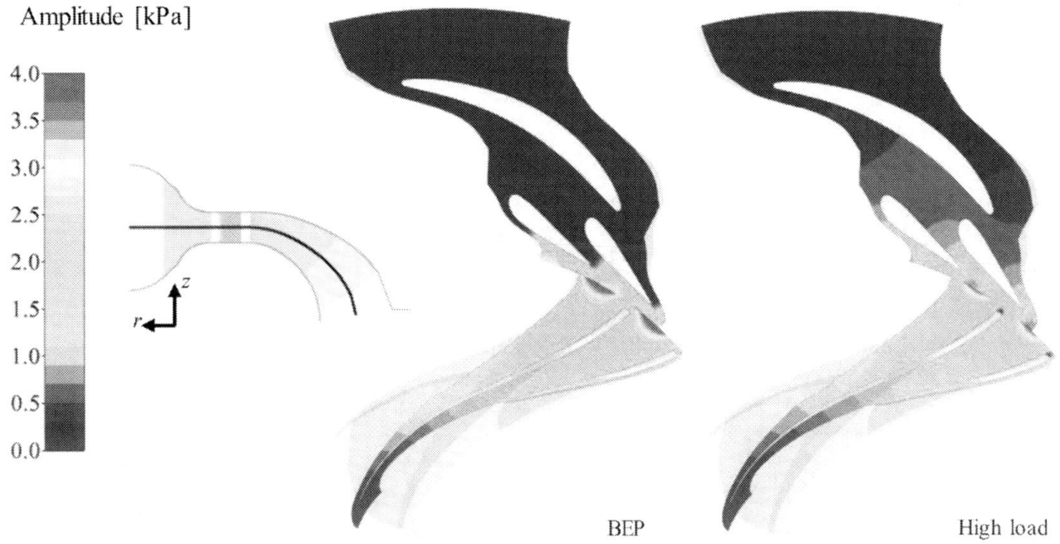

Figure 13. Pressure amplitude of the second harmonic at midspan in the distributor and runner seen from above. *Left*: BEP; *Right*: High load.

6. Conclusion

This paper focused on the methodology used to apply the nonlinear harmonic method to the Francis-99 test case and on preliminary results from intermediate grid level simulations. It was shown from steady and nonlinear harmonic simulations that the grid level used and the computation method does not modify the mean pressure values. NLH simulations on intermediate grid level tend to overestimate the pressure fluctuation amplitudes corresponding to the blade passing frequency. No major differences in the predictions were found between the simulations with two or four harmonics. At the best efficiency point, some similarities are found between the experimental and NLH pressure signals, but better agreement with experimental data is found at high load. In this particular case, simulations on the finer grid level might not be required to predict adequately the main RSI amplitude.

Further investigations are underway to assess the nonlinear harmonic method ability to approximate the unsteady flow in the Francis-99 model. These include simulations on the finest grid level for the three operating points as well as comparisons with full URANS simulations. The effect of the turbulence and boundary layer modeling is also being evaluated.

Acknowledgments
The authors would like to thank the organizers of the Francis-99 workshops for providing the test case data. They would also like to acknowledge the financial support from the Canadian Natural Sciences and Engineering Research Council and Fonds de recherche Nature et technologies du Québec, as well as the computation resources from Compute Canada and great support from the NUMECA team.

Nomenclature

P, P_s	Static pressure [kPa]
U_i	Mean velocity components [m/s]
D	Runner diameter [m]
F	Frequency [Hz]
F_{runner}	Runner frequency [Hz]
$\dot{m}$	Mass flow rate [kg/s]
N	Number of perturbations
n	Number of simulated harmonics
α	Inlet flow angle [°]
τ_{ij}	Reynolds stresses
$\nu,\ \nu_t$	Kinematic, turbulent viscosity [m^2/s]
$\phi(\vec{x},t)$	Conservative variable [m/s, kPa]
$\tilde{\phi}(\vec{x})$	Conservative variable amplitude [m/s, kPa]
$\phi'(\vec{x},t)$	Conservative variable perturbation

References

[1] Celik I B, Ghia U, Roache P J, Freitas C J, Coleman H and Raad P E 2008 Procedure for Estimation and Reporting of Uncertainty due to Discretization in CFD Applications *ASME J. Fluids Eng.* **130** 078001.

[2] Chirag T, Cervantes M J, Gandhi B K and Dahlhaug O G 2013 Experimental and numerical studies for a high head Francis turbine at several operating points *ASME J. Fluids Eng.* **135** 111102

[3] Chirag T, Cervantes M J, Gandhi B K and Dahlhaug O G 2014 Pressure measurements on a high-head Francis turbine during load acceptance and rejection *Journal of Hydraulic Research* **52**(2) pp 283-297

[4] He L and Ning W 1998 Efficient approach for analysis of unsteady viscous flows in turbomachines *AIAA J.* **36**(11) pp 2005-2012

[5] Lipej A, Čelič D, Tartinville B, Mezine M and Hirsch C 2012 Reduction of CPU time for CFD analysis of hydraulic machinery development process *26th IAHR Symp. On Hydraulic Machinery and Systems*, Beijing, China

[6] Luleå University of Technology 2014 Francis-99 Test Case *Luleå University of Technology website* http://www.ltu.se/research/subjects/Stromningslara/Konferenser/Francis-99/Test-Case-1.111520

[7] NUMECA International 2013 Dedicated Turbomachinery Models *Fine/Turbo v9.0 Theoretical Manual* Chapter 7

[8] NUMECA International 2013 Flow model *Fine/Turbo v9.0 User Manual* Chapter 4

[9] Vilmin S, Lorrain E, Hirsch Ch and Swoboda M 2006 Unsteady flow modeling across the rotor/stator interface using the nonlinear harmonic method *ASME Turbo Expo 2006* GT2006-90210

[10] Wang L, Liu B, Yang X and Cao Z 2013 Unsteady simulation of a counter-rotating research compressor using the nonlinear harmonic method *Proceedings of the Institution of Mechanical Engineers Part G (J. Aerospace Eng.)*, **227**(3) pp 415-424

Preliminary Measurements of the Radial Velocity in the Francis-99 Draft Tube Cone

Carl Bergan[1], **Kaveh Amiri**[2], **Michel J Cervantes**[1,2] **and Ole G Dahlhaug**[1]

[1] Department of Energy and Process Engineering, Waterpower Laboratory, Norwegian University of Science and Technology, Trondheim, Norway

[2] Department of Engineering Science and Mathematics, Luleå University of Technology, Luleå, Sweden

E-mail: carl.w.bergan@ntnu.no

Abstract. Two-dimensional particle image velocimetry (PIV) measurements in the draft tube cone of the Francis-99 model have been performed to complete the actual experimental data set with radial velocity data. The velocity profiles obtained presented some variation, which reason has not yet been identified. The presented results are therefore presented as preliminary until the reason is assessed. The axial velocity profiles corroborate well with the ones previously measured with laser Doppler velocimetry (LDV) for all operating points investigated. The radial velocity measured is small in magnitude for all operating points compared to the axial velocity. A gyroscopic effect induced by the swirl leaving the runner and the draft tube bend seems to induce an asymmetry in the draft tube cone.

1. Introduction

Francis-99 is a series of upcoming workshops aiming at determining the state-of-the-art in simulation of high head Francis turbines. For that purpose, the geometry, meshes and experimental results of a high Francis model, the Tokke turbine, were made available to the public at three different operating points: best efficiency point (BEP), part load (PL) and high load (HL), see www.francis-99.org and Chirag et al [1] for more information. The goal of the workshop is to promote the development of numerical tools for the design of high head Francis turbines. The experimental results made available to validate the numerical results comported efficiency, time dependent pressure measurements and velocity measurements. The velocity measurements were acquired in the draft cone at two different locations with laser Doppler velocimetry (LDV), see Figure 3. The measurements comported the axial and the tangential velocities, Sundström et al. [2]. The radial velocity component in high head draft tube cone is small but none the less of importance to assess correctly eventual separation regions. Its determination with optical instrument such as LDV and particle image velocimetry (PIV) is delicate.

PIV and LDV are well-established methods for velocity measurement and have been applied in several turbine-related applications in recent years. PIV is not yet as widespread as LDV, due to its high demand for optical access. PIV has some substantial advantages over LDV, especially for transient phenomena, due to its ability to simultaneously capture a large velocity field with a high data rate. Furthermore, the radial velocity is easily obtained with PIV compared to

LDV. Several set-up configurations of 3D spectroscopic PIV have been investigated by Ciocan [3], with extensive investigation of the data to correct for optical distortion, overlapping of two cameras and set-up for data synchronization. Inter-channel flows in propeller turbines have been investigated by Aeschlimann et al. [4] by the use of stereoscopic PIV in order to obtain 3-component velocity fields. The flow in a Francis turbine draft tube has been investigated by Iliescu et al. [5] using a combination of 3D PIV and LDV.

In the present paper, PIV measurements performed on the Francis-99 test case are presented to complete the actual data set available to validate the numerical simulations. The objectives are to determine the radial velocity in the draft tube cone at the location where the axial and tangential velocities were determined by Sundström et al. [2] with the help of LDV. The axial velocity is also measured with the radial velocity allowing investigating the repeatability of the measurements. Furthermore, contour plots of the meridional velocity are made available for qualitative comparisons.

2. Experimental Setup

The experiment was carried out on a 1:5.1 model Francis turbine of a prototype found at a Norwegian hydropower plant called Tokke. The prototype runner has an inlet diameter of 3.22 m, outlet diameter of 1.78 m, a nominal head of 377 m, and a nominal power output of 110 MW. The distributor is composed of a spiral casing with 14 stay vanes and 28 guide vanes. The runner has 15 full-length blades, and 15 half-blades, also known as splitter blades.

2.1. Test rig

The test rig was operated in a closed-loop configuration, where the pressure and the flow rate are controlled by setting the speed of the pump. The system is fitted with a high-pressure water reservoir upstream the turbine, and a low-pressure reservoir downstream the draft tube. The gauge pressure in the low-pressure reservoir can be independently controlled from 0 down to approximately −9 m, to trigger or prevent cavitation. The experiments were performed under cavitation free conditions. A sketch of the test rig is shown in Figure 1

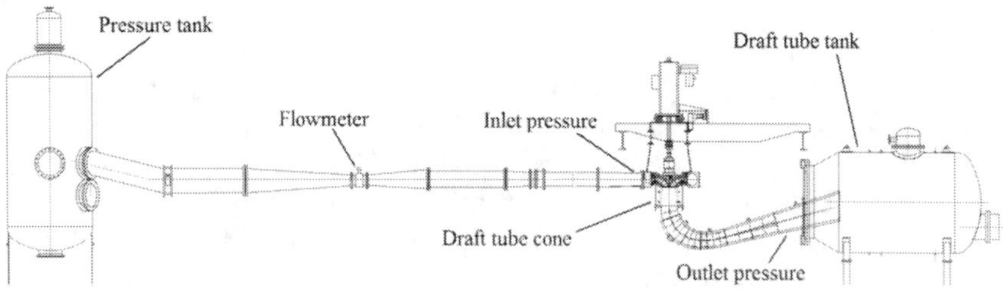

Figure 1. The Francis test rig when operated in closed loop configuration.

The measurements were performed in the draft tube cone. This section is made of Plexiglas for optical access. The draft tube cone was fitted with an index-matching box made of glass and filled with water, to reduce optical distortion.

The operating conditions investigated by Trivedi [1] are listed in Table 1. The specific speed and the specific flow listed in Table 1 are dimensionless numbers defined by Equation 1 and Equation 2.

$$n_{ED} = \frac{n \cdot D}{\sqrt{g \cdot H}} \tag{1}$$

$$Q_{ED} = \frac{Q}{D^2 \cdot \sqrt{g \cdot H}} \tag{2}$$

Table 1. Operating conditions as provided by Trivedi [1].

Operating Point	Head	Flow Rate	Runner frequency	n_{ed}	Q_{ed}	Hydraulic efficiency
-	m	m^3/s	Hz	-	-	-
BEP	11.91	0.203	5.59	0.18	0.15	92.6
HL	11.84	0.221	6.16	0.20	0.17	90.6
PL	12.29	0.071	6.77	0.22	0.05	71.7

However, upon reaching the proposed operating points, a small but clearly visible instability in the generator torque was detected. The operating conditions were slightly shifted while maintaining the values for n_{ed} and Q_{ed}, see Sundström et al. [2], giving the operating conditions listed in Table 2.

Table 2. Operating conditions used in this paper.

Operating Point	Head	Flow Rate	Runner frequency	n_{ed}	Q_{ed}	Hydraulic efficiency
-	m	m^3/s	Hz	-	-	-
BEP	12.77	0.208	5.74	0.18	0.15	92.4
HL	12.61	0.228	6.34	0.20	0.17	91.0
PL	12.30	0.072	6.77	0.22	0.05	72.5

In order to assess the variation on the velocity profiles of these new operating conditions, a sensitivity analysis was performed, by measuring the time-averaged velocity profiles for the given BEP and the shifted BEP (see Sundström et al. [2]). The velocities are scaled with the bulk velocity leaving the runner, for comparison purposes. These are presented in Figure 2. The velocity difference between the two operating heads is considered relatively small.

2.2. LDV and PIV setup

The LDV used is composed of a Spectra-Physics Model 177G, equipped with a Burst Spectrum Analyzer (BSA) from Dantec Dynamics. The LDV probe was mounted on a traverse table with the probe perpendicular to the glass wall of the index matching box [2]. The perpendicularity was checked with optical methods with an accuracy of 0.2°. The measurement positions are presented in Figure 3. The front lens had a focal length of 310 mm. The seeding particles used were of the type Expancel 46 WU 20 with an average diameter of 6 µm. The axial end tangential velocities were recorded at each coordinate for 720 s, resulting in 70 000-2 000 000 samples, depending on the measurement point location and the operating point. Each velocity profile was investigated with 16 measurement points along the radius for BEP and HL, and at 26 points for PL. The large number of measurement points at PL was necessary to capture the high velocity gradients. The recording time was reduced to 600 s for the last measurements.

A 2D PIV system from TSI was used to measure the axial and radial velocity components in the draft tube cone. Pulse light sheets with a thickness of about 3 mm were generated by

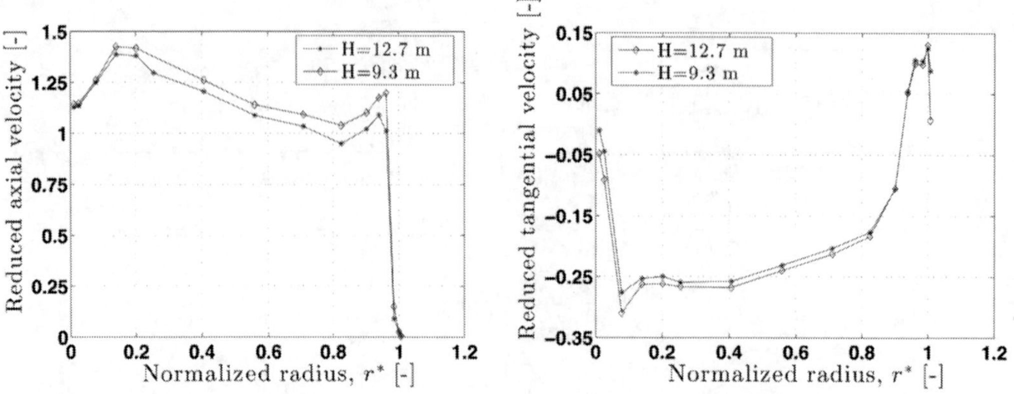

Figure 2. Sensitivity analysis for axial velocity (left) and tangential velocity (right). The velocities are reduced by the bulk velocity leaving the runner.

a Litron Laser NANO L100-50PIV. The illuminated field was recorded by a 4 Mpixel camera (VC-4MC-M180). TSI seeding particles, with a density of 1.016 g/cc, refractive index of 1.52 and mean diameter of 55 µm were used during the measurements. The PIV measurements were performed with an acquisition frequency of 40 Hz. 750 paired images with a time difference of 200 µs were recorded at each measurement section. The velocity profiles measured by PIV were measured in a near-perpendicular direction to the measurement direction of the LDV system, as shown in Figure 3. The axial velocity profiles measured by the PIV and LDV systems can therefore be used to evaluate the flow asymmetry. The PIV system was calibrated using a 2D calibration target. The calibration was performed ex-situ due to the practical limitations associated with the in-situ calibrations.

The velocity profiles measured with the LDV system were measured at two axial positions, as indicated in Figure 3. The PIV system measured velocity in a zone including these two axial coordinates. Note that the measurement axis of the LDV measurements is not parallel with the inlet pipe axis, see Figure 3. The repeatability of the PIV measurements were checked along the radius at a third axial position, indicated as section III in Figure 3.

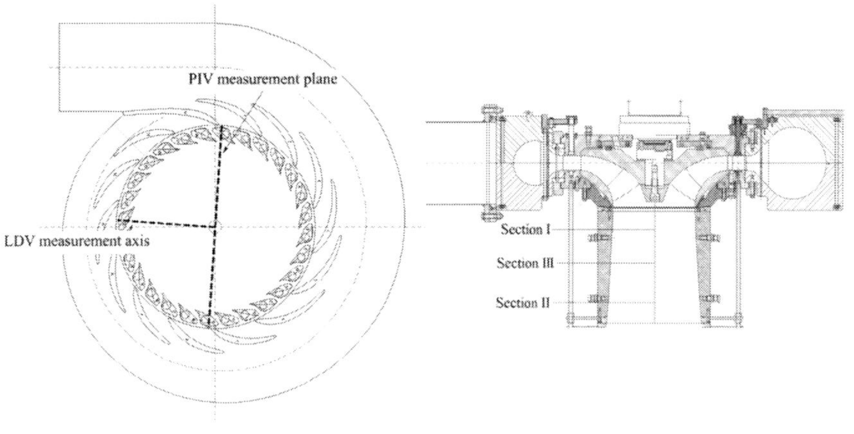

Figure 3. Measurement positions for PIV and LDV as seen from above (left) and from the side (right). Note that the LDV measurement axis is not parallel to the inlet pipe axis. In the right figure, section I, II and III are taken at 57, 377, and 188 mm below the draft tube inlet, respectively. Section I and II are the axial positions of the LDV measurements, while section III is used to test the repeatability of the PIV measurements.

3. Data Analysis

The velocity measurements from the LDV were time-averaged in each measurement point to give a velocity profile. The velocity was then made dimensionless with the bulk velocity downstream the runner for the given operating point, for comparison purposes. The values for the bulk velocity were 0.75 m/s for PL, 2.18 m/s for BEP, and 2.39 m/s for HL. The radial coordinate was made dimensionless with the runner outlet diameter. For the contour plot, the y-coordinate was made dimensionless with the runner outlet diameter, and the zero value was set to the runner outlet. In the results, the radial velocity (u_r^*) is positive in the direction away from the draft tube center, and the axial velocity (u_z^*) is positive in the stream wise direction.

The commercial PIV software, DynamicsStudio, from Dantec dynamics was used for image processing and PIV data analysis. Adaptive correlation scheme with two refinement steps and 50% overlap between the adjacent windows was applied on the acquired data after performing the 2D calibration. The averaged velocity vector maps at each operating point are presented in this paper. The velocities for the PIV were measured in two overlapping sections. In reconstructing these, each data set was re-sampled at uniform intervals in the non-overlapping sections. The overlapping sections were re-sampled using data from both set, before they were weight-averaged using a sine half-wave distribution weighing, giving a transition from the upper window to the lower window. In doing this, the data from the upper window will dominate the overlap close to the upper window, while the data from the lower window will dominate the overlap close to the lower window. The weighing method was chosen on qualitative assessment of the velocity contour plots.

4. Repeatability of the measurements

In order to evaluate the repeatability of the measurements, axial and radial velocity profiles in section III, at the overlapping region of the two measurement windows, are investigated. The velocity profiles at the BEP are presented in Figure 4. The measurements were performed three times at each window. Some discrepancy exists between the upper and lower window measurements, both in axial and radial velocities. The measurements for each window were performed on different days while the test rig was operating continuously. The highest level of discrepancies in both components is found close to the central part of the draft tube where the velocity is higher than close to the wall region. The measurements performed by Sundström et al. [2] indicate the presence of a free vortex in the central region rotating in the opposite direction of the runner. The vortex seems sensible to initial conditions.

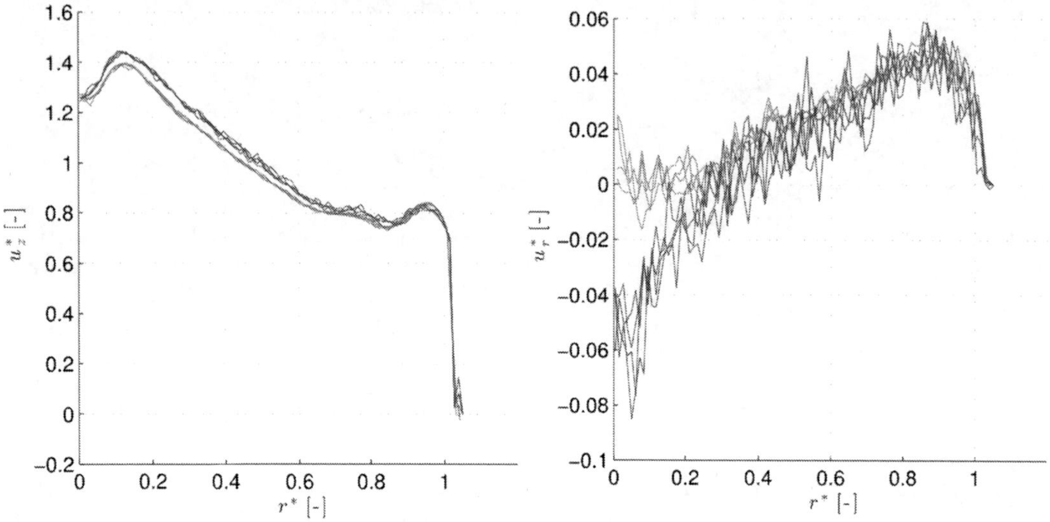

Figure 4. Mean axial (left) and radial (right) velocity profiles at section III for BEP operation. The red lines are data from the lower measurement window and the blue lines are data from the upper measurement window.

Figure 5 presents the velocity profiles at section III for HL operation. The velocity profiles are similar to the BEP case. A larger discrepancy appears in the central region of the draft tube. At this operation point, the swirl leaving the runner is larger than at the BEP point, see Sundström et al. [2].

Figure 6 presents the velocity profiles at section III for PL operation. In this figure, the measurements obtained with the full measuring window are included. At this operating point, a large recirculation region appears below the runner in the central region of the draft tube. The axial velocity, as well as the radial velocity, are subject to a large gradient in the radial direction. Furthermore, the magnitude of the radial velocity is significantly larger than for the two other operating points. The discrepancies appear in the region of high axial velocity coinciding with region of large swirl. At this operation condition, the ratio of the tangential to the axial velocity is above 1.5, see Sundström et al. [2].

The difference in measured velocities can stem from either a difference in operating conditions, errors in the PIV calibration, or a physical phenomenon. The experimental conditions are controlled by the flow rate (Q), the guide van opening angle (α), the turbine rotational speed, and the turbine head (H). Dimensionless parameters n_{ED} and Q_{ED} are used to characterize the turbine operation. The maximum deviation of these parameters is shown in Table 3. The deviations are considered to be too small to cause the observed differences in the velocity profiles. The complete results from the comparison given in Table 3 can be found in tabular form in appendix A.

An index matching box filled with water was used to decrease the light aberration during the PIV measurements. A 2D calibration target was used to compensate for the light aberration

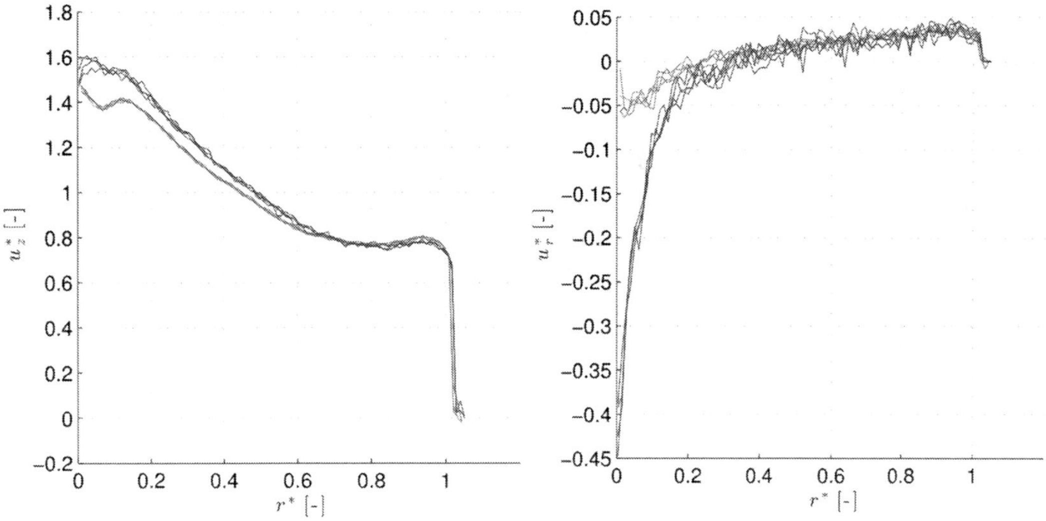

Figure 5. Mean axial (left) and radial (right) velocity profiles at section III for HL operation. The red lines are data from the lower measurement window and the blue lines are data from the upper measurement window.

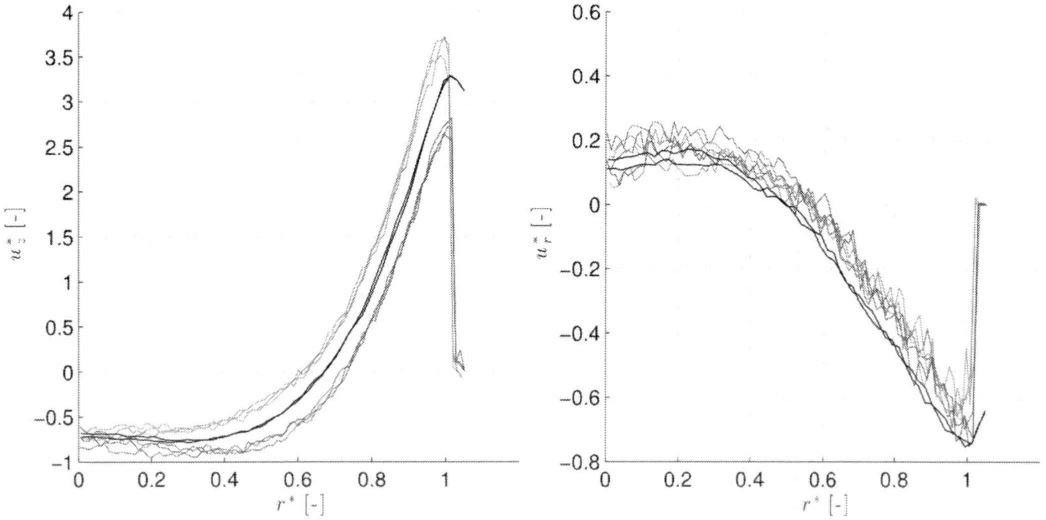

Figure 6. Mean axial (left) and radial (right) velocity profiles at section III for PL operation. The red lines are data from the lower measurement window, the blue lines are data from the upper measurement window, and the black lines are data from the full window measurement.

which still exists close to the areas of high curvature, i.e., close to the draft tube wall. Figure 7 presents a picture of the calibration target in the Plexiglas draft tube cone. Seen in the figure, there is a distance between the draft tube wall and the first calibration point. Hence, the calibration results are extrapolated close to the wall. Moreover, the picture is warped

Table 3. Maximum deviations in operating conditions. The deviations are given as maximum percentage variation from the mean value over the different measurements of the given condition.

	PL	BEP	HL
Q_{ed}	0.37 %	0.14 %	0.14 %
n_{ed}	0.03 %	0.01 %	0.03 %
H	0.05 %	0.05 %	0.05 %
Q	0.36 %	0.12 %	0.12 %
α	0.00 %	0.00 %	0.00 %

significantly close to the wall but unaffected along the draft tube cone centerline. Thus, the calibration matrix calculated may be subject to some uncertainties close to the wall, i.e., in the region $r^* > 0.85$. Consequently, the axial velocity profiles cannot be affected by the calibration process all along the radius. The calibration uncertainty in the radial direction may result in variation in the radial velocity magnitude and not its direction. However, the results showed that the uncertainty in the radial direction is too small to cause the observed differences in the velocity profiles. If calibration has some effects on the velocity profiles, it should be restricted near the wall region and just in the radial velocity profile, while, as shown in Figure 4, the discrepancies can be seen in the high velocity region close to the draft tube center in this case.

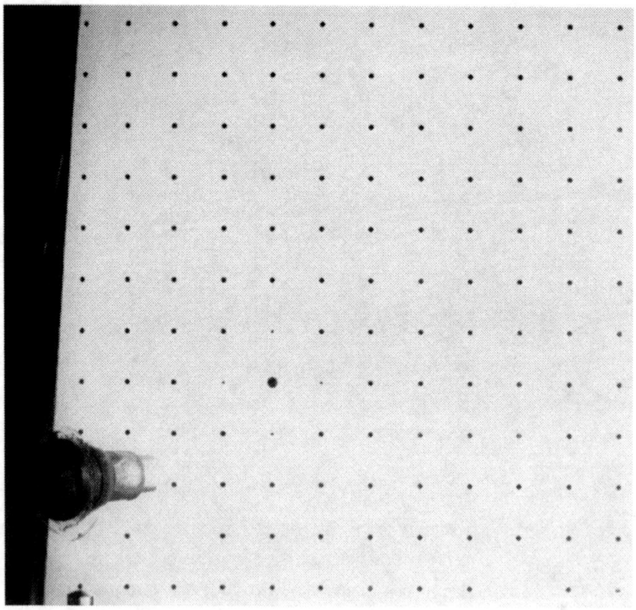

Figure 7. 2D calibration target inside the draft tube cone.

The discrepancies appear only between profiles obtained after restarting the test rig. The discrepancies in the measured profiles are mostly significant in region of high swirl and axial velocities. It may be argued that the swirl encountered in the draft tube cone is sensitive to initial conditions. As a matter of fact, the swirl living the runner creates an asymmetrical flow in the draft tube after the elbow through gyroscopic effects. The flow is divided into a low and high velocity region which extend may be sensible to initial condition. The question needs to be further addressed and the following measurements are considered as preliminary as long as the matter is not completely addressed.

5. Results and discussion

The axial and radial velocity profiles obtained at section I and II with the PIV system are presented in this section. The axial velocity is compared with the LDV measurements of Sundström et al. [2]. Velocity contours at each operating point are also included.

5.1. BEP results

Figure 8 shows the dimensionless axial and radial velocity profiles at section I for BEP operation. The trend captured by the LDV and PIV measurements is similar. There are some differences between the results especially in the wake region of the runner hub. A probable asymmetry at the runner outlet may be the reason, since the PIV and LDV measurements were not performed at the same azimuthal position as presented in Figure 3. Such an asymmetry may stem from the asymmetric flow distribution at the spiral casing distributor as reported by some researchers, see Amiri [6], but also from the gyroscopic effect as previously mentioned. The PIV measurements show slightly lower axial velocities in the middle part of the graph ($0.2 < r^* < 0.8$) and higher velocities in the region affected by the runner hub wake. An integration of the velocity profiles results in comparable flow rate for the two measurements assuming a symmetrical flow at the runner outlet, in agreement with the flow rate measurements. The axial velocity shows a linear decrease with the radius in the middle part of the graph ($0.2 < r^* < 0.8$). The axial velocity decreases again towards the centerline of the draft tube due to the runner hub wake. The axial velocity slightly increases close to the wall. This may be due to the design criteria to energize the boundary layer on the draft tube cone or may be related to the tip clearance. Illustrated in Figure 8, the radial velocity is increasing with the radius at the draft tube inlet. The results show that the radial velocity is positive close to the draft tube wall as expected for a diffuser. The flow at the draft tube cone centerline is expected to be purely axial for symmetric flow at the runner outlet. However, the negative radial velocity indicates an asymmetrical flow that may also be the source of the discrepancies between the PIV and LDV measurement results, as discussed before.

Figure 9 presents the velocity distribution at section II. The agreement between the LDV and the PIV results is much better. The axial velocity profile is similar to the velocity profile captured at section I. Surprisingly, the radial velocity is somewhat lower close to the centerline. The negative radial velocity at the draft tube center will move water into the other side of the draft tube, causing a higher velocity in that window, than in the one presented here. This is due to the gyroscopic effect induced by the draft tube bend. The LDV measurement results presented by Sundström et al. [2] showed that the flow at the runner outlet counter-rotates with respect to the runner rotation at the BEP. The rotating flow leaving the runner together with the draft tube bend produces a gyroscopic force resulting in an asymmetric flow in the draft tube cone. The asymmetry effect may propagate upstream resulting in some asymmetry at section I as presented in Figure 8.

Figure 10 presents the contour plot of the meridional velocity inside the draft tube cone. The results indicate a well-distributed flow in the draft tube cone. The figure shows the flow development along the draft tube cone. The velocity smoothly decreases in the stream wise direction without separation. The presence of the hub is seen in the upper center of the draft tube cone, where the velocities are lower. The hub effect can also be seen in Figure 9, where the radial velocity shows a sharp change at $r^* = 0.19$, the radial position of the runner hub cone. A visible boundary layer can also be seen in the axial velocity component in Figure 9, where the

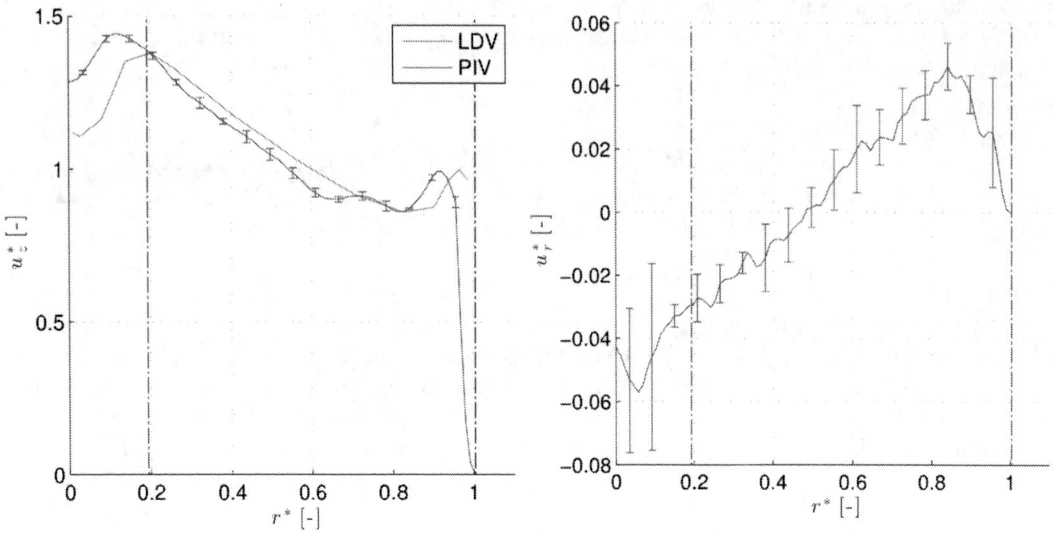

Figure 8. Axial (left) and radial (right) velocities at section I for BEP operation. The red line shows the LDV result obtained by Sundström et al. [2] and the blue line shows the mean PIV results. The error bars indicate a 95% confidence interval. The dashed lines at $r^* = 0.19$ and $r^* = 1.00$ indicate the radial position of the runner hub corner and the draft tube entrance, respectively. Note the difference in scale on the ordinate axis.

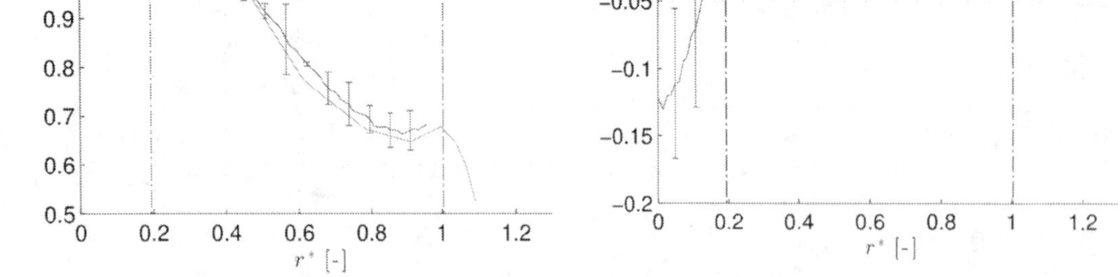

Figure 9. Axial (left) and radial (right) velocities at section II for BEP operation. The red line shows the LDV result obtained by Sundström et al. [2] and the blue line shows the mean PIV results. The error bars indicate a 95% confidence interval. The dashed lines at $r^* = 0.19$ and $r^* = 1.00$ indicate the radial position of the runner hub corner and the draft tube entrance, respectively. Note the difference in scale on the ordinate axis.

velocity shows a visible reduction at $r^* = 1$. This is only visible in the LDV measurements, since one of the lower pressure bolts prevents optical access.

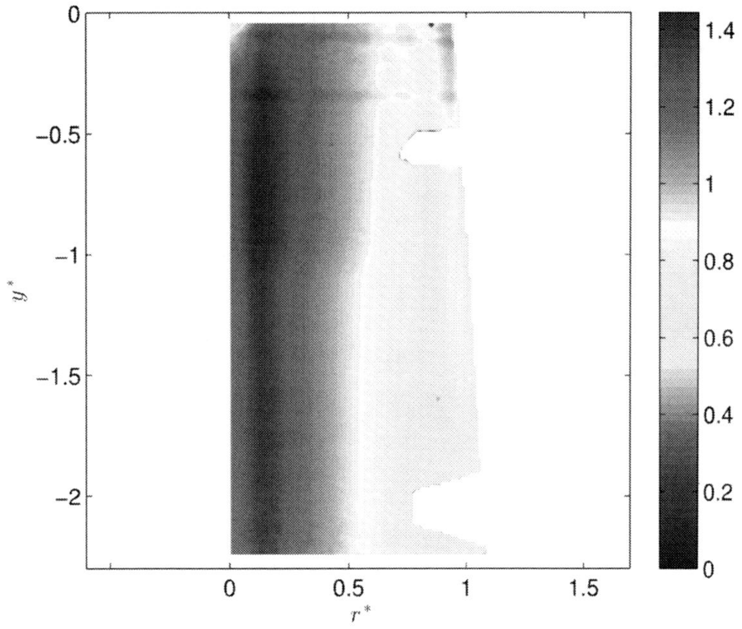

Figure 10. Contour plot of the meridional velocity within the PIV plane at BEP. Pressure bolts in the draft tube cone are seen in both the upper and lower part of the cone. The maximum velocity is obtained at $r^* = 0.15$

5.2. HL results

Axial and radial velocity profiles at section I are presented in Figure 11 while the turbine operates at HL. The low velocity region in the wake of the runner hub, presented in the BEP case, does not exist anymore and the axial velocity decreases with the radius all along the draft tube. The radial velocity shows the same profile as the one at the BEP: monotonically increasing with the radius. The flow direction is towards the draft tube wall at $r^* > 0.5$ and towards the draft tube centerline at $r^* < 0.5$. The radial velocity shows a sharp decrease in the middle of the draft tube, at the radial position of the runner hub corner. Sundström et al. [2] showed that the swirl exiting the runner at the HL is higher than the one at the BEP. Hence, the gyroscopic effect is expected to be higher for HL, resulting in a larger decrease in the radial velocity in the middle of the draft tube cone. The radial velocity at section II (Figure 12) is three times larger close to the draft tube bend compared with section I, supporting the hypothesis of an asymmetry imparted by the gyroscopic effect induced by the draft tube bend. The axial velocity profile at section II is similar to the profile captured at section I, but with a lower magnitude, as expected. The draft tube cone decreases the axial velocity, allowing for the recovery of pressure.

Figure 13 shows the contour plot of the meridional velocity at the HL operating point. The contour plot is altered along two horizontal lines of $y^* = -0.09$ and $y^* = -0.34$. It may be due to the change in draft tube angle at those positions resulting in different light aberration at the position that the angle is changed.

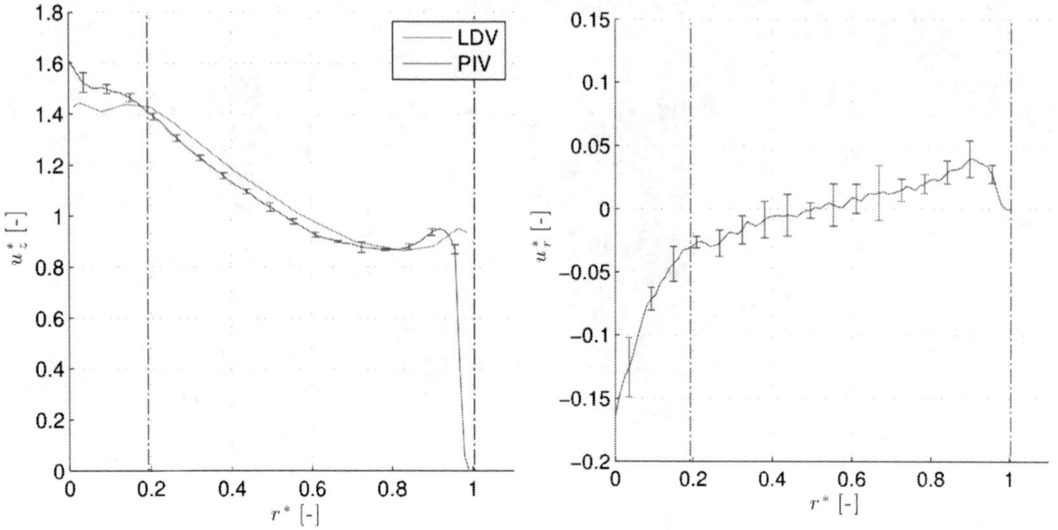

Figure 11. Axial (left) and radial (right) velocities at section I for HL operation. The red line shows the LDV result obtained by Sundström et al. [2] and the blue line shows the mean PIV results. The error bars indicate a 95% confidence interval. The dashed lines at $r^* = 0.19$ and $r^* = 1.00$ indicate the radial position of the runner hub corner and the draft tube entrance, respectively. Note the difference in scale on the ordinate axis.

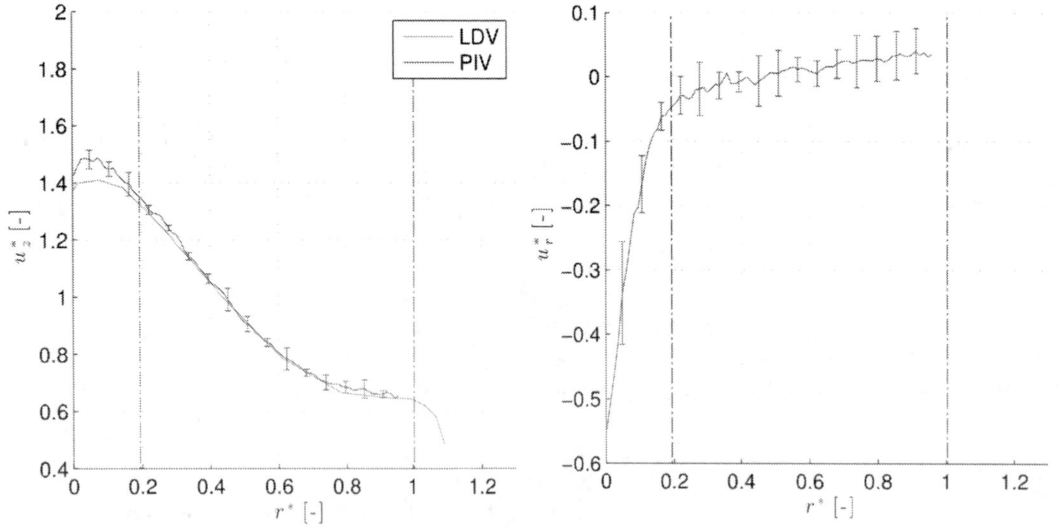

Figure 12. Axial (left) and radial (right) velocities at section II for HL operation. The red line shows the LDV result obtained by Sundström et al. [2] and the blue line shows the mean PIV results. The error bars indicate a 95% confidence interval. The dashed lines at $r^* = 0.19$ and $r^* = 1.00$ indicate the radial position of the runner hub corner and the draft tube entrance, respectively. Note the difference in scale on the ordinate axis.

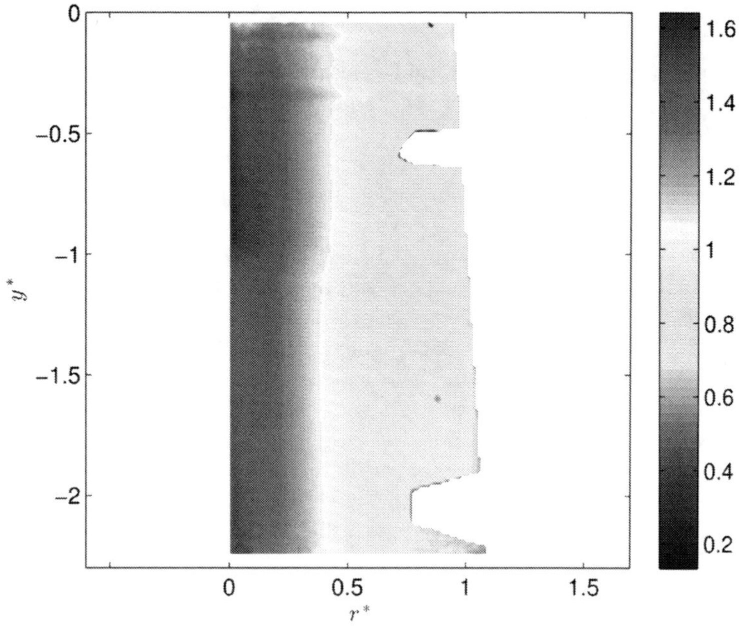

Figure 13. Contour plot of the meridional velocity within the PIV plane at HL. Pressure bolts in the draft tube cone are seen in both the upper and lower part of the cone.

5.3. PL results

Figure 14 shows the velocity profiles captured at section I while the turbine operates at the PL. The larger error-bars compared to the other operating points is attributed to the presence of the rotating vortex rope (RVR) formed in the draft tube at this operating point. The axial velocity profile at section I shows backward flow for $r^* < 0.7$. This is due to the recirculation bubble formed in the middle of the draft tube cone at this operating point. The low guide vane angle and the runner inability to extract the swirl generated by the guide vanes result in high swirl flow exiting the runner, see Sundström et al. [2]. The swirl results in a centrifugal force exerted on the flow. Hence, the axial velocity is significantly increased in the outer part of the draft tube cone for $r^* > 0.7$. The radial velocity at section I decreases with the radius. The positive radial velocity for $r^* < 0.5$ can be explained by the shape of the recirculation bubble region formed in the draft tube cone. The fact that the radial velocity is negative at $r^* > 0.5$ indicates that the flow does not follow the draft tube wall. The large pressure gradient in the radial direction close to the wall cause a negative radial velocity at this position, but it should be symmetric, i.e., the negative radial velocity should be seen close to the wall on the other side of the draft tube. This is not the case, as seen in Figure 17. This asymmetry could be caused by a local flow separation, which will propagate due to the high tangential velocity, see Sundström et al. [2]. Such a localized flow separation was not measured, but it could be the cause of the asymmetry in the radial velocity.

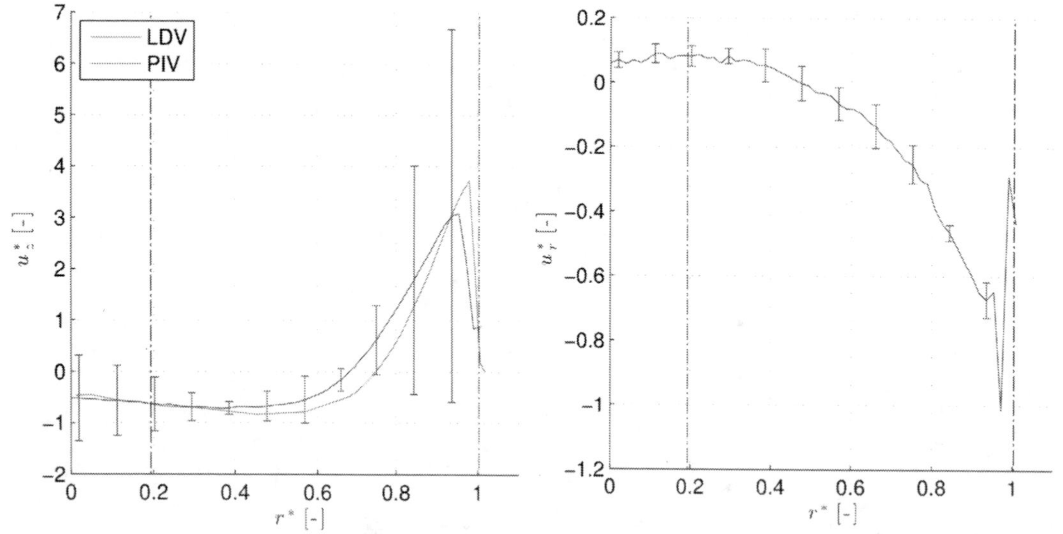

Figure 14. Axial (left) and radial (right) velocities at section I for PL operation. The red line shows the LDV result obtained by Sundström et al. [2] and the blue line shows the mean PIV results. The error bars indicate a 95% confidence interval. The dashed lines at $r^* = 0.19$ and $r^* = 1.00$ indicate the radial position of the runner hub corner and the draft tube entrance, respectively. Note the difference in scale on the ordinate axis.

The axial and radial velocity profiles at section II are presented in Figure 15. The main features of the velocity profiles are similar to the ones captured at section I. The axial velocity is monotonically increasing with the radius while the flow direction is downward for $r^* > 0.7$ and is upward for $r^* < 0.7$. The radial velocity decreases with the radius. Although the radial velocity is not measured close to the draft tube walls, the curve shows that the velocity should

Journal of Physics: Conference Series **579** (2015) 012014 doi:10.1088/1742-6596/579/1/012014

be negative close to the wall, i.e., the flow does not follow the draft tube wall. It is expected that this will vary in the azimuthal position, as seen in Figure 17.

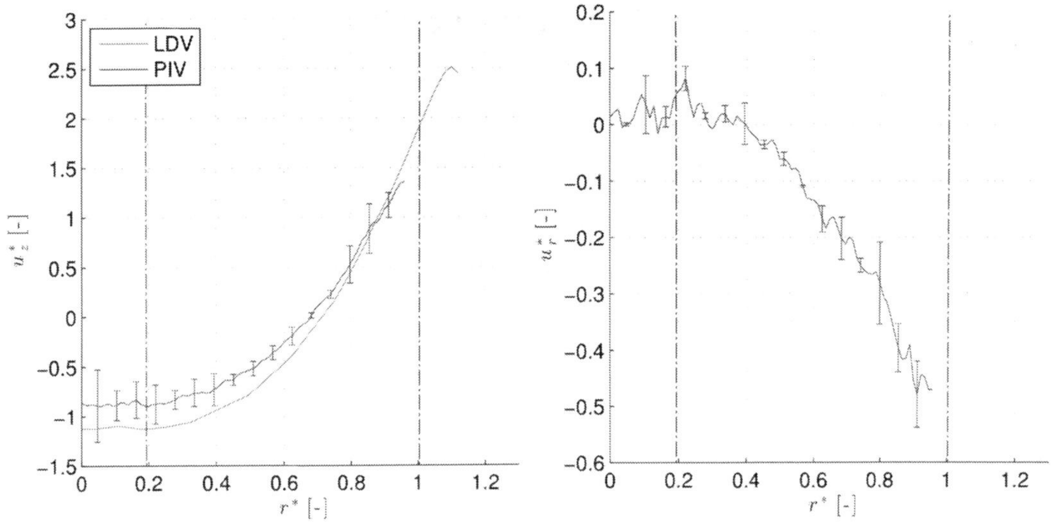

Figure 15. Axial (left) and radial (right) velocities at section II for PL operation. The red line shows the LDV result obtained by Sundström et al. [2] and the blue line shows the mean PIV results. The error bars indicate a 95% confidence interval. The dashed lines at $r^* = 0.19$ and $r^* = 1.00$ indicate the radial position of the runner hub corner and the draft tube entrance, respectively. Note the difference in scale on the ordinate axis.

Figure 16 presents the meridional velocity in the draft tube cone. Note the distinct dark blue lines of zero velocity at $r^* = 0.6$. Since the meridional velocity in this figure is a scalar value, the velocity will not be negative, but the zero-lines give an indication of the recirculation bubble in the draft tube cone.

Figure 17 presents contour of the axial velocity at PL. The figure shows the large recirculation region formed inside the draft tube cone. This recirculation region decreases the effective area of the draft tube cone affecting the turbines overall efficiency. Presented in Table 2, the turbine efficiency is decreased by about 20% at this operating point compared with the one at the BEP. The flow rate is about 1/3 of the one at BEP.

Figure 18 presents the contour plot of the radial velocity in the draft tube cone. The figure shows the asymmetry in the radial velocity distribution inside the draft tube cone. Moreover, the flow does not follow the right wall of the draft tube all along the draft tube cone as discussed before. The clear vertical line visualizes the asymmetry of the flow.

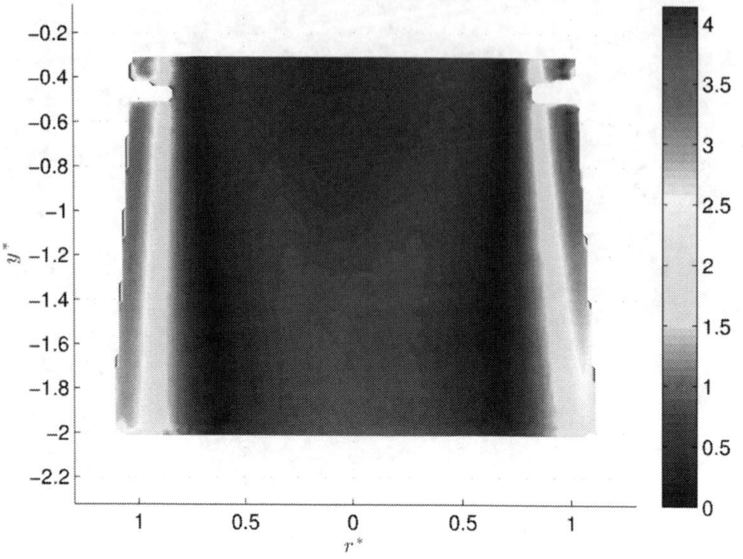

Figure 16. Contour plot of the meridional velocity within the PIV plane at PL operation. Pressure bolts in the draft tube cone are seen in both the upper part of the cone.

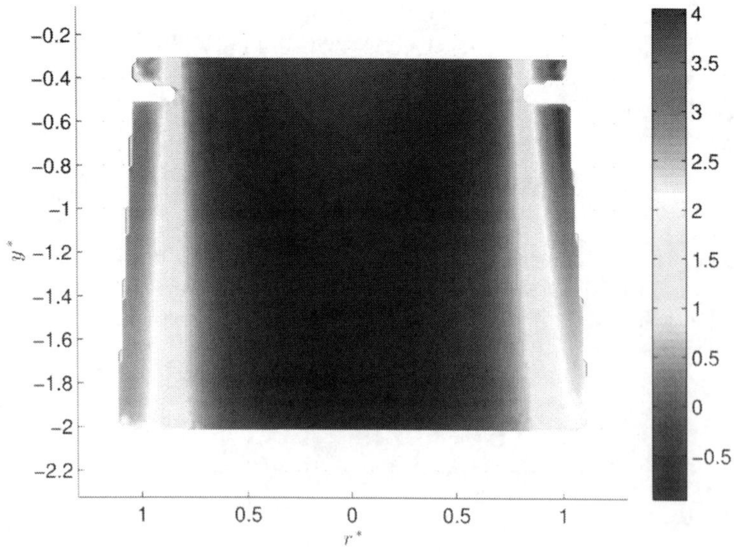

Figure 17. Contour plot of the axial velocity within the PIV plane at PL operation. Pressure bolts in the draft tube cone are seen in both the upper part of the cone.

Journal of Physics: Conference Series **579** (2015) 012014 doi:10.1088/1742-6596/579/1/012014

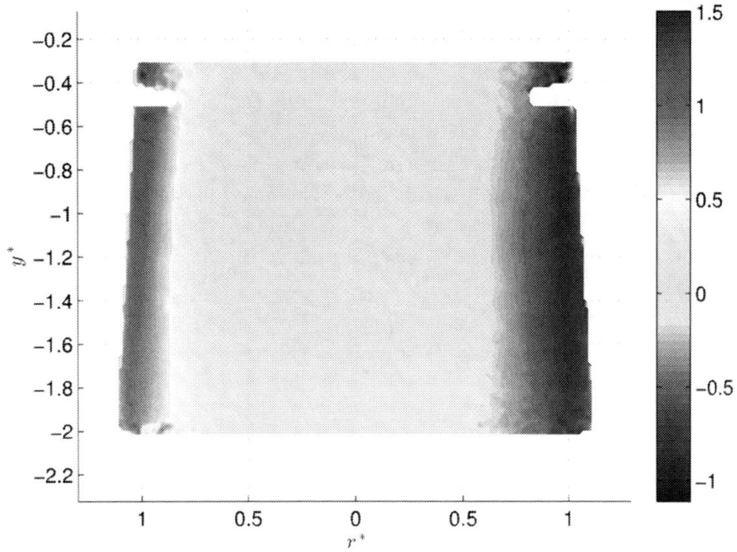

Figure 18. Contour plot of the radial velocity within the PIV plane at PL operation. Pressure bolts in the draft tube cone are seen in both the upper part of the cone.

6. Conclusions

PIV measurements have been performed on the francis-99 model turbine at the Waterpower Laboratory at the Norwegian University of Science and Technology. The results show good agreement with previously performed LDV measurements.

The measurements show discrepancies in the overlapping region between the measurement windows. Further measurements in the draft tube are needed in order to fully understand the reason. The non-zero radial velocity at the centerline of the draft tube cone indicate an asymmetry in the flow certainly due to a gyroscopic effect induced by the draft tube elbow on the swirl leaving the runner.

Journal of Physics: Conference Series **579** (2015) 012014 doi:10.1088/1742-6596/579/1/012014

Appendix A. Operating point repeatability

Table A1. Maximum deviations in operating parameters for all operating points investigated.

	PL	**BEP**	**HL**
Axial Force	1.15 %	0.82 %	1.80 %
Axial force prototype	1.16 %	0.82 %	1.79 %
Differential Pressure	0.04 %	0.05 %	0.05 %
Dimensionless Axial force	1.16 %	0.80 %	1.77 %
Dissolved Oxygen	0.00 %	0.03 %	0.00 %
Flow	0.36 %	0.12 %	0.12 %
Flow prototype	0.37 %	0.14 %	0.13 %
Friction Torque	35.87 %	27.63 %	28.14 %
Generator Speed	0.01 %	0.02 %	0.01 %
Generator Torque	1.27 %	0.22 %	0.43 %
Guide Vane Opening Degree	0.00 %	0.00 %	0.00 %
Hydraulic Efficiency	0.25 %	0.14 %	0.17 %
Hydraulic efficiency prototype	0.24 %	0.14 %	0.17 %
Hydraulic Energy	0.38 %	0.08 %	0.10 %
Kinematic Viscosity	0.00 %	0.00 %	0.00 %
Mechanical Energy	0.13 %	0.12 %	0.21 %
Mechanical Effect prototype	0.12 %	0.12 %	0.20 %
n_{11}	0.03 %	0.01 %	0.03 %
n_{ED}	0.03 %	0.01 %	0.03 %
Nominal hydraulic efficiency model	0.25 %	0.14 %	0.17 %
NPSE	2.37 %	3.31 %	1.72 %
P_{11}	0.12 %	0.08 %	0.18 %
P_{ED}	0.13 %	0.08 %	0.18 %
Pressure Inlet	0.69 %	0.93 %	0.46 %
Pressure Inlet Sensor	0.63 %	0.79 %	0.52 %
Pressure inlet sensor abs	0.62 %	0.84 %	0.41 %
Pressure Outlet	1.95 %	2.75 %	1.38 %
Prototype Head	0.05 %	0.02 %	0.05 %
Q_{11}	0.37 %	0.15 %	0.14 %
Q_{ED}	0.37 %	0.14 %	0.14 %
Reynolds Number Model	0.63 %	0.63 %	0.65 %
Scalable loss	0.00 %	0.00 %	0.00 %
T_{11}	0.12 %	0.09 %	0.17 %
T_{ED}	0.13 %	0.10 %	0.18 %
Thoma number	2.38 %	3.30 %	1.75 %
Vapor Pressure	1.60 %	1.58 %	1.64 %
Water Density Inlet	0.01 %	0.01 %	0.01 %
Water Density Mean	0.01 %	0.01 %	0.01 %
Water Density Outlet	0.01 %	0.01 %	0.01 %
Water Head Model	0.05 %	0.05 %	0.05 %
Water Temperature	1.34 %	1.34 %	1.38 %

References

[1] Trivedi C, Cervantes M J, Gandhi B K and Dahlhaug O G 2013 *Journal of Fluids Engineering* **135** 111102–111102 ISSN 0098-2202

[2] Sundstrom J, Amiri K, Bergan C, Cervantes M J and Dahlhaug O G 2014 LDA measurements in the francis-99 draft tube cone *Proceedings of IAHR2014*

[3] Ciocan G D and Iliescu M S 2012 PIV measurements applied to hydraulic machinery: Cavitating and cavitation-free flows *The Particle Image Velocimetry - Characteristics, Limits and Possible Applications* ed Cavazzini G (InTech) ISBN 978-953-51-0625-8

[4] Aeschlimann V, Beaulieu S, Houde S, Ciocan G D and Deschênes C 2013 *European Journal of Mechanics - B/Fluids* **42** 121–128 ISSN 09977546

[5] Iliescu M S, Ciocan G D and Avellan F 2002 3d PIV and LDV measurements at the outlet of a francis turbine draft tube *ASME 2002 Joint US-European Fluids Engineering Division Conference* (American Society of Mechanical Engineers) pp 311–316

[6] Amiri K 2014 *An experimental investigation of flow in a Kaplan runner steady-state and transient* Licentiate Luleå University of Technology

AUTHOR INDEX